WCDMA FOR UMTS

Radio Access For Third Generation
Mobile Communications

Revised edition

Edited by
Harri Holma and Antti Toskala
Both of Nokia, Finland

JOHN WILEY & SONS, LTD
Chichester • New York • Weinheim • Brisbane • Singapore • Toronto

Other Wiley Editorial Offices

John Wiley & Sons, Inc., 605 Third Avenue,
New York, NY 10158-0012, USA

WILEY-VCH Verlag GmbH
Pappelallee 3, D-69469 Weinheim, Germany

John Wiley & Sons Australia, 33 Park Road, Milton,
Queensland 4064, Australia

John Wiley & Sons (Asia) Pte Ltd, 2 Clementi Loop #02-01,
Jin Xing Distripark, Singapore 129809

John Wiley & Sons (Canada) Ltd, 22 Worcester Road
Rexdale, Ontario, M9W 1L1, Canada

British Library Cataloguing in Publication Data

A catalogue record for this book is available from the British Library

ISBN 0 471 48687 6

Typeset by Laser Words, Madras, India
Printed and bound in Great Britain by Antony Rowe Ltd, Chippenham, Wiltshire.
This book is printed on acid-free paper responsibly manufactured from sustainable forestry, in which at least two trees are planted for each one used for paper production.

Contents

Preface

Second generation telecommunication systems, such as GSM, enabled voice traffic to go wireless: the number of mobile phones exceeds the number of landline phones and the mobile phone penetration exceeds 70% in countries with the most advanced wireless markets. The data handling capabilities of second generation systems are limited, however, and third generation systems are needed to provide the high bit rate services that enable high quality images and video to be transmitted and received, and to provide access to the web with high data rates. These third generation mobile communication systems are referred to in this book as UMTS (Universal Mobile Telecommunication System). WCDMA (Wideband Code Division Multiple Access) is the main third generation air interface in the world and will be deployed in Europe and Asia, including Japan and Korea, in the same frequency band, around 2 GHz. The large market for WCDMA and its flexible multimedia capabilities will create new business opportunities for manufacturers, operators, and the providers of content and applications. This book gives a detailed description of the WCDMA air interface and its utilisation. The contents are summarised in Figure 1

Chapter 1 introduces the third generation air interfaces, the spectrum allocation, the time schedule, and the main differences from second generation air interfaces. Chapter 2 presents example UMTS applications, concept phones and the quality of service classes. Chapter 3 introduces the principles of the WCDMA air interface, including spreading, Rake receiver, power control and handovers. Chapter 4 presents the background to WCDMA, the global harmonisation process and the standardisation. Chapters 5–7 give a detailed presentation of the WCDMA standard, while Chapters 8–11 cover the utilisation of the standard and its performance. Chapter 5 describes the architecture of the radio access network, interfaces within the radio access network between base stations and radio network controllers (RNC), and the interface between the radio access network and the core network. Chapter 6 covers the physical layer (layer 1), including spreading, modulation, user data and signalling transmission, and the main physical layer procedures of power control, paging, transmission diversity and handover measurements. Chapter 7 introduces the radio interface protocols, consisting of the data link layer (layer 2) and the network layer (layer 3). Chapter 8 presents the guidelines for radio network dimensioning, gives an example of detailed capacity and coverage planning, and covers GSM co-planning. Chapter 9 covers the radio resource management algorithms that guarantee the efficient utilisation of the air interface resources and the quality of service. These algorithms are power control, handovers, admission and load control. Chapter 10 depicts packet access and verifies the approach presented in dynamic system simulations. Chapter 11 analyses the coverage and capacity of

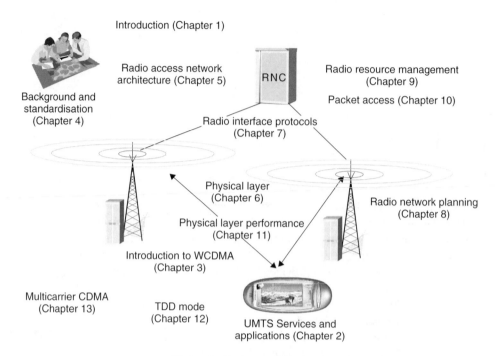

Figure 1. Contents of this book

the WCDMA air interface with bit rates up to 2 Mbps. Chapter 12 introduces the time division duplex (TDD) mode of the WCDMA air interface and its differences from the frequency division duplex (FDD) mode. In addition to WCDMA, third generation services can also be provided with EDGE or with multicarrier CDMA. EDGE is the evolution of GSM for high data rates within the GSM carrier spacing. Multicarrier CDMA is the evolution of IS-95 for high data rates using three IS-95 carriers, and is introduced in Chapter 13.

This reprint of the book includes the key modifications of 3GPP specification done since the official completion of Release'99 until December 2000.

This book is aimed at operators, network and terminal manufacturers, service providers, university students and frequency regulators. A deep understanding of the WCDMA air interface, its capabilities and its optimal usage is the key to success in the UMTS business.

This book represents the views and opinions of the authors, and does not necessarily represent the views of their employers.

Acknowledgements

The editors would like to acknowledge the time and effort put in by their colleagues in contributing to this book. Besides the editors, the contributors were Zhi-Chun Honkasalo, Seppo Hämäläinen, Markku Juntti, Janne Laakso, Jaana Laiho, Ukko Lappalainen, Otto Lehtinen, Fabio Longoni, Atte Länsisalmi, Peter Muszynski, Mika Raitola, Oscar Salonaho, Jouni Salonen, Kari Sipilä, Jukka Vialen, Heli Väätäjä, Achim Wacker and Juha Ylitalo.

While we were developing this book, many of our colleagues from different Nokia sites in three continents offered their help in suggesting improvements and finding errors. Also, a number of colleagues from other companies have helped us in improving the quality of the book. The editors are grateful for the comments received from Heikki Ahava, David Astely, Erkki Autio, Kai Heikkinen, Kari Heiska, Kimmo Hiltunen, Kaisu Iisakkila, Ann-Louise Johansson, Ilkka Keskitalo, Pasi Kinnunen, Tero Kola, Petri Komulainen, Lauri Laitinen, Anne Leino, Arto Leppisaari, Pertti Lukander, Esko Luttinen, Jonathan Moss, Olli Nurminen, Tero Ojanperä, Lauri Oksanen, Kari Pehkonen, Mika Rinne, David Soldani, Rauno Ruismäki, Kimmo Terävä, Mitch Tseng, Antti Tölli and Veli Voipio.

The team at John Wiley & Sons participating in the production of this book provided excellent support and worked hard to keep the demanding schedule. The editors especially would like to thank Sarah Lock for co-ordinating the chapter submission process, and Patrick Bonham, the copy-editor, for his efforts in smoothing out the engineering approach to the English language expressions.

We are extremely grateful to our families, as well as the families of all the authors, for their patience and support, especially during the late night and weekend editing sessions near different production milestones.

Special thanks are due to our employer, Nokia Networks, for supporting and encouraging such an effort and for providing some of the illustrations in this book.

Finally, we would like to acknowledge the efforts of our colleagues in the wireless industry for the great work done within the 3rd Generation Partnership Project to produce the global WCDMA standard in merely a year and thus to create the framework for this book. Without such an initiative this book would never have been possible.

The editors and authors welcome any comments and suggestions for improvements or changes that could be implemented in forthcoming editions of this book. The email address for gathering such information is *wcdma.for.umts@pp.nic.fi*.

Espoo, Finland   Harri Holma & Antti Toskala

Abbreviations

3GPP	3rd Generation partnership project (produces WCDMA standard)
3GPP2	3rd Generation partnership project 2 (produced cdma2000 standard)
AAL2	ATM Adaptation Layer type 2
AAL5	ATM Adaptation Layer type 5
ACELP	Algebraic code excitation linear prediction
ACIR	Adjacent channel interference ratio, caused by the transmitter non-idealities and imperfect receiver filtering
ACLR	Adjacent channel leakage ratio, caused by the transmitter non-idealities, the effect of receiver filtering is not included
ACTS	Advanced communication technologies and systems, EU research projects framework
AICH	Acquisition indicatio channel
ALCAP	Access link control application part
AM	Acknowledged mode
AMD	Acknowledged mode data
AMR	Adaptive multirate (speech codec)
ARIB	Association of radio industries and businesses (Japan)
ARQ	Automatic repeat request
ASC	Access service class
ASN.1	Abstract syntax notation one
ATM	Asynchronous transfer mode
AWGN	Additive white Gaussian noise
BB SS7	Broad band signalling system #7
BCCH	Broadcast control channel (logical channel)
BCFE	Broadcast control functional entity
BCH	Broadcast channel (transport channel)
BER	Bit error ratio
BLER	Block error ratio
BMC	Broadcast/multicast control protocol
BoD	Bandwidth on demand
BPSK	Binary phase shift keying
BS	Base station
BSS	Base station subsystem
BSC	Base station controller

CA-ICH	Channel assignment indication channel
CB	Cell broadcast
CBC	Cell broadcast center
CBS	Cell broadcast service
CCCH	Common control channel (logical channel)
CCH	Common transport channel
CCH	Control channel
CD-ICH	Collision detection indication channel
CDF	Cumulative distribution function
CDMA	Code division multiple access
CFN	Connection frame number
CIR	Carrier to interference ratio
CM	Connection management
CN	Core network
C-NBAP	Common NBAP
CODIT	Code division test bed, EU research project
CPCH	Common packet channel
CPICH	Common pilot channel
CRC	Cyclic redundancy check
CRNC	Controlling RNC
C-RNTI	Cell-RNTI, radio network temporary identity
CS	Circuit switched
CSICH	CPCH status indication channel
CTCH	Common traffic channel
CWTS	China wireless telecommunications standard group
DCA	Dynamic channel allocation
DCCH	Dedicated control channel (logical channel)
DCFE	Dedicated control functional entity
DCH	Dedicated channel (transport channel)
DECT	Digital enhanced cordless telephone
DF	Decision feedback
DL	Downlink
D-NBAP	Dedicated NBAP
DPCCH	Dedicated physical control channel
DPDCH	Dedicated physical data channel
DRNC	Drift RNC
DRX	Discontinuous reception
DS-CDMA	Direct spread code division multiple access
DSCH	Downlink shared channel
DTCH	Dedicated traffic channel
DTX	Discontinuous transmission
EDGE	Enhanced data rates for GSM evolution
EFR	Enhanced full rate speech codec
EIRP	Equivalent isotropic radiated power
EP	Elementary procedure
ETSI	European telecommunications standards institute

FACH	Forward access channel
FBI	Feedback information
FDD	Frequency division duplex
FDMA	Frequency division multiple access
FER	Frame error ratio
FP	Frame protocol
FPLMTS	Future public land mobile telecommunications system
FRAMES	Future radio wideband multiple access system, EU research project
FTP	File transfer protocol
GGSN	Gateway GPRS support node
GMSC	Gateway MSC
GPRS	General packet radio system
GPS	Global positioning system
GSIC	Groupwise serial interference cancellation
GSM	Global system for mobile communications
GTP-U	User plane part of GPRS tunnelling protocol
HLR	Home location register
IC	Interference cancellation
ID	Identity
IETF	Internet engineering task force
IMSI	International mobile subscriber identity
IMT-2000	International mobile telephony, 3rd generation networks are referred as IMT-2000 within ITU
IN	Intelligent network
IP	Internet protocol
IPI	Inter-path interference
IRC	Interference rejection combining
IS-2000	IS-95 evolution standard, (cdma2000)
IS-136	US-TDMA, one of the 2nd generation systems, mainly in Americas
IS-95	cdmaOne, one of the 2nd generation systems, mainly in Americas and in Korea
ISDN	Integrated services digital network
ISI	Inter-symbol interference
ITU	International telecommunications union
ITUN	SS7 ISUP Tunnelling
L2	Layer 2
LAI	Location area identity
LAN	Local area network
LCS	Location services
LP	Low pass
MA	Midamble
MAC	Medium access control
MAI	Multiple access interference
MAP	Maximum a posteriori
MCU	Multipoint control unit
ME	Mobile equipment

MF   Matched filter
MLSD   Maximum likelihood sequence detection
MM   Mobility management
MMSE   Minimum mean square error
MPEG   Motion picture experts group
MR-ACELP   Multirate ACELP
MS   Mobile station
MSC/VLR   Mobile services switching centre/visitor location register
MT   Mobile termination
MTP3b   Message transfer part (broadband)
MUD   Multiuser detection
NAS   Non access stratum
NBAP   Node B application part
NRT   Non-real time
ODMA   Opportunity driven multiple access
O&M   Operation and maintenance
OVSF   Orthogonal variable spreading factor
PAD   Padding
PC   Power control
PCCC   Parallel concatenated convolutional coder
PCCCH   Physical common control channel
PCCH   Paging channel (logical channel)
PCCPCH   Primary common control physical channel
PCH   Paging channel (transport channel)
PCPCH   Physical common packet channel
PCS   Personal communication systems, 2nd generation cellular systems mainly
              in Americas, operating partly on IMT-2000 band
PDC   Personal digital cellular, 2nd generation system in Japan
PDCP   Packet data converge protocol
PDP   Packet data protocol
PDSCH   Physical downlink shared channel
PDU   Protocol data unit
PER   Packed encoding rules
PHS   Personal handy phone system
PHY   Physical layer
PI   Page indicator
PIC   Parallel interference cancellation
PICH   Paging indicator channel
PLMN   Public land mobile network
PNFE   Paging and notification control function entity
PRACH   Physical random access channel
PS   Packet switched
PSCH   Physical shared channel
PSTN   Public switched telephone network
P-TMSI   Packet-TMSI
PU   Payload unit

PVC	Pre-defined virtual connection
QoS	Quality of service
QPSK	Quadrature phase shift keying
RAB	Radio access bearer
RACH	Random access channel
RAI	Routing area identity
RAN	Radio access network
RANAP	RAN application part
RB	Radio bearer
RF	Radio frequency
RLC	Radio link control
RNC	Radio network controller
RNS	Radio network sub-system
RNSAP	RNS application part
RNTI	Radio network temporary identity
RRC	Radio resource control
RRM	Radio resource management
RSSI	Received signal strength indicator
RSVP	Resource reservation protocol
RT	Real time
RTCP	Real time transport control protocol
RTP	Real time protocol
RTSP	Real time streaming protocol
RU	Resource unit
SAAL-NNI	Signalling ATM adaptation layer for network to network interfaces
SAAL-UNI	Signalling ATM adaptation layer for user to network interfaces
SAP	Service access point
SAP	Session announcement protocol
SCCP	Signalling connection control part
SCCPCH	Secondary common control physical channel
SCH	Synchronisation channel
SCTP	Simple control transmission protocol
SDD	Space division duplex
SDP	Session description protocol
SDU	Service data unit
SF	Spreading factor
SFN	System frame number
SGSN	Serving GPRS support node
SHO	Soft handover
SIB	System information block
SIC	Successive interference cancellation
SID	Silence indicator
SINR	Signal-to-noise ratio where noise includes both thermal noise and interference
SIP	Session initiation protocol
SIR	Signal to interference ratio

SM	Session management
SMS	Short message service
SN	Sequence number
SNR	Signal to noise ratio
SRB	Signalling radio bearer
SRNC	Serving RNC
SRNS	Serving RNS
SS7	Signalling system #7
SSCF	Service specific co-ordination function
SSCOP	Service specific connection oriented protocol
STD	Switched transmit diversity
STTD	Space time transmit diversity
TCH	Traffic channel
TCP	Transport control protocol
TCTF	Target channel type field
TD/CDMA	Time division CDMA, combined TDMA and CDMA
TDD	Time division duplex
TDMA	Time division multiple access
TE	Terminal equipment
TF	Transport format
TFCI	Transport format combination indicator
TFCS	Transport format combination set
TFI	Transport format indicator
TMSI	Temporary mobile subscriber identity
TPC	Transmission power control
TR	Transparent mode
TS	Technical specification
TSTD	Time switched transmit diversity
TTA	Telecommunications technology association (Korea)
TTC	Telecommunication technology commission (Japan)
TxAA	Transmit adaptive antennas
UDP	User datagram protocol
UE	User equipment
UL	Uplink
UM	Unacknowledged mode
UMTS	Universal mobile telecommunication system
URA	UTRAN registration area
URL	Universal resource locator
U-RNTI	UTRAN RNTI
USCH	Uplink shared channel
USIM	UMTS Subscriber identity module
US-TDMA	IS-136, one of the 2nd generation systems mainly in USA
UTRA	UMTS Terrestrial radio access (ETSI)
UTRA	Universal Terrestrial radio access (3GPP)
UTRAN	UMTS Terrestrial radio access network
VAD	Voice activation detection

VoIP	Voice over IP
WARC	World administrative radio conference
WCDMA	Wideband CDMA, Code division multiple access
WLL	Wireless local loop
WWW	World wide web
ZF	Zero forcing

1

Introduction

Harri Holma, Antti Toskala and Ukko Lappalainen

1.1 WCDMA in Third Generation Systems

Analog cellular systems are commonly referred to as first generation systems. The digital systems currently in use, such as GSM, PDC, cdmaOne (IS-95) and US-TDMA (IS-136), are second generation systems. These systems have enabled voice communications to go wireless in many of the leading markets, and customers are increasingly finding value also in other services such as text messaging and access to data networks, which are starting to grow rapidly.

Third generation systems are designed for multimedia communication: with them person-to-person communication can be enhanced with high quality images and video, and access to information and services on public and private networks will be enhanced by the higher data rates and new flexible communication capabilities of third generation systems. This, together with the continuing evolution of the second generation systems, will create new business opportunities not only for manufacturers and operators, but also for the providers of content and applications using these networks.

In the standardisation forums, WCDMA technology has emerged as the most widely adopted third generation air interface. Its specification has been created in 3GPP (the 3rd Generation Partnership Project), which is the joint standardisation project of the standardisation bodies from Europe, Japan, Korea, the USA and China. Within 3GPP, WCDMA is called UTRA (Universal Terrestrial Radio Access) FDD (Frequency Division Duplex) and TDD (Time Division Duplex), the name WCDMA being used to cover both FDD and TDD operation.

Throughout this book, the chapters related to specifications use the 3GPP terms UTRA FDD and TDD, the others using the term WCDMA. This book focuses on the WCDMA FDD technology. The WCDMA TDD mode and its differences from the WCDMA FDD mode are presented in Chapter 12.

WCDMA for UMTS, edited by Harri Holma and Antti Toskala
© 2001 John Wiley & Sons, Ltd

1.2   Air Interfaces and Spectrum Allocations
for Third Generation Systems

Work to develop third generation mobile systems started when the World Administrative
Radio Conference (WARC) of the ITU (International Telecommunications Union), at its
1992 meeting, identified the frequencies around 2 GHz that were available for use by future
third generation mobile systems, both terrestrial and satellite. Within the ITU these third
generation systems are called International Mobile Telephony 2000 (IMT-2000). Within
the IMT-2000 framework, several different air interfaces are defined for third generation
systems, based on either CDMA or TDMA technology, as described in Chapter 3. The
original target of the third generation process was a single common global IMT-2000 air
interface. Third generation systems are closer to this target than were second generation
systems: the same air interface—WCDMA—is to be used in Europe and Asia, including
Japan and Korea, using the frequency bands that WARC-92 allocated for the third generation
IMT-2000 system at around 2 GHz. In North America, however, that spectrum has already
been auctioned for operators using second generation systems, and no new spectrum is
available for IMT-2000. Thus, third generation services there must be implemented within
the existing bands by replacing part of the spectrum with third generation systems. This
approach is referred to as refarming. The global IMT-2000 spectrum is not available in
countries that follow the US PCS spectrum allocation.

In addition to WCDMA, the other air interfaces that can be used to provide third generation services are EDGE and multicarrier CDMA (cdma2000). EDGE (Enhanced Data Rates
for GSM Evolution) can provide third generation services with bit rates up to 500 kbps
within a GSM carrier spacing of 200 kHz [1]. EDGE includes advanced features that are
not part of GSM to improve spectrum efficiency and to support the new services. The multi-
carrier CDMA can be used as an upgrade solution for the existing IS-95 operators and will
be presented in more detail in Chapter 13.

The expected frequency bands and geographical areas where these different air interfaces
are likely to be applied are shown in Figure 1.1. Within each region there are local exceptions
in places where multiple technologies are already being deployed.

The spectrum allocation in Europe, Japan, Korea and the USA is shown in Figure 1.2.
In Europe and in most of Asia the IMT-2000 bands of 2×60 MHz (1920–1980 MHz
plus 2110–2170 MHz) will be available for WCDMA FDD. The availability of the TDD
spectrum varies: in Europe it is expected that 25 MHz will be available for licensed TDD
use in the 1900–1920 MHz and 2020–2025 MHz bands. The rest of the unpaired spectrum
is expected to be used for unlicensed TDD applications (SPA: Self Provided Applications)
in the 2010–2020 MHz band. FDD systems use different frequency bands for uplink and for
downlink, separated by the duplex distance, while TDD systems utilise the same frequency
for both uplink and downlink.

In Japan and Korea, the IMT-2000 FDD band is the same as in the rest of Asia and
in Europe. Japan has deployed PDC as a second generation system, while in Korea IS-
95 is used for both cellular and PCS operation. The PCS spectrum allocation in Korea
is different from the US PCS spectrum allocation, leaving the IMT-2000 spectrum fully
available in Korea. In Japan, part of the IMT-2000 TDD spectrum is used by PHS, the
cordless telephone system.

In China, there are reservations for PCS or WLL (Wireless Local Loop) use on one
part of the IMT-2000 spectrum, though these have not been allocated to any operators.

Figure 1.1. Expected air interfaces and spectrums for providing third generation services

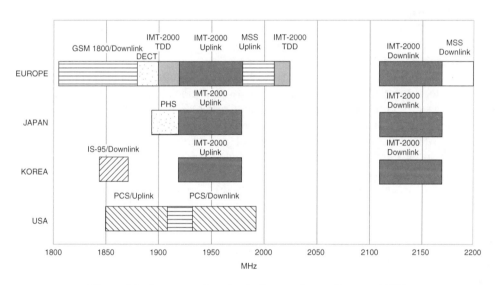

Figure 1.2. Spectrum allocation in Europe, Japan, Korea and USA

Depending on the regulation decisions, up to 2×60 MHz of the IMT-2000 spectrum will be available for WCDMA FDD use in China. The TDD spectrum is also available in China.

In the USA no new spectrum has yet been made available for third generation systems. Third generation services can be implemented by refarming third generation systems within the existing PCS spectrum. This will require replacing part of the existing second generation frequencies with third generation systems. For the US PCS band, all third generation alternatives can be considered, but EDGE has an advantage as a narrowband system. With EDGE less spectrum will need to be cleared to deploy third generation services. Multicarrier CDMA and WCDMA can also be considered for refarming.

Table 1.1. UMTS licenses by December 2000

Country	Number of operators	Number of FDD carriers (2 × 5 MHz) per operator	Number of TDD carriers (1 × 5 MHz) per operator
Finland	4	3	1
Japan	3	4	0
Spain	4	3	1
UK	5	2–3	0–1
Germany	6	2	0–1
Netherlands	5	2–3	0–1
Italy	5	2	1
Austria	6	2	0–2
Sweden	4	3	1
Norway	4	3	1

EDGE can be deployed within the existing GSM900 and GSM1800 frequencies where those frequencies are in use. These GSM frequencies are not available in Korea and Japan. The total band available for GSM900 operation is 2×25 MHz plus EGSM 2×10 MHz, and for GSM1800 operation 2×75 MHz. EGSM refers to the extension of the GSM900 band. The total GSM band is not available in all countries using the GSM system. Later, it will also be possible to refarm WCDMA to the GSM bands, but initially EDGE is the solution to providing third generation services within the GSM bands.

Licensing of the IMT-2000 spectrum is under way. The first IMT-2000 licences were granted in Finland in March 1999, and followed by Spain in March 2000. No auction was conducted in Finland or in Spain. Also, Sweden granted the licenses without auction in December 2000. However, in other countries, such as the UK, Germany and Italy, an auction similar to the US PCS spectrum auctions was conducted.

The UMTS licenses are shown in Table 1.1 in Japan and in those European countries where the licenses have been awarded by December 2000. The number of UMTS operators per country is between 4 and 6.

More frequencies have been identified for IMT-2000 in addition to the frequencies mentioned above. At the WARC-2000 meeting of the ITU in May 2000 the following frequency bands were identified for IMT-2000 use:

- 1710–1885 MHz
- 2500–2690 MHz
- 806–960 MHz

It is worth noting that some of the bands listed, especially below 2 GHz are partly used with systems like GSM. What shall be the exact duplexing arrangements etc. is under discussion at the moment.

1.3   Schedule for Third Generation Systems

European research work on WCDMA was initiated in the European Union research projects CODIT [2] and FRAMES [3], and also within large European wireless communications companies, at the start of the 1990s [4]. Those projects also produced WCDMA trial systems to evaluate link performance [5] and generated the basic understanding of WCDMA

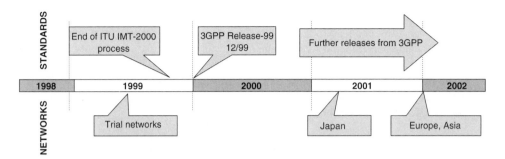

Figure 1.3. Standardisation and commercial operation schedule for WCDMA

necessary for standardisation. In January 1998 the European standardisation body ETSI decided upon WCDMA as the third generation air interface [6]. Detailed standardisation work has been carried out as part of the 3GPP standardisation process. The first full specification was completed at the end of 1999.

Commercial networks are scheduled to open in Japan during 2001, and in Europe and elsewhere in Asia at the beginning of 2002. The expected schedule is presented in Figure 1.3. This schedule relates to FDD mode operation. The TDD mode is expected to follow slightly later, and the first TDD networks will probably be based on the 3GPP Release-2000 version of the specifications. In Japan the schedule for TDD operation is also unclear, due to the unavailability of the TDD spectrum.

Looking back at the history of GSM, we note that since the opening of the first GSM network in July 1991 (Radiolinja, Finland) several countries have reached 50% cellular phone penetration. In some countries as much as 70% penetration has been reached. Second generation systems could already enable voice traffic to go wireless; now third generation systems face the challenge of making a new set of data services go wireless as well.

1.4   Differences between WCDMA and Second Generation Air Interfaces

In this section the main differences between the third and second generation air interfaces are described. GSM and IS-95 (the standard for cdmaOne systems) are the second generation air interfaces considered here. Other second generation air interfaces are PDC in Japan and US-TDMA mainly in the Americas; these are based on TDMA (time division multiple access) and have more similarities with GSM than with IS-95. The second generation systems were built mainly to provide speech services in macro cells. To understand the background to the differences between second and third generations systems, we need to look at the new requirements of the third generation systems which are listed below:

- Bit rates up to 2 Mbps
- Variable bit rate to offer bandwidth on demand
- Multiplexing of services with different quality requirements on a single connection, e.g. speech, video and packet data
- Delay requirements from delay-sensitive real-time traffic to flexible best-effort packet data

- Quality requirements from 10% frame error rate to 10^{-6} bit error rate
- Coexistence of second and third generation systems and inter-system handovers for coverage enhancements and load balancing
- Support of asymmetric uplink and downlink traffic, e.g. web browsing causes more loading to downlink than to uplink
- High spectrum efficiency
- Coexistence of FDD and TDD modes

Table 1.1 lists the main differences between WCDMA and GSM, and Table 1.2 those between WCDMA and IS-95. In this comparison only the air interface is considered. GSM also covers services and core network aspects, and this GSM platform will be used together with the WCDMA air interface: see the next section regarding core networks.

The differences in the air interface reflect the new requirements of the third generation systems. For example, the larger bandwidth of 5 MHz is needed to support higher bit rates. Transmit diversity is included in WCDMA to improve the downlink capacity to support the asymmetric capacity requirements between downlink and uplink. Transmit diversity is not supported by the second generation standards. The mixture of different bit rates, services and quality requirements in third generation systems requires advanced radio resource management algorithms to guarantee quality of service and to maximise system throughput. Also, efficient support of non-real-time packet data is important for the new services.

The main differences between WCDMA and IS-95 are discussed below. Both WCDMA and IS-95 utilise direct sequence CDMA. The higher chip rate of 3.84 Mcps in WCDMA gives more multipath diversity than the chip rate of 1.2288 Mcps, especially in small urban cells. The importance of diversity for system performance is discussed in Sections 9.2.1.2 and 11.2.1.3. Most importantly, increased multipath diversity improves the coverage. The higher chip rate also gives a higher trunking gain, especially for high bit rates, than do narrowband second generation systems.

WCDMA has fast closed-loop power control in both uplink and downlink, while IS-95 uses fast power control only in uplink. The downlink fast power control improves link performance and enhances downlink capacity. It requires new functionalities in the mobile, such as SIR estimation and outer loop power control, that are not needed in IS-95 mobiles.

Table 1.2. Main differences between WCDMA and GSM air interfaces

	WCDMA	GSM
Carrier spacing	5 MHz	200 kHz
Frequency reuse factor	1	1–18
Power control frequency	1500 Hz	2 Hz or lower
Quality control	Radio resource management algorithms	Network planning (frequency planning)
Frequency diversity	5 MHz bandwidth gives multipath diversity with Rake receiver	Frequency hopping
Packet data	Load-based packet scheduling	Time slot based scheduling with GPRS
Downlink transmit diversity	Supported for improving downlink capacity	Not supported by the standard, but can be applied

Table 1.3. Main differences between WCDMA and IS-95 air interfaces

	WCDMA	IS-95
Carrier spacing	5 MHz	1.25 MHz
Chip rate	3.84 Mcps	1.2288 Mcps
Power control frequency	1500 Hz, both uplink and downlink	Uplink: 800 Hz, downlink: slow power control
Base station synchronisation	Not needed	Yes, typically obtained via GPS
Inter-frequency handovers	Yes, measurements with slotted mode	Possible, but measurement method not specified
Efficient radio resource management algorithms	Yes, provides required quality of service	Not needed for speech only networks
Packet data	Load-based packet scheduling	Packet data transmitted as short circuit switched calls
Downlink transmit diversity	Supported for improving downlink capacity	Not supported by the standard

The IS-95 system was targeted mainly for macro cellular applications. The macro cell base stations are located on masts or rooftops where the GPS signal can be easily received. IS-95 base stations need to be synchronised and this synchronisation is typically obtained via GPS. The need for a GPS signal makes the deployment of the indoor and micro cells more problematic, since GPS reception is difficult without line-of-sight connection to the GPS satellites. Therefore, WCDMA is designed to operate with asynchronous base stations where no synchronisation from GPS is needed. The asynchronous base stations make the WCDMA handover slightly different from that of IS-95.

Inter-frequency handovers are considered important in WCDMA, to maximise the use of several carriers per base station. In IS-95 inter-frequency measurements are not specified, making inter-frequency handovers more difficult.

Experience from second generation air interfaces has been important in the development of the third generation interface, but there are many differences, as listed above. In order to make the fullest use of the capabilities of WCDMA, a deep understanding of the WCDMA air interface is needed, from the physical layer to network planning and performance optimisation.

1.5 Core Networks

There are three basic solutions for the core network to which WCDMA radio access networks can be connected. The basis of the second generation has been either the GSM core network or one based on IS-41. Both will naturally be important options in third generation systems. An emerging alternative is GPRS with an all-IP-based core network. The most typical connections between the core networks and the air interfaces are illustrated in Figure 1.4. Other connections are also possible and are expected to appear in the standardisation forums in due course.

The market needs will determine which combinations will be used by the operators. It is expected that operators will remain with their second generation core network for voice services and will then add packet data functionalities on top of that. Later, it will be possible to use IP-based core networks for all services.

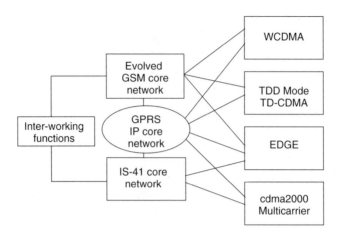

Figure 1.4. Core network relation to the third generation air interface alternatives

Because of the different technologies and frequency allocations, global roaming will continue to require specific arrangements between operators, such as multi-mode and multi-band handset and roaming gateways between the different core networks. To the end user the operator arrangements will not be visible, and global roaming terminals will probably emerge for those consumers willing to pay for global service.

References

[1] Pirhonen, R., Rautava, T. and Penttinen, J., 'TDMA Convergence for Packet Data Services', *IEEE Personal Communications Magazine*, June 1999, Vol. 6, No. 3, pp. 68–73.
[2] Andermo, P.-G. (ed.), 'UMTS Code Division Testbed (CODIT)', CODIT Final Review Report, September 1995.
[3] Nikula, E., Toskala, A., Dahlman, E., Girard, L. and Klein, A., 'FRAMES Multiple Access for UMTS and IMT-2000', *IEEE Personal Communications Magazine*, April 1998, pp. 16–24.
[4] Ojanperä, T., Rikkinen, K., Häkkinen, H., Pehkonen, K., Hottinen, A. and Lilleberg, J., 'Design of a 3rd Generation Multirate CDMA System with Multiuser Detection, MUD-CDMA', *Proc. ISSSTA'96*, Mainz, Germany, September 1996, pp. 334–338.
[5] Pajukoski, K. and Savusalo, J., 'Wideband CDMA Test System', *Proc. IEEE Int. Conf. on Personal Indoor and Mobile Radio Communications*, PIMRC'97, Helsinki, Finland, 1–4 September 1997, pp. 669–672.
[6] Holma, H., Toskala, A. and Latva-aho, M., 'Asynchronous Wideband CDMA for IMT-2000', *SK Telecom Journal,* South Korea, Vol. 8, No. 6, 1998, pp. 1007–1021.

2

UMTS Services and Applications

Jouni Salonen and Antti Toskala

2.1 Introduction

The best known new feature of UMTS is higher user bit rates: on circuit-switched connections 384 kbps, and on packet-switched connections up to 2 Mbps, can be reached. Higher bit rates naturally facilitate some new services, such as video telephony and quick downloading of data. If there is to be a killer application, it is most likely to be quick access to information and its filtering appropriate to the location of a user: see Figure 2.1. Often the requested information is on the Internet, which calls for effective handling of TCP/UDP/IP traffic in the UMTS network. At the start of the UMTS era almost all traffic will be voice, but later the share of data will increase. It is, however, difficult to predict the pace at which the share of data will start to dominate the overall traffic volume. At the same time that transition from voice to data occurs, traffic will move from circuit-switched connections to packet-switched connections. At the start of UMTS service not all of the Quality of Service (QoS) functions will be implemented, and therefore delay-critical applications such as speech and video telephony will be carried on circuit-switched bearers. Later, it will be possible to support delay-critical services as packet data with QoS functions.

Compared to GSM and other existing mobile networks, UMTS provides a new and important feature, namely it allows negotiation of the properties of a radio bearer. Attributes that define the characteristics of the transfer may include throughput, transfer delay and data error rate. To be a successful system, UMTS has to support a wide range of applications that possess different quality of service (QoS) requirements. At present it is not possible to predict the nature and usage of many of these applications. Therefore it is neither possible nor sensible to optimise UMTS to only one set of applications. UMTS bearers have to be generic by nature, to allow good support for existing applications and to facilitate the evolution of new applications. Since most of the telecommunications applications today are Internet or N-ISDN applications, it is natural that these applications and services dictate primarily the procedures for bearer handling.

WCDMA for UMTS, edited by Harri Holma and Antti Toskala
© 2001 John Wiley & Sons, Ltd

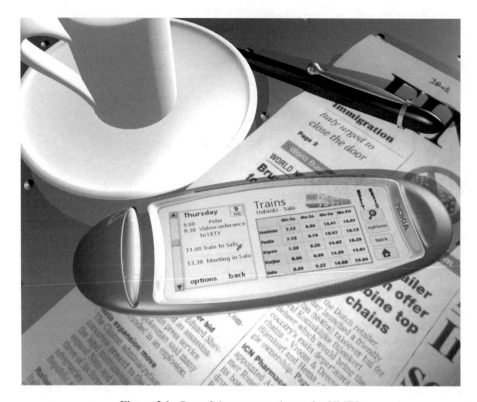

Figure 2.1. One of the concept phones for UMTS

2.2 UMTS Bearer Service

UMTS allows a user/application to negotiate bearer characteristics that are most appropriate for carrying information. It is also possible to change bearer properties via a bearer renegotiation procedure in the course of an active connection. Bearer negotiation is initiated by an application, while renegotiation may be initiated either by the application or by the network (e.g. in handover situations). An application-initiated negotiation is basically similar to a negotiation that occurs in the bearer establishment phase: the application requests a bearer depending on its needs, and the network checks the available resources and the user's subscription and then responds. The user either accepts or rejects the offer. The properties of a bearer affect directly the price of the service.

The bearer class, bearer parameters and parameter values are directly related to an application as well as to the networks that lie between the sender and the receiver. The set of parameters should be selected so that negotiation and renegotiation procedures are simple and unambiguous. In addition, parameters should allow easy policing and monitoring. The format and semantics will take into account the existing reservation protocols such as RSVP and those used in GPRS. Furthermore, the QoS concept should be flexible and versatile enough to allow bearer negotiation in the future with as yet unknown applications.

The layered architecture of a UMTS bearer service is depicted in Figure 2.2; each bearer service on a specific layer offers its individual services using those provided by the layers

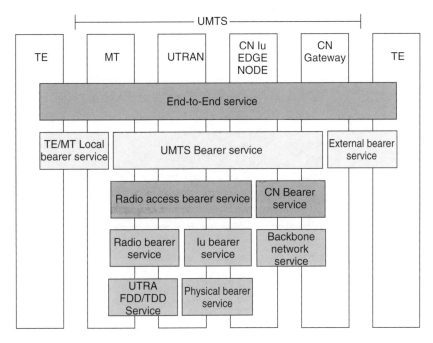

Figure 2.2. Architecture of a UMTS bearer service

below. As can be seen from the figure, the UMTS bearer service plays a major role in end-to-end service provisioning [1].

2.3   UMTS QoS Classes

In general, applications and services can be divided into different groups, depending on how they are considered. Like new packet-switched protocols, UMTS attempts to fulfil QoS requests from the application or the user. In UMTS four traffic classes have been identified:

— conversational,
— streaming,
— interactive, and
— background classes.

The main distinguishing factor between these classes is how delay-sensitive the traffic is: the conversational class is meant for very delay-sensitive traffic, while the background class is the most delay-insensitive. The UMTS QoS classes are summarised in Table 2.1.

In the initial phase of UMTS the conversational and streaming classes will be transmitted as real-time connections over the WCDMA air interface, while the interactive and background classes are transmitted as scheduled non-real-time packet data. The transmission of non-real-time packet data in WCDMA is described in detail in Chapter 10.

Table 2.1. UMTS QoS classes

Traffic class	Conversational class	Streaming class	Interactive class	Background
Fundamental characteristics	Preserve time relation (variation) between information entities of the stream Conversational pattern (stringent and low delay)	Preserve time relation (variation) between information entities of the stream	Request response pattern Preserve data integrity	Destination is not expecting the data within a certain time Preserve data integrity
Example of the application	Voice, videotelephony, video games	Streaming multimedia	Web browsing, network games	Background download of emails

2.3.1 Conversational Class

The best-known application of this class is speech service over circuit-switched bearers. With Internet and multimedia, a number of new applications will require this type, for example voice over IP and video telephony. Real-time conversation is always performed between peers (or groups) of live (human) end-users. This is the only type of the four where the required characteristics are strictly imposed by human perception.

Real-time conversation is characterised by the fact that the end-to-end delay is low and the traffic is symmetric or nearly symmetric. The maximum end-to-end delay is given by the human perception of video and audio conversation: subjective evaluations have shown that the end-to-end delay has to be less than 400 ms. Therefore the limit for acceptable delay is strict, as failure to provide sufficiently low delay will result in unacceptable quality.

2.3.1.1 AMR Speech Service

The speech codec in UMTS will employ the Adaptive Multi-rate (AMR) technique. The multi-rate speech coder is a single integrated speech codec with eight source rates: 12.2 (GSM-EFR), 10.2, 7.95, 7.40 (IS-641), 6.70 (PDC-EFR), 5.90, 5.15 and 4.75 kbps. The AMR bit rates are controlled by the radio access network and do not depend on the speech activity. To facilitate interoperability with existing cellular networks, some of the modes are the same as in existing cellular networks. The 12.2 kbps AMR speech codec is equal to the GSM EFR codec, 7.4 kbps is equal to the US-TDMA speech codec, and 6.7 kbps is equal to the Japanese PDC codec. The AMR speech coder is capable of switching its bit rate every 20 ms speech frame upon command. For AMR mode switching there are two candidates: in-band signalling or dedicated channel.

The AMR coder operates on speech frames of 20 ms corresponding to 160 samples at the sampling frequency of 8000 samples per second. The coding scheme for the multi-rate coding modes is the so-called Algebraic Code Excited Linear Prediction Coder (ACELP). The multi-rate ACELP coder is referred to as MR-ACELP. Every 160 speech samples, the speech signal is analysed to extract the parameters of the CELP model (LP filter coefficients, adaptive and fixed codebooks' indices and gains). The speech parameter bits delivered by the speech encoder are rearranged according to their subjective importance before they are sent to the network. The rearranged bits are further sorted based on their sensitivity to errors and are divided into three classes of importance: A, B and C. Class A is the most sensitive, and the strongest channel coding is used for class A bits in the air interface.

During a normal telephone conversation, the participants alternate so that, on the average, each direction of transmission is occupied about 50% of the time. The AMR has three basic functions to utilise effectively discontinuous activity:

- Voice Activity Detector (VAD) on the TX side.
- Evaluation of the background acoustic noise on the TX side, in order to transmit characteristic parameters to the RX side.
- The transmission of comfort noise information to the RX side is achieved by means of a Silence Descriptor (SID) frame, which is sent at regular intervals.
- Generation of comfort noise on the RX side during periods when no normal speech frames are received.

DTX has some obvious positive implications: in the user terminal battery life will be prolonged or a smaller battery could be used for a given operational duration. From the network point of view, the average required bit rate is reduced, leading to a lower interference level and hence increased capacity.

The AMR specification also contains error concealment. The purpose of frame substitution is to conceal the effect of lost AMR speech frames. The purpose of muting the output in the case of several lost frames is to indicate the breakdown of the channel to the user and to avoid generating possibly annoying sounds as a result of the frame substitution procedure [2] [3]. The AMR speech codec can tolerate about a 1% frame error rate (FER) of class A bits without any deterioration of speech quality. For class B and C bits a higher FER is allowed. The corresponding bit error rate (BER) of class A bits will be about 10^{-4}.

The bit rate of the AMR speech connection can be controlled by the radio access network depending on the air interface loading and the quality of the speech connections. During high loading, such as during busy hours, it is possible to use lower AMR bit rates to offer higher capacity while providing slightly lower speech quality. Also, if the mobile is running out of the cell coverage area and using its maximum transmission power, a lower AMR bit rate can be used to extend the cell coverage area. The uplink coverage of the AMR speech codec is discussed in Section 11.2.1. With the AMR speech codec it is possible to achieve a trade-off between the network's capacity, coverage and speech quality according to the operator's requirements.

2.3.1.2 Video Telephony

Video telephony has similar delay requirements to speech service. Due to the nature of video compression, the BER requirement is more stringent than that of speech. UMTS has specified that ITU-T Rec. H.324M should be used for video telephony in circuit-switched connections [4]. At the moment there are two video telephony candidates for packet-switched connections: ITU-T Rec. H.323 and IETF SIP.

ITU-T Rec. H.324

Originally Rec. H.324 was intended for multimedia communication over a fixed telephone network, i.e. PSTN. It is specified that for PSTN connections, a synchronous V.34 modem is used. Later on, when wireless networks evolved, mobile extensions were added to the

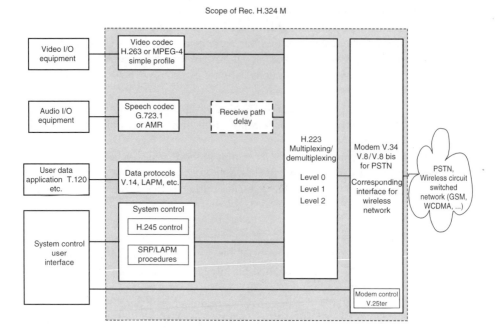

Figure 2.3. Scope of ITU Rec. H.324

specification to make the system more robust against transmission errors. The overall picture of the H.324 system is shown in Figure 2.3.

H.324 consists of the following mandatory elements: H.223 for multiplexing and H.245 for control. Elements that are optional but are typically employed are H.263 video codec, G.723.1 speech codec, and V.8bis. Later, MPEG-4 video and AMR were added as optional codecs into the system. The recommendation defines the seven phases of a call: set-up, speech only, modem training, initialisation, message, end, and clearing. Level 0 of H.223 multiplexing is exactly the same as that of H.324, thus providing backward compatibility with older H.324 terminals. With a standardised negotiation procedure the terminal can adapt to the prevailing radio link conditions by selecting the appropriate error resiliency level.

V.8bis contains procedures for the identification and selection of common modes of operation between data circuit-terminating equipment (DCE) and between data terminal equipment (DTE) over general switched telephone network and leased point-to-point telephone types. The basic features of V.8bis are as follows:

- It allows a desired communication mode to be selected by either the calling or the answering station.
- It allows terminals to automatically identify common operating modes (applications).
- It enables automatic selection between multiple terminals that share a common telephone circuit.
- It provides user-friendly switching from normal voice telephony to a modem-based communication mode.

The capabilities exchange feature of V.8bis permits a list of communication modes, as well as software applications, to be exchanged between terminals. Each terminal is therefore able to establish the modes of operation it shares with the remote station. A capability exchange between stations thus ensures, *a priori*, that a selected communication mode is possible. Attempts to establish incompatible modes of operation are thus avoided, which speeds up the application level connection.

As with the mode selection procedure, a capabilities exchange may be performed either at call set-up, automatically under the control of either the calling or the answering station, or during the course of telephony. In the latter case, on completion of the information exchange, the communication link may be configured either to return to voice telephony mode or to adopt immediately one of the common modes of communication.

V.8bis has been designed so that, when a capabilities exchange takes place in telephony mode, and the capabilities exchanged are limited to standard features, the interruption in voice communications is short (less than approximately 2 seconds) and as unobtrusive as possible.

In order to guarantee seamless data services between UMTS and PSTN, the call control mechanism of UMTS should take the V.8bis messages into account. V.8bis messages should be interpreted and converted into UMTS messages and vice versa.

One of the recent developments of H.324 is an operating mode that makes it possible to use an H.324 terminal over ISDN links. This mode of operation is defined in Annex D of the H.324 recommendation and is also referred to as H.324/I. H.324/I terminals use the I.400 series ISDN user-network interface in place of the V.34 modem. The output of the H.223 multiplex is applied directly to each bit of the digital channel, in the order defined by H.223. Operating modes are defined bit rates ranging from 56 kbps to 1920 kbps, so that H.324/I allows the use of several 56 or 64 kbps links at the same time.

H.324/I provides direct interoperability with H.320 terminals, H.324 terminals on the GSTN (using GSTN modems), H.324 terminals operating on ISDN through user substitution of I.400 series ISDN interfaces for V.34 modems, and voice telephones (both GSTN and ISDN). H.324/I terminals support H.324/Annex F (= V.140) which is for establishing communication between two multiprotocol audio-visual terminals using digital channels at a multiple of 64 or 56 kbps [5].

ITU-T Rec. H.323

In H.323 (see Figure 2.4), logical channels are multiplexed at the destination port transport address level. The transport address is the combination of a network address and the port that identifies a transport level endpoint, for example an IP address and a UDP port. Packets are transmitted from a source transport address to a destination transport address. For example, each logical channel for H.245, T.120 data, audio, video, and RTCP is sent to a separate destination transport address. Packets of different payload types are sent to different transport addresses, eliminating the need for a separate multiplexing /demultiplexing layer in H.225.0. If a participant agrees to the H.245 control protocol to open a logical audio and/or video channel, the receiver terminal sends over the H.245 control channel the port where it wishes to receive the corresponding bitstream. The data stream will be transmitted to that port as UDP datagrams [6].

It should be noted that the current specification of GPRS does not support more than one PDP context per IP address. In order to fully exploit the nuances of UMTS, port-level

Scope of Rec. H.323

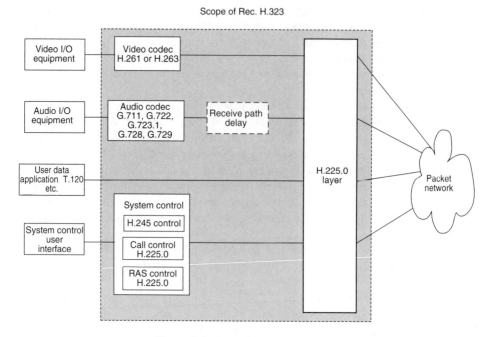

Figure 2.4. Scope of ITU Rec. H.323

context should be supported. This is to make it possible to specify portwise QoS, i.e. dedicated bearer for audio and for video.

H.225.0 makes use of RTP/RTCP (real-time transport protocol/real-time transport control protocol) for media stream packetisation and synchronisation for all underlying LANs. The usage of RTP/RTCP is tied to the usage of UDP/TCP/IP.

It is assumed that bit errors are detected in the lower layers, and erroneous packets are not sent up to H.225.0. H.225.0 terminals shall be capable of sending and receiving audio and video on separate transport addresses using separate instances of RTP to allow for media-specific frame sequence numbers and separate QoS treatment for each medium.

If both audio and video media are used in a conference, they are transmitted as separate RTP sessions; RTCP packets are transmitted for each medium using two different UDP port pairs and/or multicast addresses. There is no direct coupling at the RTP level between audio and video sessions, except that a user participating in both sessions should use the same distinctive name in the RTCP packets for both so that the sessions can be associated. Despite the separation, synchronised playback of a source's audio and video can be achieved using timing information carried in the RTCP packets for both sessions.

Establishment of a point-to-point H.323 conference requires two TCP connections between the two terminals, one for call setup and the other for conference control and capability exchange. The initial connection is made from the caller to a well-known port on the callee. This connection carries the call setup messages defined in H.225.0, and is commonly called the Q.931 channel. Upon receipt of the incoming call, the callee listens to a TCP connection on a dynamic port; the callee communicates this port in the acceptance message. The caller then establishes the second TCP connection to that port. The second connection carries the conference control messages defined in H.245. Once the H.245 channel is established, the first connection is no longer necessary (in a simple conference environment) and may be closed by either endpoint.

The H.245 channel is used by the terminals to exchange audio and video capabilities and perform master/slave determination. It is then used to signal opening logical channels for audio and video, which causes RTP sessions to be created for the media streams. The H.245 channel remains open for the duration of the conference. It is used to signal the end of the conference.

IETF Multimedia Architecture

Current IP video telephony visions are based primarily on the ITU-T H.323 standard. However, this is not necessarily the best solution to be used on the Internet and particularly on wireless links due to its complex control signalling. H.323 employs H.245 control signalling and ASN.1/PER binary coding. Instead, IETF, the main body defining standards for the Internet, has been using text-based signalling protocols. Now these simple IETF conferencing *de facto* standards and experiences have led to a vision called IETF Multimedia Architecture (see Figure 2.5), which covers several areas:

- SIP (Session Initiation Protocol): Signalling protocol to be used instead of H.323/H.245.
- SAP (Session Announcement Protocol): Multicast announcement protocol (advertises Internet A/V sessions such as pop concerts, lectures, etc.). Current MBone is based on this.

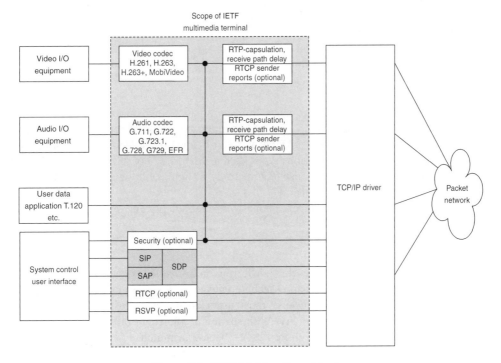

Figure 2.5. IETF Multimedia architecture

- SDP (Session Description Protocol): Text-based syntax to describe sessions (replaces ASN.1/BER in H.323).
- RTSP (Real Time Streaming Protocol): Protocol for controlling remote servers (e.g. VOD servers to play a file).

RTP is used for media encapsulation and RTCP for control information delivery and lip sync purposes. Multiparty application sharing is also possible: at least shared workspace and a network text editor have been developed. Both approaches are based on reliable multicast. SIP provides the necessary protocol mechanisms so that end systems and proxy servers can provide services:

— call forwarding, call-forwarding no answer;
— call-forwarding busy;
— call-forwarding unconditional;
— other address translation services;
— callee and calling number delivery, where numbers can be any (preferably a unique) naming scheme;
— personal mobility, i.e., the ability to reach a called party under a single, location-independent address even when the user changes terminals;
— terminal-type negotiation and selection.

Callers can be given a choice how to reach the party, e.g., via Internet telephony, mobile phone, an answering service, etc.:

— terminal capability negotiation;
— caller and callee authentication;
— blind and supervised call transfer;
— invitations to multicast conferences.

Extensions of SIP are available to allow third-party signalling, for example for click-to-dial services, fully meshed conferences and connections to multipoint control units (MCUs), as well as mixed modes and the transition between them. SIP is addressing-neutral, with addresses expressed as URLs of various types such as SIP, H.323 or telephone (E.164). SIP is independent of the packet layer and requires only an unreliable datagram service, as it provides its own reliability mechanism [7].

Figure 2.6 shows one of the concept phones for video telephony.

2.3.2   Streaming Class

Multimedia streaming is a technique for transferring data such that it can be processed as a steady and continuous stream. Streaming technologies are becoming increasingly important with the growth of the Internet because most users do not have fast enough access to download large multimedia files quickly. With streaming, the client browser or plug-in can start displaying the data before the entire file has been transmitted.

For streaming to work, the client side receiving the data must be able to collect the data and send it as a steady stream to the application that is processing the data and converting

Figure 2.6. 3G concept phone for video telephony

it to sound or pictures. Streaming applications are very asymmetric and therefore typically withstand more delay than more symmetric conversational services. This also means that they tolerate more jitter in transmission. Jitter can be easily smoothed out by buffering.

Internet video products and the accompanying media industry as a whole is clearly divided into two different target areas: (1) Web broadcast and (2) video streaming on-demand. Web broadcast providers usually target very large audiences that connect to a highly performance-optimised media server (or choose from a multitude of servers) via the actual Internet, which at present is very slow. The on-demand services are more often used by big corporations that wish to store video clips or lectures to a server connected to a higher bandwidth local intranet—these on-demand lectures are seldom used simultaneously by more than hundreds of people.

Both application types use basically similar core video compression technology, but the coding bandwidths, level of tuning within network protocol use, and robustness of server technology needed for broadcast servers differ from the technology used in on-demand, smaller-scale systems. This has led to a situation where the few major companies developing and marketing video streaming products have specialised their end-user products to meet the needs of these two target groups. Basically, they have optimised their core products differently: those directed to the '28.8 kbps market' for bandwidth variation-sensitive streaming over the Internet and those for the 100–7300 kbps intranet market.

At the receiver the streaming data or video clip is played by a suitable independent media player application or a browser plug-in. Plug-ins can be downloaded from the Web, usually free of charge, or may be readily bundled to a browser. This depends largely on the browser and its version in use—new browsers tend to have integrated plug-ins for the most popular streaming video players.

In conclusion, a client player implementation in a mobile system seems to lead to an application-level module that could handle video streams independently (with independent connection and playback activation) or in parallel with the browser application when the service is activated from the browser. The module would interface directly to the socket interface of applied packet network protocol layers, here most likely UDP/IP or TCP/IP [8].

2.3.3   Interactive Class

When the end-user, either a machine or a human, is on line requesting data from remote equipment (e.g. a server), this scheme applies. Examples of human interaction with the remote equipment are Web browsing, database retrieval, and server access. Examples of machine interaction with remote equipment are polling for measurement records and automatic database enquiries (tele-machines).

Interactive traffic is the other classical data communication scheme that is broadly characterised by the request response pattern of the end-user. At the message destination there is an entity expecting the message (response) within a certain time. Round-trip delay time is therefore one of the key attributes. Another characteristic is that the content of the packets must be transparently transferred (with low bit error rate).

2.3.3.1   Location-based Services

It is easy to predict that location-based services and applications will become one of the new dimensions in UMTS. A location-based service is provided either by a teleoperator or by a third party service provider that utilises available information on the terminal location. The service is either push (e.g. automatic distribution of local information) or pull type (e.g. localisation of emergency calls). Other possible location-based services are discount calls in a certain area, broadcasting of a service over a limited number of sites (broadcasting video on demand), and retrieval and display of location-based information, such as the location of the nearest gas stations, hotels, restaurants, and so on. Figure 2.7 shows an example. Depending on the service, the data may be retrieved interactively or as background. For instance, before travelling to an unknown city abroad one may request night-time download of certain points of interest from the city. The downloaded information typically contains a map and other data to be displayed on top of the map. By clicking the icon on the map, one gets information from the point. Information to be downloaded background or interactively can be limited by certain criteria and personal interest.

The location information can be input by the user or detected by the network or mobile station. Release-99 of UMTS specifies the following positioning methods:

— the cell coverage-based positioning method,
— Observed Time Difference Of Arrival—Idle Period DownLink (OTDOA-IPDL),
— network-assisted GPS methods.

These methods are complementary rather than competing, and are suited for different purposes. A requested location is given according to the requirements set by a number of attributes, for instance horizontal and vertical accuracy, response time and priority, as well

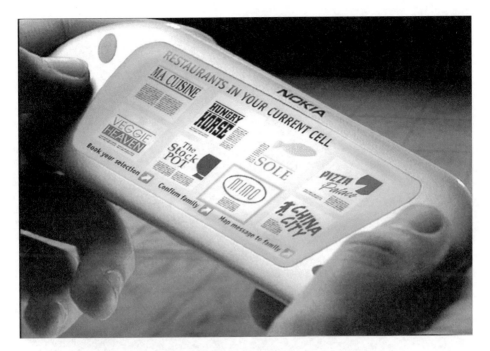

Figure 2.7. 3G concept phone showing location-based service

as security. Accuracy is obviously the most important. The measurement of position is a statistical process and not all measurements of the same location will yield the same result. The overall system accuracy reported involves a statistical measure of many operations at many times and at many locations through the UTRAN coverage area. The accuracy reported for an individual measurement may vary considerably from the overall system performance statistics. At the moment UMTS specifies that it will provide location information for a mobile station to an accuracy of 50 m [9][10].

2.3.3.2   Computer Games

Playing a computer game interactively across the network is one example of applications that can be seen to be part of the interactive class. However, depending on the nature of a game, i.e. how intensive data transfer is, it may rather belong to the conversational class due to high requirements for the maximum allowed end-to-end delay.

2.3.4   Background Class

Data traffic of applications such as e-mail delivery, SMS, downloading of databases and reception of measurement records can be delivered background since such applications do not require immediate action. The delay may be seconds, tens of seconds or even minutes.

Background traffic is one of the classical data communication schemes that is broadly characterised by the fact that the destination is not expecting the data within a certain time. It is thus more or less insensitive to delivery time. Another characteristic is that the content

Figure 2.8. 3G concept phone showing electronic postcard application

of the packets does not need to be transparently transferred. Data to be transmitted has to be received error free.

The electronic postcard (see Figure 2.8) is one example of new applications that are gradually becoming more and more common. It is easy to predict that once terminals have built-in cameras and large colour displays become small enough, the electronic postcard will soon take off.

2.4   Service Capabilities with Different Terminal Classes

In WCDMA the same principle as with GSM with Terminal class mark is not used. WCDMA terminals shall tell the network, upon connection set up, larger set of parameters indicating the radio access capabilities of the particular terminal. These capabilities determine e.g. what is the maximum user data rate supported in particular radio configuration, given independently for the uplink and downlink directions. To provide guidance on which capabilities should be applied together, reference terminal radio access capability combinations have been specified in 3GPP standardisation, see [11]. The following reference combinations have been defined for Release'99:

- 32 kbps class: This is intended to provide basic speech service, including AMR speech as well as some limited data rate capabilities up to 32 kbps

- 64 kbps class: This is intended to provide speech and data service, with also simultaneous data and AMR speech capability
- 144 kbps class: This class has the air interface capability to provide for example video telephony or then various other data services
- 384 kbps class is being further enhanced from 144 kbps and has for example multicode capability which points toward support of advanced packet data methods provided in WCDMA
- 768 kbps class has been defined as an intermediate step between 384 kbps and 2 Mbps class
- 2 Mbps class: This is the state of the art class and has been defined for downlink direction only

These classes are defined so that a higher class has all the capabilities covered by a lower class. It should be noted that terminals may deviate from these classes when giving their parameters to the network, thus 2 Mbps is possible for the uplink also though not covered by any of the classes directly.

2.5   Concluding Remarks

In this chapter we have briefly looked at UMTS from the perspective of services and applications. The list is by no means complete, but hopefully it helps readers to understand the variety of different services and gives some flavour of what we will see just a few years from now. UMTS provides high bit rates for both circuit-switched and packet-switched connections, effective bearer handling, multicall, and many other new features to make it possible to create new applications in a cost-efficient manner.

References

[1] 3GPP, Technical Specification Group Services and System Aspects, QoS Concept (3G TR 23.907 version 1.3.0), 1999.
[2] 3GPP, Mandatory Speech Codec Speech Processing Functions, AMR Speech Codec: General Description (3G TS 26.071 version 3.0.1), 1999.
[3] 3GPP, Mandatory Speech Codec Speech Processing Functions, AMR Speech Codec: Frame Structure General Description (3G TS 26.101 version 1.4.0), 1999.
[4] 3GPP, Technical Specification Group Services and System Aspects, Codec for Circuit Switched Multimedia Telephony Service: General Description (3G TS 26.110 version 3.0.1), 1999.
[5] ITU-T H.324, Terminal for Low Bitrate Multimedia Communication, 1998.
[6] ITU-T H.323v2, Packet Based Multimedia Communications Systems, 1998.
[7] Handley, M., et al., SIP: Session Initiation Protocol, RFC2543, IETF, 1999.
[8] Honko, H., Internet Video Prestudy, 1997.
[9] 3GPP, Technical Specification Group (TSG) RAN, Working Group 2 (WG2), Stage 2 Functional Specification of Location Services in URAN (3G TR 25.923 version 1.4.0), 1999.
[10] 3GPP, Technical Specification Group Services and System Aspects, Location Services (LCS), Service description, Stage 1 (3G TS 22.071 version 3.1.0), 1999.
[11] 3GPP, Technical Specification Group (TSG) RAN, Working Group 2 (WG2), UE Radio Access Capabilities, 3G TS 25.306 version 3.0.0, 2000.

He is Alive!

A Special Easter Week-end Featuring:

❖ **Fire-Proof** (A Film Show and a Couples' Night out with the Lord -Good Friday March 29, 2013 at 7.30 pm)

❖ **Resurrection Power** (Special Family Service, Easter Sunday, March 31, 2013 at 10 am)

❖ **Tabitha & Royal Priesthood** (Singles' Outreach Visit to the Hilcrest Nursing Home, Easter Monday April 1, 2013 at 2 pm)

At The Garden of the Lord Parish Unit 1, Clydesmuir Industrial Estate, Tremorfa, Cardiff, CF24 2QS

3

Introduction to WCDMA

Peter Muszynski and Harri Holma

3.1 Introduction

This chapter introduces the principles of the WCDMA air interface. Special attention is drawn to those features by which WCDMA differs from GSM and IS-95. The main parameters of the WCDMA physical layer are introduced in Section 3.2. The concept of spreading and despreading is described in Section 3.3, followed by a presentation of the multipath radio channel and Rake receiver in Section 3.4. Other key elements of the WCDMA air interface discussed in this chapter are power control and soft and softer handovers. The need for power control and its implementation are described in Section 3.5, and soft and softer handover in Section 3.6.

3.2 Summary of Main Parameters in WCDMA

We present the main system design parameters of WCDMA in this section and give brief explanations for most of them. Table 3.1 summarises the main parameters related to the WCDMA air interface. Here we highlight some of the items that characterise WCDMA.

- WCDMA is a wideband Direct-Sequence Code Division Multiple Access (DS-CDMA) system, i.e. user information bits are spread over a wide bandwidth by multiplying the user data with quasi-random bits (called chips) derived from CDMA spreading codes. In order to support very high bit rates (up to 2 Mbps), the use of a variable spreading factor and multicode connections is supported. An example of this arrangement is shown in Figure 3.1.

- The chip rate of 3.84 Mcps used leads to a carrier bandwidth of approximately 5 MHz. DS-CDMA systems with a bandwidth of about 1 MHz, such as IS-95, are commonly referred to as narrowband CDMA systems. The inherently wide carrier bandwidth of WCDMA supports high user data rates and also has certain performance benefits, such as increased multipath diversity. Subject to his operating licence, the network operator can deploy multiple such 5 MHz carriers to increase capacity, possibly in the form of

WCDMA for UMTS, edited by Harri Holma and Antti Toskala
© 2001 John Wiley & Sons, Ltd

Table 3.1. Main WCDMA parameters

Multiple access method	DS-CDMA
Duplexing method	Frequency division duplex/time division duplex
Base station synchronisation	Asynchronous operation
Chip rate	3.84 Mcps
Frame length	10 ms
Service multiplexing	Multiple services with different quality of service requirements multiplexed on one connection
Multirate concept	Variable spreading factor and multicode
Detection	Coherent using pilot symbols or common pilot
Multiuser detection, smart antennas	Supported by the standard, optional in the implementation

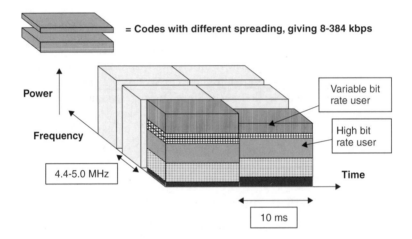

Figure 3.1. Allocation of bandwidth in WCDMA in the time-frequency-code space

hierarchical cell layers. Figure 3.1 also shows this feature. The actual carrier spacing can be selected on a 200 kHz grid between approximately 4.4 and 5 MHz, depending on interference between the carriers.

- WCDMA supports highly variable user data rates, in other words the concept of obtaining Bandwidth on Demand (BoD) is well supported. Each user is allocated frames of 10 ms duration, during which the user data rate is kept constant. However, the data capacity among the users can change from frame to frame. Figure 3.1 also shows an example of this feature. This fast radio capacity allocation will typically be controlled by the network to achieve optimum throughput for packet data services.

- WCDMA supports two basic modes of operation: Frequency Division Duplex (FDD) and Time Division Duplex (TDD). In the FDD mode, separate 5 MHz carrier frequencies are used for the uplink and downlink respectively, whereas in TDD only one 5 MHz is time-shared between uplink and downlink. Uplink is the connection from the mobile to the base station, and downlink is that from the base station to the mobile. The TDD mode is based heavily on FDD mode concepts and was added in order to leverage the basic WCDMA system also for the unpaired spectrum allocations of the ITU for the IMT-2000 systems. The TDD mode is described in detail in Chapter 12.

- WCDMA supports the operation of asynchronous base stations, so that unlike in the synchronous IS-95 system there is no need for a global time reference, such as a GPS. Deployment of indoor and micro base stations is easier when no GPS signal needs to be received.

- WCDMA employs coherent detection on uplink and downlink based on the use of pilot symbols or common pilot. While already used on the downlink in IS-95, the use of coherent detection on the uplink is new for public CDMA systems and will result in an overall increase of coverage and capacity on the uplink.

- The WCDMA air interface has been crafted in such a way that advanced CDMA receiver concepts, such as multiuser detection and smart adaptive antennas, can be deployed by the network operator as a system option to increase capacity and/or coverage. In most second generation systems no provision has been made for such receiver concepts and as a result they are either not applicable or can be applied only under severe constraints with limited increases in performance.

- WCDMA is designed to be deployed in conjunction with GSM. Therefore, handovers between GSM and WCDMA are supported in order to be able to leverage the GSM coverage for the introduction of WCDMA.

In the following sections of this chapter we will briefly review the generic principles of CDMA operation. In the subsequent chapters, the above mentioned aspects specific to the WCDMA standard will be presented and explained in more detail. The basic CDMA principles are also described in references [1], [2], [3] and [4].

3.3   Spreading and Despreading

Figure 3.2 depicts the basic operations of spreading and despreading for a DS-CDMA system.

User data is here assumed to be a BPSK-modulated bit sequence of rate R, the user data bits assuming the values of ±1. The spreading operation, in this example, is the multiplication of each user data bit with a sequence of 8 code bits, called chips. We assume this also for the BPSK spreading modulation. We see that the resulting spread data is at a rate of $8 \times R$ and has the same random (pseudo-noise-like) appearance as the spreading code. In this case we would say that we used a spreading factor of 8. This wideband signal would then be transmitted across a wireless channel to the receiving end.

During despreading we multiply the spread user data/chip sequence, bit duration by bit duration, with the very same 8 code chips as we used during the spreading of these bits. As shown, the original user bit sequence has been recovered perfectly, provided we have (as shown in Figure 3.2) also perfect synchronisation between the spread user signal and the (de)spreading code.

The increase of the signalling rate by a factor of 8 corresponds to a widening (by a factor of 8) of the occupied spectrum of the spread user data signal. Due to this virtue,

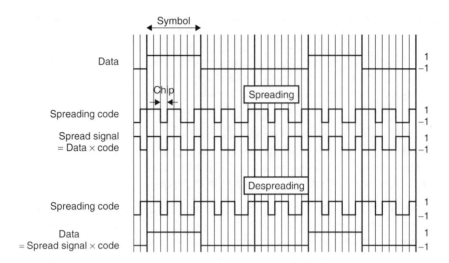

Figure 3.2. Spreading and despreading in DS-CDMA

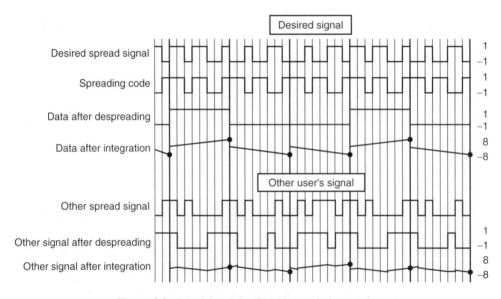

Figure 3.3. Principle of the CDMA correlation receiver

CDMA systems are more generally called spread spectrum systems. Despreading restores a bandwidth proportional to R for the signal.

The basic operation of the correlation receiver for CDMA is shown in Figure 3.3. The upper half of the figure shows the reception of the desired own signal. As in Figure 3.2, we see the despreading operation with a perfectly synchronised code. Then, the correlation receiver integrates (i.e. sums) the resulting products (data × code) for each user bit.

The lower half of Figure 3.3 shows the effect of the despreading operation when applied to the CDMA signal of another user whose signal is assumed to have been spread with a different spreading code. The result of multiplying the interfering signal with the own code and integrating the resulting products leads to interfering signal values lingering around 0.

As can be seen, the amplitude of the own signal increases on average by a factor of 8 relative to that of the user of the other interfering system, i.e. the correlation detection has raised the desired user signal by the spreading factor, here 8, from the interference present in the CDMA system. This effect is termed 'processing gain' and is a fundamental aspect of all CDMA systems, and in general of all spread spectrum systems. Processing gain is what gives CDMA systems the robustness against self-interference that is necessary in order to reuse the available 5 MHz carrier frequencies over geographically close distances. Let's take an example with real WCDMA parameters. Speech service with a bit rate of 12.2 kbps has a processing gain of 25 dB $= 10 \times \log_{10}(3.84e6/12.2e3)$. After despreading, the signal power needs to be typically a few decibels above the interference and noise power. The required power density over the interference power density after despreading is designated as E_b/N_0 in this book, where E_b is the energy, or power density, per user bit and N_0 is the interference and noise power density. For speech service E_b/N_0 is typically in the order of 5.0 dB, and the required wideband signal-to-interference ratio is therefore 5.0 dB minus the processing gain $= -20.0$ dB. In other words, the signal power can be 20 dB under the interference or thermal noise power, and the WCDMA receiver can still detect the signal. The wideband signal-to-interference ratio is also called the carrier-to-interference ratio C/I. Due to spreading and despreading, C/I can be lower in WCDMA than, for example, in GSM. A good quality speech connection in GSM requires $C/I = 9 - 12$ dB.

Since the wideband signal can be below the thermal noise level, its detection is difficult without knowledge of the spreading sequence. For this reason, the spread spectrum systems have their origin in military applications where the wideband nature of the signal allows it to be hidden below the omnipresent thermal noise.

Note that within any given channel bandwidth (chip rate) we will have a higher processing gain for lower user data bit rates than for high bit rates. In particular, for user data bit rates of 2 Mbps, the processing gain is less than $2(= 3.84$ Mcps $\div 2$ Mbps $= 1.92$ which corresponds to 2.8 dB) and some of the robustness of the WCDMA waveform against interference is clearly compromised. The performance of high bit rates with WCDMA is presented in Section 11.4.

Both base stations as well as mobiles for WCDMA use essentially this type of correlation receiver. However, due to multipath propagation (and possibly multiple receive antennas), it is necessary to use multiple correlation receivers in order to recover the energy from all paths and/or antennas. Such a collection of correlation receivers, termed 'fingers', is what comprises the CDMA Rake receiver. We will describe the operation of the CDMA Rake receiver in further detail in the following section, but before doing so, we make some final remarks regarding the transformation of spreading/despreading when used for wireless systems.

It is important to understand that spreading/despreading by itself does not provide any signal enhancement for wireless applications. Indeed, the processing gain comes at the price of an increased transmission bandwidth (by the amount of the processing gain).

All the WCDMA benefits come rather 'through the back door' by the wideband properties of the signals when examined at the system level, rather than the level of an individual radio link:

1. The processing gain, together with the wideband nature, suggest a frequency reuse
    of 1 between different cells of a wireless system (i.e. a frequency is reused in every
    cell/sector). This feature can be used to obtain high spectral efficiency.

2. Having many users share the same wideband carrier for their communications provides
    interferer diversity, i.e. the multiple access interference from many system users is
    averaged out, and this again will boost capacity compared to systems where one has to
    plan for the worst-case interference.

3. However, both the above benefits require the use of tight power control and soft
    handover to avoid one user's signal blocking the other's communications. Power control
    and soft handover will be explained later in this chapter.

4. With a wideband signal, the different propagation paths of a wireless radio signal can
    be resolved at higher accuracy than with signals at a lower bandwidth. This results in
    a higher diversity content against fading, and thus improved performance.

3.4   Multipath Radio Channels and Rake Reception

Radio propagation in the land mobile channel is characterised by multiple reflections, diffrac-
tions and attenuation of the signal energy. These are caused by natural obstacles such as
buildings, hills, and so on, resulting in so-called multipath propagation. There are two effects
resulting from multipath propagation that we are concerned with in this section:

1. The signal energy (pertaining, for example, to a single chip of a CDMA waveform) may
    arrive at the receiver across clearly distinguishable time instants. The arriving energy
    is 'smeared' into a certain multipath delay profile: see Figure 3.4, for example. The
    delay profile extends typically from 1 to 2 μs in urban and suburban areas, although
    in some cases delays as long as 20 μs or more with significant signal energy have
    been observed in hilly areas. The chip duration at 3.84 Mcps is 0.26 μs. If the time
    difference of the multipath components is at least 0.26 μs, the WCDMA receiver can

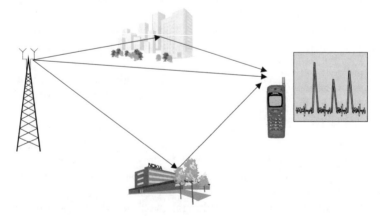

Figure 3.4. Multipath propagation leads to a multipath delay profile

separate those multipath components and combine them coherently to obtain multipath diversity. The 0.26 μs delay can be obtained if the difference in path lengths is at least 78 m (= speed of light ÷ chip rate = $3.0 \cdot 10^8$ m s^{-1} ÷ 3.84 Mcps). With a chip rate of about 1 Mcps, the difference in the path lengths of the multipath components must be about 300 m, which cannot be obtained in small cells. Therefore, it is easy to see that the 5 MHz WCDMA can provide multipath diversity in small cells, which is not possible with IS-95.

2. Also, for a certain time delay position there are usually many paths nearly equal in length along which the radio signal travels. For example, paths with a length difference of half a wavelength (at 2 GHz this is approximately 7 cm) arrive at virtually the same instant when compared to the duration of a single chip, which is 78 m at 3.84 Mcps. As a result, signal cancellation, called fast fading, takes place as the receiver moves across even short distances. Signal cancellation is best understood as a summation of several weighted phasors that describe the phase shift (usually modulo radio wavelength) and attenuation along a certain path at a certain time instant.

Figure 3.5 shows an exemplary fast fading pattern as would be discerned for the arriving signal energy at a particular delay position as the receiver moves. We see that the received signal power can drop considerably (by 20–30 dB) when phase cancellation of multipath reflections occurs. Because of the underlying geometry causing the fading and dispersion phenomena, signal variations due to fast fading occur several orders of magnitude more

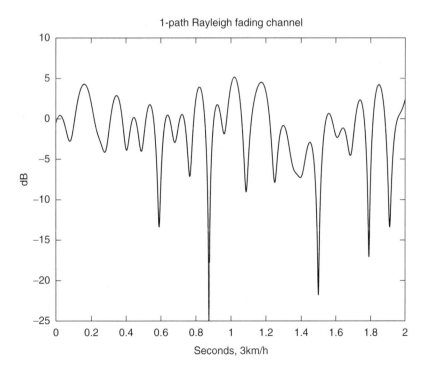

Figure 3.5. Fast Rayleigh fading as caused by multipath propagation

frequently than changes in the average multipath delay profile. The statistics of the received signal energy for a short-term average are usually well described by the Rayleigh distribution: see, e.g., [5] and [6]. These fading dips make error-free reception of data bits very difficult, and countermeasures are needed in WCDMA. The countermeasures against fading in WCDMA are shown below.

1.   The delay dispersive energy is combined by utilising multiple Rake fingers (correlation receivers) allocated to those delay positions on which significant energy arrives.

2.   Fast power control and the inherent diversity reception of the Rake receiver are used to mitigate the problem of fading signal power.

3.   Strong coding and interleaving and retransmission protocols are used to add redundancy and time diversity to the signal and thus help the receiver in recovering the user bits across fades.

The dynamics of the radio propagation suggest the following operating principle for the CDMA signal reception:

1.   Identify the time delay positions at which significant energy arrives and allocate correlation receivers, i.e. Rake fingers, to those peaks. The measurement grid for acquiring the multipath delay profile is in the order of one chip duration (typically within the range of $\frac{1}{4}-\frac{1}{2}$ chip duration) with an update rate in the order of some tens of milliseconds.

2.   Within each correlation receiver, track the fast-changing phase and amplitude values originating from the fast fading process and remove them. This tracking process has to be very fast, with an update rate in the order of 1 ms or less.

3.   Combine the demodulated and phase-adjusted symbols across all active fingers and present them to the decoder for further processing.

Figure 3.6 illustrates points 2 and 3 by depicting modulation symbols (BPSK or QPSK) as well as the instantaneous channel state as weighted complex phasors. To facilitate point 2, WCDMA uses known pilot symbols that are used to sound the channel and provide an estimate of the momentary channel state (value of the weighted phasor) for a particular finger. Then the received symbol is rotated back, so as to undo the phase rotation caused

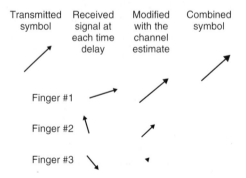

Figure 3.6. The principle of maximal ratio combining within the CDMA Rake receiver

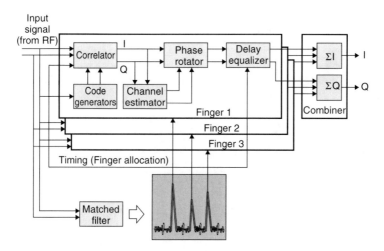

Figure 3.7. Block diagram of the CDMA Rake receiver

by the channel. Such channel-compensated symbols can then be simply summed together to recover the energy across all delay positions. This processing is also called Maximal Ratio Combining (MRC).

Figure 3.7 shows a block diagram of a Rake receiver with three fingers according to these principles. Digitised input samples are received from the RF front-end circuitry in the form of I and Q branches (i.e. in complex low-pass number format). Code generators and correlator perform the despreading and integration to user data symbols. The channel estimator uses the pilot symbols for estimating the channel state which will then be removed by the phase rotator from the received symbols. The delay is compensated for the difference in the arrival times of the symbols in each finger. The Rake combiner then sums the channel-compensated symbols, thereby providing multipath diversity against fading. Also shown is a matched filter used for determining and updating the current multipath delay profile of the channel. This measured and possibly averaged multipath delay profile is then used to assign the Rake fingers to the largest peaks.

In typical implementations of the Rake receiver, processing at the chip rate (correlator, code generator, matched filter) is done in ASICs, whereas symbol-level processing (channel estimator, phase rotator, combiner) is implemented by a DSP. Although there are several differences between the WCDMA Rake receiver in the mobile and the base station, all the basic principles presented here are the same.

Finally, we note that multiple receive antennas can be accommodated in the same way as multiple paths received from a single antenna: by just adding additional Rake fingers to the antennas, we can then receive all the energy from multiple paths *and* antennas. From the Rake receiver's perspective, there is essentially no difference between these two forms of diversity reception.

3.5 Power Control

Tight and fast power control is perhaps the most important aspect in WCDMA, in particular on the uplink. Without it, a single overpowered mobile could block a whole cell.

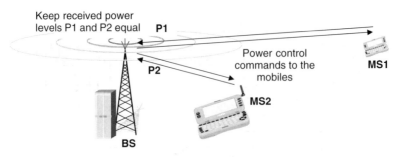

Figure 3.8. Closed-loop power control in CDMA

Figure 3.8 depicts the problem and the solution in the form of closed-loop transmission power control.

Mobile stations MS1 and MS2 operate within the same frequency, separable at the base station only by their respective spreading codes. It may happen that MS1 at the cell edge suffers a path loss, say 70 dB above that of MS2 which is near the base station BS. If there were no mechanism for MS1 and MS2 to be power-controlled to the same level at the base station, MS2 could easily overshout MS1 and thus block a large part of the cell, giving rise to the so-called near–far problem of CDMA. The optimum strategy in the sense of maximising capacity is to equalise the received power per bit of all mobile stations at all times.

While one can conceive open-loop power control mechanisms that attempt to make a rough estimate of path loss by means of a downlink beacon signal, such a method would be far too inaccurate. The prime reason for this is that the fast fading is essentially uncorrelated between uplink and downlink, due to the large frequency separation of uplink and downlink band of the WCDMA FDD mode. Open-loop power control is, however, used in WCDMA, but only to provide a coarse initial power setting of the mobile station at the beginning of a connection.

The solution to power control in WCDMA is fast closed-loop power control, also shown in Figure 3.8. In closed-loop power control in the uplink, the base station performs frequent estimates of the received Signal-to-Interference Ratio (SIR) and compares it to a target SIR. If the measured SIR is higher than the target SIR, the base station will command the mobile station to lower the power; if it is too low it will command the mobile station to increase its power. This measure–command–react cycle is executed at a rate of 1500 times per second (1.5 kHz) for each mobile station and thus operates faster than any significant change of path loss could possibly happen and, indeed, even faster than the speed of fast Rayleigh fading for low to moderate mobile speeds. Thus closed-loop power control will prevent any power imbalance among all the uplink signals received at the base station.

The same closed-loop power control technique is also used on downlink, though here the motivation is different: on the downlink there is no near–far problem due to the one-to-many scenario. All the signals within one cell originate from the one base station to all mobiles. It is, however, desirable to provide a marginal amount of additional power to mobile stations at the cell edge, as they suffer from increased other-cell interference. Also on the downlink a method of enhancing weak signals caused by Rayleigh fading with additional power is needed at low speeds when other error-correcting methods based on interleaving and error correcting codes do not yet work effectively.

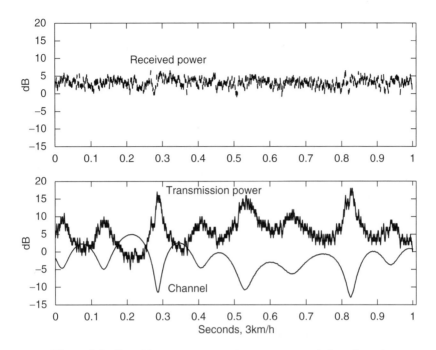

Figure 3.9. Closed-loop power control compensates a fading channel

Figure 3.9 shows how uplink closed-loop power control works on a fading channel at low speed. Closed-loop power control commands the mobile station to use a transmit power proportional to the inverse of the received power (or SIR). Provided the mobile station has enough headroom to ramp the power up, only very little residual fading is left and the channel becomes an essentially non-fading channel as seen from the base station receiver.

While this fading removal is highly desirable from the receiver point of view, it comes at the expense of increased average transmit power at the transmitting end. This means that a mobile station in a deep fade, i.e. using a large transmission power, will cause increased interference to other cells. Figure 3.9 illustrates this point. The gain from the fast power control is discussed in more detail in Section 9.2.1.1.

Before leaving the area of closed-loop power control, we mention one more related control loop connected with it: outer loop power control. Outer loop power control adjusts the target SIR setpoint in the base station according to the needs of the individual radio link and aims at a constant quality, usually defined as a certain target bit error rate (BER) or block error rate (BLER). Why should there be a need for changing the target SIR setpoint? The required SIR (there exists a proportional E_b/N_0 requirement) for, say, BLER = 1% depends on the mobile speed and the multipath profile. Now, if one were to set the target SIR setpoint for the worst case, i.e. high mobile speeds, one would waste much capacity for those connections at low speeds. Thus, the best strategy is to let the target SIR setpoint float around the minimum value that just fulfils the required target quality. The target SIR setpoint will change over time, as shown in the graph in Figure 3.10, as the speed and propagation environment changes. The gain of outer loop power control is discussed in detail in Section 9.2.2.1.

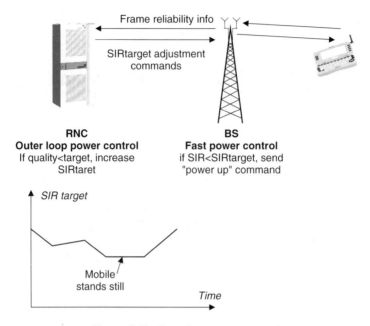

Figure 3.10. Outer loop power control

Outer loop control is typically implemented by having the base station tag each uplink user data frame with a frame reliability indicator, such as a CRC check result obtained during decoding of that particular user data frame. Should the frame quality indicator indicate to the Radio Network Controller (RNC) that the transmission quality is decreasing, the RNC in turn will command the base station to increase the target SIR setpoint by a certain amount. The reason for having outer loop control reside in the RNC is that this function should be performed after a possible soft handover combining. Soft handover will be presented in the next section.

3.6 Softer and Soft Handovers

During softer handover, a mobile station is in the overlapping cell coverage area of two adjacent sectors of a base station. The communications between mobile station and base station take place concurrently via *two* air interface channels, one for each sector separately. This requires the use of two separate codes in the downlink direction, so that the mobile station can distinguish the signals. The two signals are received in the mobile station by means of Rake processing, very similar to multipath reception, except that the fingers need to generate the respective code for each sector for the appropriate despreading operation. Figure 3.11 shows the softer handover scenario.

In the uplink direction a similar process takes place at the base station: the code channel of the mobile station is received in each sector, then routed to the same baseband Rake receiver and the maximal ratio combined there in the usual way. During softer handover only one power control loop per connection is active. Softer handover typically occurs in about 5–15% of connections.

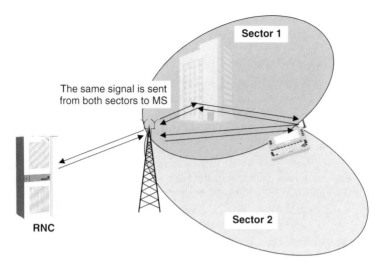

Figure 3.11. Softer handover

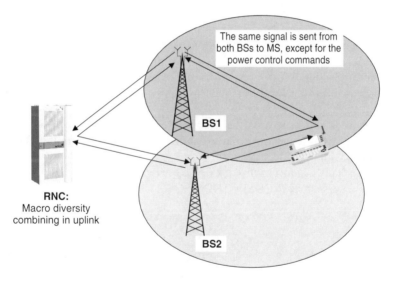

Figure 3.12. Soft handover

Figure 3.12 shows soft handover. During soft handover, a mobile station is in the overlapping cell coverage area of two sectors belonging to different base stations. As in softer handover, the communications between mobile station and base station take place concurrently via two air interface channels from each base station separately. As in softer handover, both channels (signals) are received at the mobile station by maximal ratio combining Rake processing. Seen from the mobile station, there are very few differences between softer and soft handover.

However, in the uplink direction soft handover differs significantly from softer handover: the code channel of the mobile station is received from both base stations, but the received data is then routed to the RNC for combining. This is typically done so that the same frame reliability indicator as provided for outer loop power control is used to select the better frame between the two possible candidates within the RNC. This selection takes place after each interleaving period, i.e. every 10–80 ms.

Note that during soft handover two power control loops per connection are active, one for each base station. Power control in soft handover is discussed in Section 9.2.1.3.

Soft handover occurs in about 20–40% of connections. To cater for soft handover connections, the following additional resources need to be provided by the system and must be considered in the planning phase:

- Additional Rake receiver channels in the base stations
- Additional transmission links between base station and RNC
- Additional Rake fingers in the mobile stations

We also note that soft and softer handover can take place in combination with each other.

Why are these CDMA-specific handover types needed? They are needed for similar reasons as closed-loop power control: without soft/softer handover there would be near–far scenarios of a mobile station penetrating from one cell deeply into an adjacent cell without being power-controlled by the latter. Very fast and frequent hard handovers could largely avoid this problem; however, they can be executed only with certain delays during which the near–far problem could develop. So, as with fast power control, soft/softer handovers are an essential interference-mitigating tool in WCDMA. Soft and softer handovers are described in more detail in Section 9.3.

In addition to soft/softer handover, WCDMA provides other handover types:

- Inter-frequency hard handovers that can be used, for example, to hand a mobile over from one WCDMA frequency carrier to another. One application for this is high capacity base stations with several carriers.
- Inter-system hard handovers that take place between the WCDMA FDD system and another system, such as WCDMA TDD or GSM.

References

[1]   Ojanperä, T. and Prasad, R., *Wideband CDMA for Third Generation Mobile Communications*, Artech House, 1998.
[2]   Viterbi, A., *Principles of Spread Spectrum Communication*, Addison-Wesley, 1997.
[3]   Cooper, G. and McGillem, C., *Modern Communications and Spread Spectrum*, McGraw-Hill, 1998.
[4]   Dixon, R., *Spread Spectrum Systems with Commercial Applications*, John Wiley & Sons, 1994.
[5]   Jakes, W., *Microwave Mobile Communications*, John Wiley & Sons, 1974.
[6]   Saunders, S., *Antennas and Propagation for Wireless Communication Systems*, John Wiley & Sons, 1999.

4

Background and Standardisation of WCDMA

Antti Toskala

4.1 Introduction

In the first phase of third generation standardisation, the basic process of selecting the best technology for multiple radio access was conducted in several regions. This chapter describes the selection process that took place in ETSI mainly during 1997, as well as the decisions made by the regional standardisation organisations into early 1998. Other standardisation bodies that carried out WCDMA-related work are also introduced, and then 3GPP, the common standardisation effort to create a global standard for WCDMA, is described. Finally, the developments in ITU for work on the IMT-2000 recommendations and the relationship of the ITU work to the regional activities are presented.

4.2 Background in Europe

In Europe a long period of research preceded the selection of third generation technology, as shown in Figure 4.1. The RACE I (Research of Advanced Communication Technologies in Europe) programme started the basic third generation research work in 1988. This programme was followed by RACE II, with the development of the CDMA-based CODIT (Code Division Testbed) and TDMA-based ATDMA (Advanced TDMA Mobile Access) air interfaces during 1992–95. In addition, wideband air interface proposals were studied in a number of industrial projects in Europe: see, for example, [1].

The European research programme ACTS (Advanced Communication Technologies and Services) was launched at the end of 1995 in order to support mobile communications research and development. Within ACTS the FRAMES (Future Radio Wideband Multiple Access System) project [2] was set up with the objective of defining a proposal for a UMTS radio access system. The main industrial partners in FRAMES were Nokia, Siemens, Ericsson, France Télécom and CSEM/Pro Telecom, with participation also from

WCDMA for UMTS, edited by Harri Holma and Antti Toskala
© 2001 John Wiley & Sons, Ltd

several European universities. Based on an initial proposal evaluation phase in FRAMES, a harmonised multiple access platform was defined, consisting of two modes: FMA1, a wideband TDMA [3], and FMA2, a wideband CDMA [4]. The FRAMES wideband CDMA and wideband TDMA proposals were submitted to ETSI as candidates for UMTS air interface and ITU IMT-2000 submission.

The proposals for the UMTS Terrestrial Radio Access (UTRA) air interface received by the milestone were grouped into five concept groups in ETSI in June 1997, after their submission and presentation during 1996 and early 1997.

The following groups were formed:

- Wideband CDMA (WCDMA)
- Wideband TDMA (WTDMA)
- TDMA/CDMA
- OFDMA
- ODMA

The concept groups formed in ETSI are introduced briefly in the following section. The evaluation of the proposals was based on the requirements defined in the ITU-R IMT-2000 framework (and in ETSI defined specifically in UMTS 21.01 [5] as well as on the evaluation principles and conditions covered in UMTS 30.03 [6]). The results of the evaluation were collated in UMTS 30.06 [7].

4.2.1 Wideband CDMA

The WCDMA concept group was formed around the WCDMA proposals from FRAMES/FMA2, Fujitsu, NEC and Panasonic. Several European, Japanese and US companies contributed to the development of the WCDMA concept. The physical layer of the WCDMA uplink was adopted mainly from FRAMES/FMA2, while the downlink solution was modified following the principles of the other proposals made to the WCDMA concept group.

The basic system features consisted of

- Wideband CDMA operation with 5 MHz
- Physical layer flexibility for integration of all data rates on a single carrier
- Reuse 1 operation

The enhancements covered included

- Transmit diversity
- Adaptive antenna operation
- Support for advanced receiver structures

The WCDMA concept achieved the greatest support, one of the technical motivating issues being the flexibility of the physical layer for accommodating different service types simultaneously. This was considered to be an advantage, especially with respect to low and medium bit rates. Among the drawbacks of WCDMA, it was recognised that in an unlicensed system in the TDD band, with the continuous transmit and receive operation, pure WCDMA technology does not facilitate interference avoidance techniques in cordless-like operating environments.

4.2.2   Wideband TDMA

The WTDMA concept group was formed by taking the non-spread option from the FRAMES/FMA1 proposal. FRAMES/FMA1 was basically a TDMA-based system concept with 1.6 MHz carrier spacing for wideband service implementation. The concept aimed at high capacity with the aid of interference averaging over the operator bandwidth, with fractional loading and frequency hopping.

The basic system features consisted of

- Equalisation with training sequences in TDMA bursts
- Interference averaging with frequency hopping
- Link adaptation
- Two basic burst types, $1/16^{th}$ and $1/64^{th}$ burst lengths for high and low data rates respectively
- Low reuse sizes

The enhancements covered included

- Inter-cell interference suppression
- Support of adaptive antennas
- TDD operation
- Less complex equalisers for large delay spread environments

The main limitation associated with the system was considered to be the range with respect to low bit rate services. This is due to the fact that in TDMA-based operation the slot duration is, at a minimum, only $1/64^{th}$ of the frame timing, which results in either very high peak power or a low average output power level. This means that for large ranges with, for example, speech, the WTDMA concept would not have been competitive on its own, but would have required a narrowband option as a companion.

4.2.3   Wideband TDMA/CDMA

The WTDMA/CDMA group was based on the spreading option in the FRAMES/FMA1 proposal, resulting in the hybrid CDMA/TDMA concept with 1.6 MHz carrier spacing.

The basic system features consisted of

- TDMA burst structure with midamble for channel estimation
- CDMA concept applied on top of the TDMA structure for additional flexibility
- Reduction of intra-cell interference by multi-user detection for users within a timeslot on the same carrier
- Low reuse sizes, down to 3

Enhancements covered included

- Frequency hopping
- Inter-cell interference cancellation
- Support of adaptive antennas
- Operation in TDD mode
- Dynamic Channel Allocation (DCA)

This proposal, especially the issues related to receiver complexity, led to lively discussions during the selection process.

4.2.4   OFDMA

The OFDMA group was based on OFDMA technology with inputs mainly from Telia, Sony and Lucent. The system concept was shaped by the discussions about OFDMA in other forums, such as the Japanese standardisation forum, ARIB.

The basic concept features included

- Operation with slow frequency hopping with TDMA and OFDM multiplexing
- A 100 kHz wide bandslot from the OFDM signal as the basic resource unit
- Higher rates built by allocating several bandslots, creating a wideband signal
- Diversity provided by dividing the information among several bandslots over the carrier

The enhancement techniques covered were

- Transmit diversity
- Multi-user detection for interference cancellation
- Adaptive antenna solutions

A main technical weakness of the system concept was the uplink transmission direction, where the resulting envelope variations caused concern for power amplifier design.

4.2.5   ODMA

Vodafone proposed Opportunity Driven Multiple Access (ODMA), basically a relaying protocol, not a pure multiple access as such. ODMA was later integrated in the WCDMA and WCDMA/TDMA concept groups and was not considered in the selection process as a concept on its own.

Briefly, in ODMA a terminal beyond the reach of the cell coverage would use another terminal as a relay to transmit packets to the base station. ODMA operation has been considered to be feasible with TDD operation where both reception and transmission are in the same frequency band. If it were desired to implement ODMA with FDD, this would require terminals either to receive in their normal transmission band or then to transmit in their normal reception band. Such a requirement makes ODMA undesirable for FDD from an implementation point of view.

4.2.6   ETSI Selection

All the proposed technologies were basically able to fulfil the UMTS requirements, although it was difficult to reach a consensus on issues such as system capacity, since the results of simulations can vary greatly depending on the assumptions. However, it soon became evident in the selection process that WCDMA and TDMA/CDMA were the main candidates. Also, issues such as the global potential of a technology naturally had an impact in cases where obvious technical conclusions were very limited; in this respect, the outcome of the ARIB technology selection in Japan gave support to WCDMA.

ETSI decided between the technologies in January 1998 [8], selecting WCDMA as the standard for the UTRA (UMTS Terrestrial Radio Access) air interface on the paired frequency bands, i.e. for FDD (Frequency Division Duplexing) operation, and WTDMA/CDMA for operation with unpaired spectrum allocation, i.e. for TDD (Time Division Duplexing) operation. The detailed standardisation of UTRA proceeded within ETSI until the work was handed over to the 3rd Generation Partnership Project (3GPP). The technical work was transferred to 3GPP with the contribution of UTRA in early 1999.

4.3   Background in Japan

In Japan, ARIB (the Association for Radio Industries and Businesses) evaluated possible third generation systems around three different main technologies based on WCDMA, WTDMA and OFDMA.

The WCDMA technology in Japan was very similar to that being considered in Europe in ETSI; indeed, the members of ARIB contributed their technology to ETSI's WCDMA concept group. Details of FRAMES/FMA2 were provided from Europe for consideration in the ARIB process. The other technologies considered, WTDMA and OFDMA, also had many similarities to the candidates in the ETSI selection process.

The outcome of the ARIB selection process in 1997 was WCDMA, with both FDD and TDD modes of operation. Since WCDMA had been chosen in ARIB before the process was completed in ETSI, it carried more weight in the ETSI selection as the global technological alternative. Since the creation of 3GPP for the third generation standardisation framework, ARIB have contributed their WCDMA to 3GPP, in the same way as ETSI have contributed UTRA. In Japan, work on higher layer specifications is the responsibility of TTC (the Telecommunication Technology Committee), which has also shifted the activity to 3GPP.

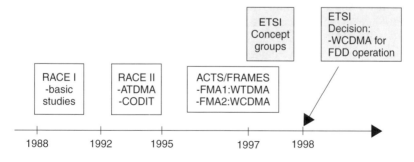

Figure 4.1. European research programmes towards third generation systems and the ETSI decision

4.4   Background in Korea

In Korea, the Telecommunications Technology Association (TTA) adopted a two-track approach to the development of third generation CDMA technology. The TTA1 and TTA2 air interface proposals (later renamed Global CDMA 1 and 2 respectively) were based on synchronous and asynchronous wideband CDMA technologies respectively. TTA1 WCDMA was similar to WCDMA technology in ETSI, ARIB and T1P1, while TTA2 was similar to cdma2000 in TR45.5.

Several technical details of the Korean technology that differed from the ETSI and ARIB solutions were submitted to the ETSI and ARIB standardisation processes, leading to a high degree of commonality between the ETSI, ARIB and TTA WCDMA solutions. The Korean standardisation efforts were later moved to 3GPP and 3GPP2 to contribute to WCDMA and cdma2000 standardisation respectively.

4.5   Background in the United States

In the US there exist several second generation technologies, the most widely distributed digital systems being those based on either GSM-1900, US-TDMA (D-AMPS) or US-CDMA (IS-95) standards. For all those technologies, a natural path of evolution towards the third generation had been defined. In addition, a third generation CDMA proposal that had no direct relation to second generation systems, namely WIMS W-CDMA, came from the TR46.1 standardisation committee.

4.5.1   W-CDMA N/A

Work on GSM-1900 related standardisation was carried out within T1P1, as with GSM standardisation in ETSI, with similar discussions concerning technology selection. As a result, W-CDMA N/A (N/A for North America) was submitted to the ITU-R IMT-2000 process. The proposals had much in common with the ETSI and ARIB WCDMA technologies, since the contributing companies had also been active in the ETSI and ARIB selection processes.

4.5.2   UWC-136

In TR45.3, discussions concerning the evolution of IS-136 (Digital AMPS) technology towards the third generation took place. The resulting selection was a combination of narrowband and wideband TDMA technologies, with the narrowband component identical to the EDGE concept, part of GSM evolution in ETSI and T1P1. The wideband part for indoor service provision up to 2 Mbps was based on the same WTDMA concept as was considered in ETSI. The selection of EDGE technology in TR45.3 will create a clear connection between TDMA technologies within TR45.3, T1P1 and ETSI. The development of TDMA-based technology is expected to accelerate around the common component in the air interface in the form of the EDGE component. It is foreseen that the work on EDGE will be covered by 3GPP in the near future.

4.5.3 cdma2000

The cdma2000 air interface proposal to ITU is the result of work in TR45.5 on the evolution of IS-95 towards the third generation. The cdma2000 proposal is based partly on IS-95 principles with respect to synchronous network operation, common pilot channels, and so on, but it is a wideband version with three times the bandwidth of IS-95. The ITU proposal contains further bandwidth options as well as the multi-carrier option for downlink. The cdma2000 proposal has a high degree of commonality with the Global CDMA 1 ITU proposal from TTA, Korea.

The cdma2000 multi-carrier option is covered in more detail in Chapter 13, as being currently standardised by 3GPP2.

4.5.4 TR46.1

The WIMS W-CDMA was not based on work derived from an existing second generation technology but was a new third generation technology proposal with no direct link to any second generation standardisation. It was based on the constant processing gain principle with a high number of multicodes in use, thus showing some fundamental differences but also a level of commonality with WCDMA technology in other forums.

4.5.5 WP-CDMA

WP-CDMA (Wideband Packet CDMA) resulted from the convergence between W-CDMA N/A of T1P1 and WIMS W-CDMA of TR46.1 in the US. The main features of the WIMS W-CDMA proposal were merged with the principles of W-CDMA N/A. The merged proposal was submitted to the ITU-R IMT-2000 process towards the end of 1998, and to the 3GPP process at the beginning of 1999. Its the most characteristic feature, compared with the other WCDMA-based proposals, was a common packet mode channel operation for the uplink direction, but there were also a few smaller differences.

4.6   Creation of 3GPP

As similar technologies were being standardised in several regions around the world, it became evident that achieving identical specifications to ensure equipment compatibility globally would be very difficult with work going on in parallel. Also, having to discuss similar issues in several places was naturally a waste of resources for the participating companies. Therefore initiatives were made to create a single forum for WCDMA standardisation for a common WCDMA specification.

The standardisation organisations involved in the creation of the 3rd Generation Partnership Project (3GPP) [9] were ARIB (Japan), ETSI (Europe), TTA (Korea), TTC (Japan) and T1P1 (USA) as shown in Figure 4.2. The partners agreed on joint efforts for the standardisation of UTRA, now standing for Universal Terrestrial Radio Access, as distinct from UTRA (UMTS Terrestrial Radio Access) from ETSI, also submitted to 3GPP. Companies such as manufacturers and operators are members of 3GPP through the respective standardisation organisation to which they belong.

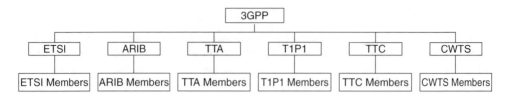

Figure 4.2. 3GPP organisational partners

Later during 1999, CWTS (the China Wireless Telecommunication Standard Group) also joined 3GPP and contributed technology from TD/SCDMA, a TDD-based CDMA third generation technology already submitted to ITU-R earlier.

3GPP also includes market representation partners: GSM Association, UMTS Forum, Global Mobile Suppliers Association, IPv6 Forum and Universal Wireless Communications Consortium (UWCC). In [9] there are up-to-date links to all participating organisations.

The work was initiated formally at the end of 1998 and the detailed technical work was started in early 1999, with the aim of having the first version of the common specification, called Release-99, ready by the end of 1999.

Within 3GPP, four different technical specification groups (TSG) were set up as follows:

- Radio Access Network TSG
- Core Network TSG
- Service and System Aspects TSG
- Terminals TSG

Within these groups the one most relevant to the WCDMA technology is the Radio Access Network TSG (RAN TSG), which has been divided into four different working groups as illustrated in Figure 4.3.

The RAN TSG will produce Release-99 of the UTRA air interface specification. The work done within the 3GPP RAN TSG working groups has been the basis of the technical description of the UTRA air interface covered in this book. Without such a global initiative, this book would have been forced to focus on a single regional specification, though with many similarities to those of other regions. Thus the references throughout this book are to the specification volumes from 3GPP.

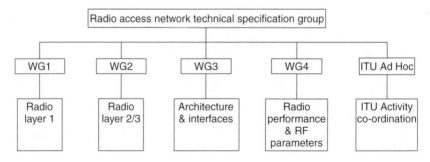

Figure 4.3. 3GPP RAN TSG working groups

During the first half of 1999 the inputs from the various participating organisations were merged in a single standard, leaving the rest of the year to finalise the detailed parameters for the first full release, Release-99, of UTRA from 3GPP. The member organisations have undertaken individually to produce standard publications based on the 3GPP specification. Thus, for example, the Release-99 UMTS specifications from ETSI are identical to the Release-99 specifications produced by 3GPP. The latest specifications can be obtained from 3GPP [9].

During 2000, further work on GSM evolution was moved from ETSI and other forums to 3GPP, including work on GPRS and EDGE. A new TSG, TSG GERAN was set up for this purpose.

4.7   Creation of 3GPP2

Work done in TR45.5 and TTA was merged to form 3GPP2, focused on the development of cdma2000 Direct-Sequence (DS) and Multi-Carrier (MC) mode for the cdma2000 third generation component. This activity has been running in parallel with the 3GPP project, with participation from ARIB, TTC and CWTS as member organisations. Recently the main concentration has been on the MC mode work, due to decisions resulting from the global harmonisation efforts.

4.8   Harmonisation Phase

During the spring of 1999 several operators and manufacturers held series of meetings to seek further harmonisation and convergence between the CDMA-based third generation solutions, WCDMA and cdma2000. For the 3GPP framework the ETSI, ARIB, TTA and T1P1 concepts had already been merged to a single specification, while cdma2000 was still on its own in TR45.5. As a result of several meetings and telephone conferences, the manufacturers and operators agreed to adopt a harmonised global third generation CDMA standard consisting of three modes: Multi-Carrier (MC), Direct Spread (DS) and Time Division Duplex (TDD). The MC mode was based on the cdma2000 multi-carrier option, the DS mode on WCDMA (UTRA FDD), and the TDD mode on UTRA TDD. The agreement was to phase in a modular approach in which both core networks could be used with all air interface alternatives, as described in Figure 1.4 in Chapter 1.

The main technical impacts of these harmonisation activities were the change of UTRA FDD and TDD mode chip rate from 4.096 Mcps to 3.84 Mcps and the inclusion of a common pilot for UTRA FDD. The work in 3GPP2 focused on the MC mode, and the DS mode from cdma2000 was abandoned. The result is that globally there is only one Direct Spread (DS) wideband CDMA standard, WCDMA.

4.9   IMT2000 Process in ITU

In the ITU, recommendations have been developed for third generation mobile communications systems, the ITU terminology being called IMT-2000 [10], formerly FPLMTS. In the ITU-R, ITU-R TG8/1 has worked on the radio-dependent aspects, while the radio-independent aspects have been covered in ITU-T SG11.

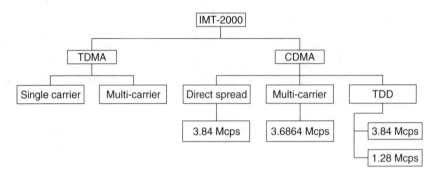

Figure 4.4. ITU-R IMT-2000 grouping

In the radio aspects, ITU-R TG8/1 received a number of different proposals during the IMT-2000 candidate submission process. In the second phase of the process, evaluation results were received from the proponent organisations as well as from the other evaluation groups that studied the technologies. During the first half of 1999 recommendation IMT.RKEY was created which describes the IMT-2000 multimode concept.

The ITU-R IMT-2000 process was finalised at the end of 1999, when the detailed specification (IMT-RSCP) was created and the radio interface specifications were approved by ITU-R [11]. The detailed implementation of IMT-2000 will continue in the regional standards bodies. The ITU-R process has been an important external motivation and timing source for IMT-2000 activities in regional standards bodies. The requirements set by ITU for an IMT-2000 technology have been reflected in the requirements in the regional standards bodies, for example in ETSI UMTS 21.01 [5], in order for the ETSI submission to fulfil the IMT-2000 requirements. The ITU-R interaction between regional standardisation bodies in the IMT-2000 process is reflected in Figure 4.5.

The ITU-R IMT-2000 grouping, with TDMA- and CDMA-based groups, is illustrated in Figure 4.4. The UTRA FDD (WCDMA) and cdma2000 are part of the CDMA interface, as CDMA Direct Spread and CDMA Multi-Carrier respectively. UWC-136 and DECT are part of the TDMA-based interface in the concept, as TDMA Single Carrier and TDMA Multi-Carrier respectively. The TDD part in CDMA consists of UTRA TDD from 3GPP and TD-SCDMA from CWTS. For the FDD part in the CDMA interface, harmonisation has been completed, and the harmonisation process is expected to continue for the CDMA TDD mode within 3GPP during 2000 and 2001.

4.10 Beyond 3GPP Release-99

Upon completion of the Release-99 specifications, work will concentrate on specifying new features as well as making the necessary corrections to Release-99. Typically such corrections arise as implementation proceeds and test systems are updated to include the latest changes in the specifications. As experience in various forums has shown, a major step forward in system capabilities with many new features requires a phasing-in period for the specifications. Fortunately, the main functions have been verified in the various test systems in operation since 1995, but only the actual implementation will reveal any errors and inconsistencies in the fine detail of the specifications.

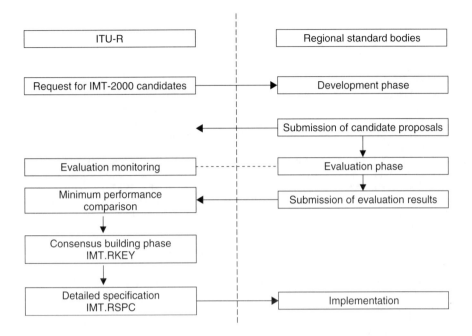

Figure 4.5. Relationship of ITU-R to the regional standards bodies

In 3GPP the next version of the specifications was originally considered as Release 2000, but since that the Release naming was adjusted so that next release is called Release 4 due to the 03/2001. Release 4 will contain only minor adjustment with respect to the Release'99. Bigger items are foreseen for inclusions in Release 5, including reaching for the data rates even up to the 10 Mbps in the downlink direction. Release 5 is due for the end of 2001. Release'99 specifications have a version number starting with 3 while Release 4 and 5 specifications have version number starting logically with 4 and 5 respectively.

During coming years, part of the process is to specify the extensions for connecting UTRA FDD to IS-41 based core networks. Work will be carried out on cdma2000 to allow connection to GSM-based core networks. This development will offer operators some degree of flexibility in selecting their third generation technology, assuming they are not aiming to launch such a service immediately, since Release-99 does not yet contain these options.

On the TDD side, further alignment is expected between the TDD mode in UTRA and the TDD mode from CWTS (China). This process is expected to continue after Release-99 is completed in 3GPP. 3GPP Release-4 is also planned to cover the lower chip rate (1.28 Mcps) TDD.

Other interesting developments are expected in the area of using IP based technology in UTRA. First step is going to be the IP based transport option in Release 4 and further releases are likely to see IP technology shaping the internal architecture of the UTRA Network (UTRAN). Also some of the protocols developed by the Internet Engineering Task Force (IETF) [12], such as robust IP header compression suitable for cellular transmission, are foreseen to be adopted as part the UTRA Rel'4 specifications.

References

[1]   Pajukoski, K. and Savusalo, J., 'Wideband CDMA Test System', *Proc. IEEE Int. Conf. on Personal Indoor and Mobile Radio Communications*, PIMRC'97, Helsinki, Finland, 1–4 September 1997, pp. 669–672.

[2]   Nikula, E., Toskala, A., Dahlman, E., Girard, L., and Klein, A., 'FRAMES Multiple Access for UMTS and IMT-2000', *IEEE Personal Communications Magazine*, April 1998, pp. 16–24.

[3]   Klein, A., Pirhonen, R., Sköld, J., and Suoranta, R., 'FRAMES Multiple Access Mode 1—Wideband TDMA with and without Spreading', *Proc. IEEE Int. Conf. on Personal Indoor and Mobile Radio Communications*, PIMRC'97, Helsinki, Finland, 1–4 September 1997, pp. 37–41.

[4]   Ovesjö, F., Dahlman, E., Ojanperä, T., Toskala, A., and Klein, A., 'FRAMES Multiple Access Mode 2—Wideband CDMA', *Proc. IEEE Int. Conf. on Personal Indoor and Mobile Radio Communications*, PIMRC'97, Helsinki, Finland, 1–4 September 1997, pp. 42–46.

[5]   Universal Mobile Telecommunications System (UMTS), Requirements for the UMTS Terrestrial Radio Access System (UTRA), ETSI Technical Report, UMTS 21.01 version 3.0.1, November 1997.

[6]   Universal Mobile Telecommunications System (UMTS), Selection Procedures for the Choice of Radio Transmission Technologies of the UMTS, ETSI Technical Report, UMTS 30.03 version 3.1.0, November 1997.

[7]   Universal Mobile Telecommunications System (UMTS), UMTS Terrestrial Radio Access System (UTRA) Concept Evaluation, ETSI Technical Report, UMTS 30.06 version 3.0.0, December 1997.

[8]   ETSI Press Release, SMG Tdoc 40/98, 'Agreement Reached on Radio Interface for Third Generation Mobile System, UMTS', Paris, France, January 1998.

[9]   http://www.3GPP.org

[10] http://www.itu.int/imt/

[11] ITU Press Release, ITU/99–22, 'IMT-2000 Radio Interface Specifications Approved in ITU Meeting in Helsinki', 5 November 1999.

[12] http://www.ietf.org

5

Radio Access Network Architecture

Fabio Longoni and Atte Länsisalmi

5.1   System Architecture

This chapter gives a wide overview of the UMTS system architecture, including an introduction to the logical network elements and the interfaces. The UMTS system utilises the same well-known architecture that has been used by all main second generation systems and even by some first generation systems. The reference list contains the related 3GPP specifications.

The UMTS system consists of a number of logical network elements that each has a defined functionality. In the standards, network elements are defined at the logical level, but this quite often results in a similar physical implementation, especially since there are a number of open interfaces (for an interface to be 'open', the requirement is that it has been defined to such a detailed level that the equipment at the endpoints can be from two different manufacturers). The network elements can be grouped based on similar functionality, or based on which sub-network they belong to.

Functionally the network elements are grouped into the Radio Access Network (RAN, UMTS Terrestrial RAN = UTRAN) that handles all radio-related functionality, and the Core Network, which is responsible for switching and routing calls and data connections to external networks. To complete the system, the User Equipment (UE) that interfaces with the user and the radio interface is defined. The high-level system architecture is shown in Figure 5.1.

From a specification and standardisation point of view, both UE and UTRAN consist of completely new protocols, the design of which is based on the needs of the new WCDMA radio technology. On the contrary, the definition of CN is adopted from GSM. This gives the system with new radio technology a global base of known and rugged CN technology that accelerates and facilitates its introduction, and enables such competitive advantages as global roaming.

WCDMA for UMTS, edited by Harri Holma and Antti Toskala
© 2001 John Wiley & Sons, Ltd

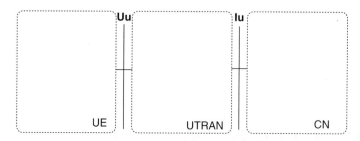

Figure 5.1. UMTS high-level system architecture

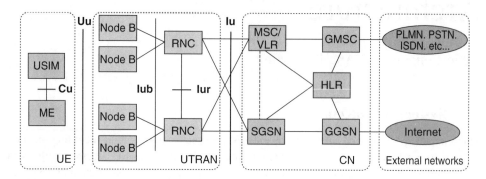

Figure 5.2. Network elements in a PLMN

Another way to group UMTS network elements is to divide them into sub-networks. The UMTS system is modular in the sense that it is possible to have several network elements of the same type. In principle, the minimum requirement for a fully featured and operational network is to have at least one logical network element of each type (note that some features and consequently some network elements are optional). The possibility of having several entities of the same type allows the division of the UMTS system into sub-networks that are operational either on their own or together with other sub-networks, and that are distinguished from each other with unique identities. Such a sub-network is called a UMTS PLMN (Public Land Mobile Network). Typically one PLMN is operated by a single operator, and is connected to other PLMNs as well as to other types of networks, such as ISDN, PSTN, the Internet, and so on. Figure 5.2 shows elements in a PLMN and, in order to illustrate the connections, also external networks.

The UTRAN architecture is presented in Section 5.2. A short introduction to all the elements is given below.

The UE consists of two parts:

- The Mobile Equipment (ME) is the radio terminal used for radio communication over the Uu interface.

- The UMTS Subscriber Identity Module (USIM) is a smartcard that holds the subscriber identity, performs authentication algorithms, and stores authentication and encryption keys and some subscription information that is needed at the terminal.

UTRAN also consists of two distinct elements:

- The Node B converts the data flow between the Iub and Uu interfaces. It also participates in radio resource management. (*Note that the term 'Node B' from the corresponding 3GPP specifications is used throughout Chapter 5. The more generic term 'Base Station' used elsewhere in this book means exactly the same thing.*)

- The Radio Network Controller (RNC) owns and controls the radio resources in its domain (the Node Bs connected to it). RNC is the service access point for all services UTRAN provides the CN, for example management of connections to the UE.

The main elements of the GSM CN (there are other entities not shown in Figure 5.2, such as those used to provide IN services) are as follows:

- HLR (Home Location Register) is a database located in the user's home system that stores the master copy of the user's service profile. The service profile consists of, for example, information on allowed services, forbidden roaming areas, and Supplementary Service information such as status of call forwarding and the call forwarding number. It is created when a new user subscribes to the system, and remains stored as long as the subscription is active. For the purpose of routing incoming transactions to the UE (e.g. calls or short messages), the HLR also stores the UE location on the level of MSC/VLR and/or SGSN, i.e. on the level of serving system.

- MSC/VLR (Mobile Services Switching Centre/Visitor Location Register) is the switch (MSC) and database (VLR) that serves the UE in its current location for Circuit Switched (CS) services. The MSC function is used to switch the CS transactions, and the VLR function holds a copy of the visiting user's service profile, as well as more precise information on the UE's location within the serving system. The part of the network that is accessed via the MSC/VLR is often referred to as the CS domain.

- GMSC (Gateway MSC) is the switch at the point where UMTS PLMN is connected to external CS networks. All incoming and outgoing CS connections go through GMSC.

- SGSN (Serving GPRS (General Packet Radio Service) Support Node) functionality is similar to that of MSC/VLR but is typically used for Packet Switched (PS) services. The part of the network that is accessed via the SGSN is often referred to as the PS domain.

- GGSN (Gateway GPRS Support Node) functionality is close to that of GMSC but is in relation to PS services.

The external networks can be divided into two groups:

- CS networks. These provide circuit-switched connections, like the existing telephony service. ISDN and PSTN are examples of CS networks.

- PS networks. These provide connections for packet data services. The Internet is one example of a PS network.

The UMTS standards are structured so that internal functionality of the network elements is not specified in detail. Instead, the interfaces between the logical network elements have been defined. The following main open interfaces are specified:

- Cu Interface. This is the electrical interface between the USIM smartcard and the ME. The interface follows a standard format for smartcards.

- Uu Interface. This is the WCDMA radio interface, which is the subject of the main part of this book. The Uu is the interface through which the UE accesses the fixed part of the system, and is therefore probably the most important open interface in UMTS. There are likely to be many more UE manufacturers than manufacturers of fixed network elements.

- Iu Interface. This connects UTRAN to the CN and is introduced in detail in Section 5.4. Similarly to the corresponding interfaces in GSM, A (Circuit Switched) and Gb (Packet Switched), the open Iu interface gives UMTS operators the possibility of acquiring UTRAN and CN from different manufacturers. The enabled competition in this area has been one of the success factors of GSM.

- Iur Interface. The open Iur interface allows soft handover between RNCs from different manufacturers, and therefore complements the open Iu interface. Iur is described in more detail in Section 5.5.1.

- Iub Interface. The Iub connects a Node B and an RNC. UMTS is the first commercial mobile telephony system where the Controller-Base Station interface is standardised as a fully open interface. Like the other open interfaces, open Iub is expected to further motivate competition between manufacturers in this area. It is likely that new manufacturers concentrating exclusively on Node Bs will enter the market.

5.2 UTRAN Architecture

UTRAN architecture is highlighted in Figure 5.3.

UTRAN consists of one or more Radio Network Sub-systems (RNS). An RNS is a sub-network within UTRAN and consists of one Radio Network Controller (RNC) and one or more Node Bs. RNCs may be connected to each other via an Iur interface. RNCs and Node Bs are connected with an Iub Interface.

Before entering into a brief description of the UTRAN network elements (in this section) and a more extensive description of UTRAN interfaces (in the following sections), we present the main characteristics of UTRAN that have also been the main requirements for

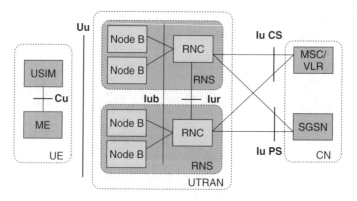

Figure 5.3. UTRAN architecture

the design of the UTRAN architecture, functions and protocols. These can be summarised in the following points:

- *Support of UTRA* and all the related functionality. In particular, the major impact on the design of UTRAN has been the requirement to support *soft handover* (one terminal connected to the network via two or more active cells) and the WCDMA-specific *Radio Resource Management* algorithms.

- Maximisation of the *commonalities in the handling of packet- switched and circuit-switched data*, with a unique air interface protocol stack and with the use of the same interface for the connection from UTRAN to both the PS and CS domains of the core network.

- Maximisation of the *commonalities with GSM*, when possible.

- Use of the *ATM transport* as the main transport mechanism in UTRAN.

5.2.1   The Radio Network Controller

The RNC (Radio Network Controller) is the network element responsible for the control of the radio resources of UTRAN. It interfaces the CN (normally to one MSC and one SGSN) and also terminates the RRC (Radio Resource Control) protocol that defines the messages and procedures between the mobile and UTRAN. It logically corresponds to the GSM BSC.

5.2.1.1   Logical Role of the RNC

The RNC controlling one Node B (i.e. terminating the Iub interface towards the Node B) is indicated as the *Controlling RNC* (CRNC) of the Node B. The Controlling RNC is responsible for the load and congestion control of its own cells, and also executes the admission control and code allocation for new radio links to be established in those cells.

In case one mobile–UTRAN connection uses resources from more than one RNS (see Figure 5.4), the RNCs involved have two separate logical roles (*with respect to this mobile–UTRAN connection*):

- *Serving RNC.* The SRNC for one mobile is the RNC that terminates both the Iu link for the transport of user data and the corresponding RANAP signalling to/from the

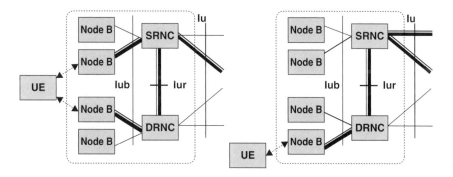

Figure 5.4. Logical role of the RNC for one UE UTRAN connection. The left-hand scenario shows one UE in inter-RNC soft handover (combining is performed in the SRNC). The right-hand scenario represents one UE using resources from one Node B only, controlled by the DRNC

network (this connection is referred to as the RANAP connection). The SRNC also terminates the Radio Resource Control Signalling, that is the signalling protocol between the UE and UTRAN. It performs the L2 processing of the data to/from the radio interface. Basic Radio Resource Management operations, such as the mapping of Radio Access Bearer parameters into air interface transport channel parameters, the handover decision, and outer loop power control, are executed in the SRNC. The SRNC may also (but not always) be the CRNC of some Node B used by the mobile for connection with UTRAN. One UE connected to UTRAN has one and only one SRNC.

- *Drift RNC.* The DRNC is any RNC, other than the SRNC, that controls cells used by the mobile. If needed, the DRNC may perform macrodiversity combining and splitting. The DRNC does not perform L2 processing of the user plane data, but routes the data transparently between the Iub and Iur interfaces, except when the UE is using a common or shared transport channel. One UE may have zero, one or more DRNCs.

Note that one physical RNC normally contains all the CRNC, SRNC and DRNC functionality.

5.2.2   The Node B (Base Station)

The main function of the Node B is to perform the air interface L1 processing (channel coding and interleaving, rate adaptation, spreading, etc.). It also performs some basic Radio Resource Management operation as the inner loop power control. It logically corresponds to the GSM Base Station. The enigmatic term 'Node B' was initially adopted as a temporary term during the standardisation process, but then never changed.

The logical model of the Node B is described in Section 5.5.2.

5.3   General Protocol Model for UTRAN Terrestrial Interfaces

5.3.1   General

Protocol structures in UTRAN terrestrial interfaces are designed according to the same general protocol model. This model is shown in Figure 5.5. The structure is based on the principle that the layers and planes are logically independent of each other, and if needed, parts of the protocol structure may be changed in the future while other parts remain intact.

5.3.2   Horizontal Layers

The protocol structure consists of two main layers, the Radio Network Layer and the Transport Network Layer. All UTRAN-related issues are visible only in the Radio Network Layer, and the Transport Network Layer represents standard transport technology that is selected to be used for UTRAN but without any UTRAN-specific changes.

5.3.3   Vertical Planes

5.3.3.1   Control Plane

The Control Plane is used for all UMTS-specific control signalling. It includes the Application Protocol (i.e. RANAP in Iu, RNSAP in Iur and NBAP in Iub), and the Signalling Bearer for transporting the Application Protocol messages.

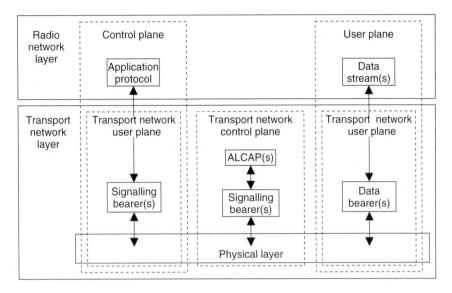

Figure 5.5. General protocol model for UTRAN terrestrial interfaces

The Application Protocol is used, among other things, for setting up bearers to the UE (i.e. the Radio Access Bearer in Iu and subsequently the Radio Link in Iur and Iub). In the three-plane structure the bearer parameters in the Application Protocol are not directly tied to the User Plane technology, but rather are general bearer parameters.

The Signalling Bearer for the Application Protocol may or may not be of the same type as the Signalling Bearer for the ALCAP. It is always set up by O&M actions.

5.3.3.2   User Plane

All information sent and received by the user, such as the coded voice in a voice call or the packets in an Internet connection, are transported via the User Plane. The User Plane includes the Data Stream(s), and the Data Bearer(s) for the Data Stream(s). Each Data Stream is characterised by one or more frame protocols specified for that interface.

5.3.3.3   Transport Network Control Plane

The Transport Network Control Plane is used for all control signalling within the Transport Layer. It does not include any Radio Network Layer information. It includes the ALCAP protocol that is needed to set up the transport bearers (Data Bearer) for the User Plane. It also includes the Signalling Bearer needed for the ALCAP.

The Transport Network Control Plane is a plane that acts between the Control Plane and the User Plane. The introduction of the Transport Network Control Plane makes it possible for the Application Protocol in the Radio Network Control Plane to be completely independent of the technology selected for the Data Bearer in the User Plane.

When the Transport Network Control Plane is used, the transport bearers for the Data Bearer in the User Plane are set up in the following fashion. First there is a signalling transaction by the Application Protocol in the Control Plane, which triggers the setup of the Data Bearer by the ALCAP protocol that is specific for the User Plane technology.

The independence of the Control Plane and the User Plane assumes that an ALCAP signalling transaction takes place. It should be noted that ALCAP might not be used for all types of Data Bearers. If there is no ALCAP signalling transaction, the Transport Network Control Plane is not needed at all. This is the case when preconfigured Data Bearers are used. It should also be noted that the ALCAP protocol(s) in the Transport Network Control Plane is/are not used for setting up the Signalling Bearer for the Application Protocol or for the ALCAP during real-time operation.

The Signalling Bearer for the ALCAP may or may not be of the same type as that for the Application Protocol. The UMTS specifications assume that the Signalling Bearer for ALCAP is always set up by O&M actions, and do not specify this in detail.

5.3.3.4 Transport Network User Plane

The Data Bearer(s) in the User Plane, and the Signalling Bearer(s) for the Application Protocol, also belong to the Transport Network User Plane. As described in the previous section, the Data Bearers in the Transport Network User Plane are directly controlled by the Transport Network Control Plane during real-time operation, but the control actions required for setting up the Signalling Bearer(s) for the Application Protocol are considered O&M actions.

5.4   Iu, the UTRAN–CN Interface

The Iu interface connects UTRAN to CN. Iu is an open interface that divides the system into radio-specific UTRAN and CN which handles switching, routing and service control. As can be seen from Figure 5.3, the Iu can have two different instances, which are Iu CS (Iu Circuit Switched) for connecting UTRAN to Circuit Switched (CS) CN, and Iu PS (Iu Packet Switched) for connecting UTRAN to Packet Switched (PS) CN. The original design goal in the standardisation was to develop only one Iu interface, but then it was realised that fully optimised User Plane transport for CS and PS services can only be achieved if different transport technologies are allowed. Consequently, the Transport Network Control Plane is different. One of the main design guidelines has still been that the Control Plane should be the same for Iu CS and Iu PS, and the differences are minor.

5.4.1   Protocol Structure for Iu CS

The Iu CS overall protocol structure is depicted in Figure 5.6. The three planes in the Iu interface share a common ATM (Asynchronous Transfer Mode) transport which is used for all planes. The physical layer is the interface to the physical medium: optical fibre, radio link or copper cable. The physical layer implementation can be selected from a variety of standard off-the-shelf transmission technologies, such as SONET, STM1, or E1.

5.4.1.1   Iu CS Control Plane Protocol Stack

The Control Plane protocol stack consists of RANAP, on top of Broad Band (BB) SS7 (Signalling System #7) protocols. The applicable layers are the Signalling Connection Control Part (SCCP), the Message Transfer Part (MTP3-b) and SAAL-NNI (Signalling ATM Adaptation Layer for Network to Network Interfaces). SAAL-NNI is further divided into Service Specific Co-ordination Function (SSCF), Service Specific Connection Oriented Protocol

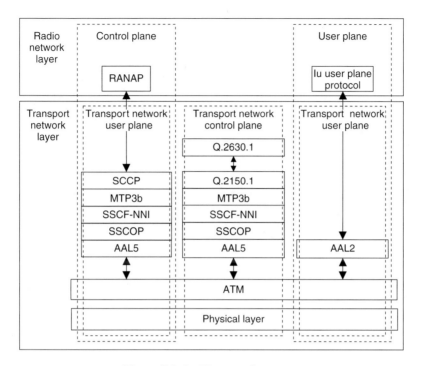

Figure 5.6. Iu CS protocol structure

(SSCOP) and ATM Adaptation Layer 5 (AAL) layers. SSCF and SSCOP layers are specifically designed for signalling transport in ATM networks, and take care of such functions as signalling connection management. AAL5 is used for segmenting the data to ATM cells.

5.4.1.2   Iu CS Transport Network Control Plane Protocol Stack

The Transport Network Control Plane protocol stack consists of the Signalling Protocol for setting up AAL2 connections (Q.2630.1 and adaptation layer Q.2150.1), on top of BB SS7 protocols. The applicable BB SS7 are those described above without the SCCP layer.

5.4.1.3   Iu CS User Plane Protocol Stack

A dedicated AAL2 connection is reserved for each individual CS service. The Iu User Plane Protocol residing directly on top of AAL2 is described in more detail in Section 5.4.4.

5.4.2   Protocol Structure for Iu PS

The Iu PS protocol structure is depicted in Figure 5.7. Again, a common ATM transport is applied for both User and Control Plane. Also the physical layer is as specified for Iu CS.

5.4.2.1   Iu PS Control Plane Protocol Stack

The Control Plane protocol stack again consists of RANAP, and the same BB SS7-based signalling bearer as described in Section 5.4.1.1. Also as an alternative, an IP-based signalling bearer is specified. The SCCP layer is also used commonly for both. The IP-based signalling bearer consists of M3UA (SS7 MTP3—User Adaptation Layer), SCTP (Simple

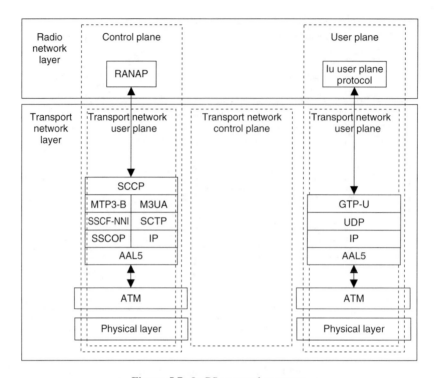

Figure 5.7. Iu PS protocol structure

Control Transmission Protocol), IP (Internet Protocol), and AAL5 which is common to both alternatives. The SCTP layer is specifically designed for signalling transport in the Internet. Specific adaptation layers are specified for different kinds of signalling protocols, such as M3UA for SS7-based signalling.

5.4.2.2   Iu PS Transport Network Control Plane Protocol Stack

The Transport Network Control Plane is not applied to Iu PS. The setting up of the GTP tunnel requires only an identifier for the tunnel, and the IP addresses for both directions, and these are already included in the RANAP RAB Assignment messages. The same information elements that are used in Iu CS for addressing and identifying the AAL2 signalling are used for the User Plane data in Iu CS.

5.4.2.3   Iu PS User Plane Protocol Stack

In the Iu PS User Plane, multiple packet data flows are multiplexed on one or several AAL5 PVCs. The GTP-U (User Plane part of the GPRS Tunnelling Protocol) is the multiplexing layer that provides identities for individual packet data flow. Each flow uses UDP connectionless transport and IP addressing.

5.4.3   RANAP Protocol

RANAP is the signalling protocol in Iu that contains all the control information specified for the Radio Network Layer. The functionality of RANAP is implemented by various RANAP Elementary Procedures. Each RANAP function may require the execution of one or more EPs. Each EP consists of either just the request message (class 2 EP), the request and response message pair (class 1 EP), or one request message and one or more response messages (class 3 EP). The following RANAP functions are defined:

- Relocation. This function handles both SRNS Relocation and Hard Handover, including intersystem case to/from GSM:

    - SRNS Relocation: the serving RNS functionality is relocated from one RNS to another without changing the radio resources and without interrupting the user data flow. The prerequisite for SRNS relocation is that all Radio Links are already in the same DRNC that is the target for the relocation.
    - Inter RNS Hard Handover: used to relocate the serving RNS functionality from one RNS to another and to change the radio resources correspondingly by a hard handover in the Uu interface. The prerequisite for Hard Handover is that the UE is at the border of the source and target cells.

- RAB (Radio Access Bearer) Management. This function combines all RAB handling:

    - RAB Setup, including the possibility for queuing the setup,
    - modification of the characteristics of an existing RAB,
    - clearing an existing RAB, including the RAN-initiated case.

- Iu Release. Releases all resources (Signalling link and U-Plane) from a given instance of Iu related to the specified UE. Also includes the RAN-initiated case.

- Reporting Unsuccessfully Transmitted Data. This function allows the CN to update its charging records with information from UTRAN if part of the data sent was not successfully sent to the UE.

- Common ID management. In this function the permanent identification of the UE is sent from the CN to UTRAN to allow paging coordination from possibly two different CN domains.

- Paging. This is used by CN to page an idle UE for a UE terminating service request, such as a voice call. A paging message is sent from the CN to UTRAN with the UE common identification (permanent Id) and the paging area. UTRAN will either use an existing signalling connection, if one exists, to send the page to the UE or broadcast the paging in the requested area.

- Management of tracing. The CN may, for operation and maintenance purposes, request UTRAN to start recording all activity related to a specific UE–UTRAN connection.

- UE–CN signalling transfer. This functionality provides transparent transfer of UE–CN signalling messages that are not interpreted by UTRAN in three cases:

    - Transfer of the first UE message from UTRAN to UE: this may be, for example, a response to paging, a request of a UE-originated call, or just registration to a new area. It also initiates the signalling connection for the Iu.
    - Direct Transfer: used for carrying all consecutive signalling messages over the Iu signalling connection in both the uplink and downlink directions.

- CN Information Broadcast: allows the CN to set system information to be broadcast repetitively to all users in a specified area.

- Security Mode Control. This is used to set the ciphering or integrity checking on or off. When ciphering is on, the signalling and user data connections in the radio interface are encrypted with a secret key algorithm. When integrity checking is on, an integrity checksum, further secured with a secret key, is added to some or all of the Radio Interface signalling messages. This ensures that the communication partner has not changed, and the content of the information has not been altered.

- Management of overload. This is used to control the load over the Iu interface against overload due, for example, to processor overload at the CN or UTRAN. A simple mechanism is applied that allows stepwise reduction of the load and its stepwise resumption, triggered by a timer.

- Reset. This is used to reset the CN or the UTRAN side of the Iu interface in error situations. One end of the Iu may indicate to the other end that it is recovering from a restart, and the other end can remove all previously established connections.

- Location Reporting. This functionality allows the CN to receive information on the location of a given UE. It includes two elementary procedures, one for controlling the location reporting in the RNC and the other to send the actual report to the CN.

5.4.4 Iu User Plane Protocol

The Iu User Plane protocol is in the Radio Network Layer of the Iu User Plane. It has been defined so that it would be, as much as possible, independent of the CN domain that it is used for. The purpose of the User Plane protocol is to carry user data related to RABs over the Iu interface. Each RAB has its own instance of the protocol. The protocol performs either a fully transparent operation, or framing for the user data segments and some basic control signalling to be used for initialisation and online control. Based on these cases, the protocol has two modes:

- Transparent Mode. In this mode of operation the protocol does not perform any framing or control. It is applied for RABs that do not require such features but that assume fully transparent operation.

- Support Mode for predefined SDU sizes. In this mode the User Plane performs framing of the user data into segments of predefined size. The SDU sizes typically correspond to AMR (Adaptive Multirate Codec) speech frames, or to the frame sizes derived from the data rate of a CS data call. Also, control procedures for initialisation and rate control are defined, and a functionality is specified for indicating the quality of the frame based, for example, on CRC from the radio interface.

5.5 UTRAN Internal Interfaces

5.5.1 RNC–RNC Interface (Iur Interface) and the RNSAP Signalling

The protocol stack of the RNC to RNC interface (Iur interface) is shown in Figure 5.8.

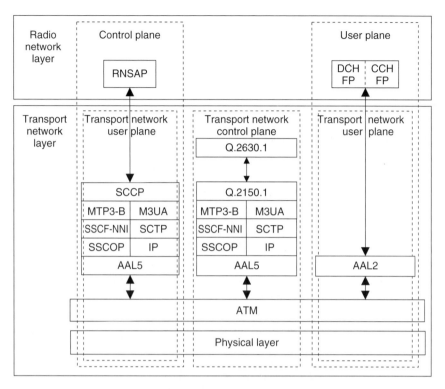

Figure 5.8. Protocol stack of the Iur interface. As for the Iu interface, two options are possible for the transport of the RNSAP signalling: the SS7 stack (SCCP and MTP3b) and the new SCTP/IP based transport. Two User Plane protocols are defined (DCH: dedicated channel; CCH: common channel)

Although this interface was initially designed in order to support the inter-RNC soft handover (shown on the left-hand side of Figure 5.4), more features were added during the development of the standard and now the Iur interface provides four distinct functions:

1. Support of Basic Inter-RNC Mobility
2. Support of Dedicated Channel Traffic
3. Support of Common Channel Traffic
4. Support of Global Resource Management

For this reason the Iur signalling protocol itself (RNSAP, *Radio Network System Application Part*) is divided into four different *modules* (to be intended as groups of procedures). In general, it is possible to implement only part of the four Iur functions between two Radio Network Controllers, according to the operator's need.

5.5.1.1 Iur1: Support of the Basic Inter-RNC Mobility

This functionality requires the *basic* module of RNSAP signalling as described in [25.423]. This first brick for the construction of the Iur interfaces provides by itself the functionality needed for the mobility of the user between the two RNCs, but does not support the exchange of any user data traffic. If this module is not implemented, the Iur interface as such does not exist, and the only way for a user connected to UTRAN via the RNS1 to utilise a cell in RNS2 is to disconnect itself temporarily from UTRAN (release the RRC connection).

The functions offered by the Iur basic module include:

- *Support of SRNC relocation*
- *Support of inter-RNC cell and UTRAN registration area update*
- *Support of inter-RNC packet paging*
- *Reporting of protocol errors*

Since this functionality does not involve user data traffic across Iur, the User Plane and the Transport Network Control Plane protocols are not needed.

5.5.1.2   Iur2: Support of Dedicated Channel Traffic

This functionality requires the *Dedicated Channel* module of RNSAP signalling and allows the dedicated channel traffic between two RNCs. Even if the initial need for this functionality is to support the inter-RNC soft handover state, it also allows the anchoring of the SRNC for all the time the user is utilising dedicated channels (dedicated resources in the Node B), commonly for as long as the user has an active connection to the circuit-switched domain.

This functionality requires also the User Plane *Frame Protocol* for the dedicated channel, plus the Transport Network Control Plane protocol (*Q.2630.1*) used for the setup of the transport connections (AAL2 connections). Each dedicated channel is conveyed over one transport connection, except the coordinated DCH used to obtain unequal error protection in the air interface.

The *Frame Protocol* for dedicated channels, in short DCH FP [25.427], defines the structure of the *data frames* carrying the user data and the *control frames* used to exchange measurements and control information. For this reason, the Frame Protocol specifies also simple messages and procedures. The user data frames are normally routed transparently through the DRNC; thus the Iur frame protocol is used also in Iub and referred to as Iur/Iub DCH FP.

The functions offered by the Iur DCH module are:

- *Establishment, modification and release of the dedicated channel in the DRNC due to hard and soft handover in the dedicated channel state*
- *Setup and release of dedicated transport connections across the Iur interface*
- *Transfer of DCH Transport Blocks between SRNC and DRNC*
- *Management of the radio links in the DRNS, via dedicated measurement report procedures and power setting procedures*

5.5.1.3   Iur3: Support of Common Channel Traffic

This functionality allows the handling of common and shared channel data streams across the Iur interface. It requires the Common Transport Channel module of the RNSAP protocol and the Iur Common Transport Channel Frame Protocol (in short, CCH FP). The Q.2630.1

signalling protocol of the Transport Network Control Plane is also needed if signalled AAL2 connections are used.

If this functionality is not implemented, every inter-RNC cell update always triggers an SRNC relocation, i.e. the serving RNC is always the RNC controlling the cell used for common or shared channel transport.

The identification of the benefits of this feature caused a long debate in the relevant standardisation body. On the one hand, this feature allows the implementation of the total anchor RNC concept, avoiding the complex SRNC relocation procedure (via the CN); on the other hand, it requires the splitting of the Medium Access Control layer functionality into two network elements, generating inefficiency in the utilisation of the resources and complexity in the Iur interface. The debate could not reach an agreement, thus the feature is supported by the standard but is not essential for the operation of the system.

The functions offered by the Iur common transport channel module are:

- *Setup and release of the transport connection across the Iur for common channel data streams*
- *Splitting of the MAC layer between the SRNC (MAC-d) and the DRNC (MAC-c and MAC-sh). The scheduling for DL data transmission is performed in the DRNC*
- *Flow control between the MAC-d and MAC-c/MAC-sh*

5.5.1.4   Iur4: Support of Global Resource Management

This functionality provides signalling to support enhanced radio resource and O&M features across the Iur interface. It is implemented via the *global module* of the RNSAP protocol, and does not require any User Plane protocol, since there is no transmission of user data across the Iur interface. The function is considered optional. This function is not present in the Release'99 UTRAN specification, but it is foreseen to be introduced in subsequent releases for the support of advanced positioning methods and Iur optimisation purposes.

The functions offered by the Iur global resource module are:

- *Transfer of cell measurements between two RNCs*
- *Transfer of Node B timing information between two RNCs*

5.5.2   RNC–Node B Interface and the NBAP Signalling

The protocol stack of the RNC–Node B interface (Iub interface) is shown, with the typical triple plane notation, in Figure 5.9.

In order to understand the structure of the interface, it is necessary to briefly introduce the logical model of the Node B, depicted in Figure 5.10. This consists of a common control port (a common signalling link) and a set of traffic termination points, each controlled by a dedicated control port (dedicated signalling link). One traffic termination point controls a number of mobiles having dedicated resources in the Node B, and the corresponding traffic is conveyed through dedicated data ports. Common data ports outside the traffic termination points are used to convey RACH, FACH and PCH traffic.

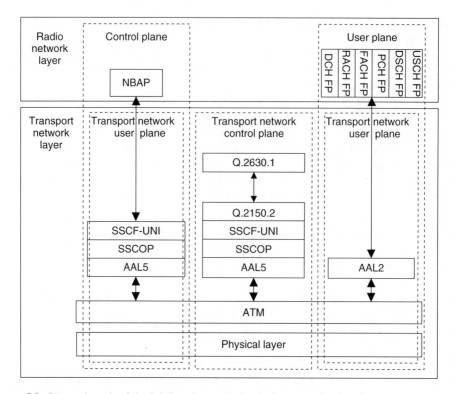

Figure 5.9. Protocol stack of the Iub interface. This is similar to the Iur interface protocol, the main difference being that in the Radio Network and Transport Network Control Planes the SS7 stack is replaced by the simpler SAAL-UNI as signalling bearer. Note also that the SCTP/IP option is not present here

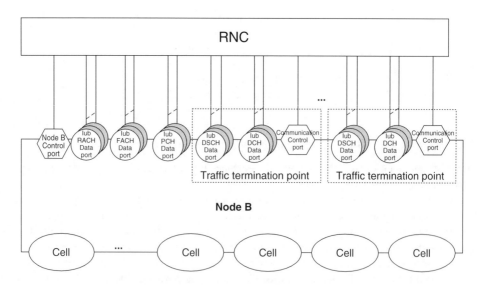

Figure 5.10. Logical model of the Node B for FDD

Note that there is no relation between the traffic termination point and the cells, i.e. one traffic termination point can control more than one cell, and one cell can be controlled by more than one traffic termination point.

The Iub interface signalling (NBAP, *Node B Application Part*) is divided into two essential components: the common NBAP, that defines the signalling procedures across the common signalling link, and the dedicated NBAP, used in the dedicated signalling link.

The User Plane Iub frame protocols define the structures of the frames and the basic in-band control procedures for every type of transport channel (i.e. for every type of data port of the model). The Q.2630.1 signalling is used for the dynamic management of the AAL2 connections used in the User Plane.

5.5.2.1 Common NBAP and the Logical O&M

The common NBAP (C-NBAP) procedures are used for the signalling that is not related to one specific UE context already existing in the Node B. In particular, the C-NBAP defines all the procedures for the logical O&M (Operation and Maintenance) of the Node B, such as configuration and fault management.

The main functions of the Common NBAP are:

- *Setup of the first RL of one UE, and selection of the traffic termination point*
- *Cell configuration*
- *Handling of the RACH/FACH and PCH channels*
- *Initialisation and reporting of Cell or Node B specific measurement*
- *Fault management*

5.5.2.2 Dedicated NBAP

When the RNC requests the first radio link for one UE via the C-NBAP *Radio Link Setup* procedure, the Node B assigns a traffic termination point for the handling of this UE context, and every subsequent signalling related to this mobile is exchanged with dedicated NBAP (D-NBAP) procedures across the dedicated control port of the given Traffic Termination Point.

The main functions of the Dedicated NBAP are:

- *Addition, release and reconfiguration of radio links for one UE context*
- *Handling of dedicated and shared channels*
- *Handling of softer combining*
- *Initialisation and reporting of radio link specific measurement*
- *Radio link fault management*

References

[1]   3GPP Technical Specification 25.401 UTRAN Overall Description.
[2]   3GPP Technical Specification 25.410 UTRAN Iu Interface: General Aspects and Principles.
[3]   3GPP Technical Specification 25.411 UTRAN Iu Interface: Layer 1.

[4]   3GPP Technical Specification 25.412 UTRAN Iu Interface: Signalling Transport.
[5]   3GPP Technical Specification 25.413 UTRAN Iu Interface: RANAP Signalling.
[6]   3GPP Technical Specification 25.414 UTRAN Iu Interface: Data transport and Transport Signalling.
[7]   3GPP Technical Specification 25.415 UTRAN Iu Interface: CN-RAN User Plane Protocol.
[8]   3GPP Technical Specification 25.420 UTRAN Iur Interface: General Aspects and Principles.
[9]   3GPP Technical Specification 25.421 UTRAN Iur Interface: Layer 1.
[10] 3GPP Technical Specification 25.422 UTRAN Iur Interface: Signalling Transport.
[11] 3GPP Technical Specification 25.423 UTRAN Iur Interface: RNSAP Signalling.
[12] 3GPP Technical Specification 25.424 UTRAN Iur Interface: Data Transport and Transport Signalling for CCH Data Streams.
[13] 3GPP Technical Specification 25.425 UTRAN Iur Interface: User Plane Protocols for CCH Data Streams.
[14] 3GPP Technical Specification 25.426 UTRAN Iur and Iub Interface Data Transport and Transport Signalling for DCH Data Streams.
[15] 3GPP Technical Specification 25.427 UTRAN Iur and Iub Interface User Plane Protocols for DCH Data Streams.
[16] 3GPP Technical Specification 25.430 UTRAN Iub Interface: General Aspects and Principles.
[17] 3GPP Technical Specification 25.431 UTRAN Iub Interface: Layer 1.
[18] 3GPP Technical Specification 25.432 UTRAN Iub Interface: Signalling Transport.
[19] 3GPP Technical Specification 25.433 UTRAN Iub Interface: NBAP Signalling.
[20] 3GPP Technical Specification 25.434 UTRAN Iub Interface: Data Transport and Transport Signalling for CCH Data Streams.
[21] 3GPP Technical Specification 25.435 UTRAN Iub Interface: User Plane Protocols for CCH Data Streams.

6

Physical Layer

Antti Toskala

6.1 Introduction

In this chapter the WCDMA (UTRA FDD) physical layer is described. The physical layer of the radio interface has been typically the main discussion topic when different cellular systems have been compared against each other. The physical layer structures naturally relate directly to the achievable performance issues, when observing a single link between a terminal station and a base station. For the overall system performance the protocols in the other layers, such as handover protocols, also have a great deal of impact. Naturally it is essential to have low Signal-to-Interference Ratio (SIR) requirements for sufficient link performance with various coding and diversity solutions in the physical layer, since the physical layer defines the fundamental capacity limits. The performance of the WCDMA physical layer is described in detail in Chapter 11.

The physical layer has a major impact on equipment complexity with respect to the required baseband processing power in the terminal station and base station equipment. As well as the diversity benefits on the performance side, the wideband nature of WCDMA also offers new challenges in its implementation. As third generation systems are wideband from the service point of view as well, the physical layer cannot be designed around only a single service, such as speech; more flexibility is needed for future service introduction. The new requirements of the third generation systems and for the air interface are summarised in Section 1.4. This chapter presents the WCDMA physical layer solutions to meet those requirements.

This chapter uses the term 'terminal' for the user equipment. In 3GPP terminology the terms User Equipment (UE) and Mobile Equipment (ME) are often used, the difference being that UE also covers the Subscriber Identification Module (SIM) as shown in Chapter 5, in which the UTRA network architecture is presented. The term 'base station' is also used throughout this chapter, though in part of the 3GPP specifications the term Node B is used to represent the parts of the base station that contain the relevant parts from the physical layer perspective. The UTRA FDD physical layer specifications are contained in references [1–5].

WCDMA for UMTS, edited by Harri Holma and Antti Toskala
© 2001 John Wiley & Sons, Ltd

₊ᵤₛ chapter has been divided as follows. First, the transport channels are described together with their mapping to different physical channels in Section 6.2. Spreading and modulation for uplink and downlink are presented in Section 6.3, and the physical channels for user data and control data are described in Sections 6.4 and 6.5. In Section 6.6 the key physical layer procedures, such as power control and handover measurements, are covered.

6.2 Transport Channels and their Mapping to the Physical Channels

In UTRA the data generated at higher layers is carried over the air with transport channels, which are mapped in the physical layer to different physical channels. The physical layer is required to support variable bit rate transport channels to offer bandwidth-on-demand services, and to be able to multiplex several services to one connection. This section presents the mapping of the transport channels to the physical channels, and how those two requirements are taken into account in the mapping.

Each transport channel is accompanied by the Transport Format Indicator (TFI) at each time event at which data is expected to arrive for the specific transport channel from the higher layers. The physical layer combines the TFI information from different transport channels to the Transport Format Combination Indicator (TFCI). The TFCI is transmitted in the physical control channel to inform the receiver which transport channels are active for the current frame; the exception to this is the use of Blind Transport Format Detection (BTFD) that will be covered in connection with the downlink dedicated channels. The TFCI is decoded appropriately in the receiver and the resulting TFI is given to higher layers for each of the transport channels that can be active for the connection. In Figure 6.1 two transport channels are mapped to a single physical channel, and also error indication is

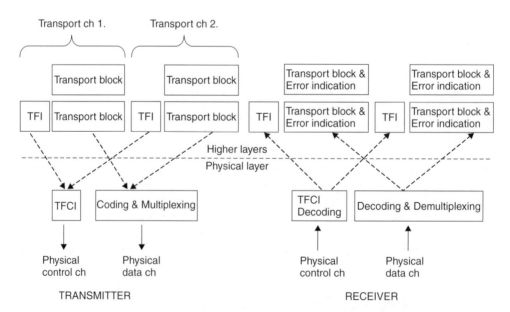

Figure 6.1. The interface between higher layers and the physical layer

provided for each transport block. The transport channels may have a different number of blocks and at any moment not all the transport channels are necessarily active.

One physical control channel and one or more physical data channels form a single Coded Composite Transport Channel (CCTrCh). There can be more than one CCTrCh on a given connection but only one physical layer control channel is transmitted in such a case.

The interface between higher layers and the physical layer is less relevant for terminal implementation, since basically everything takes place within the same equipment, thus the interfacing here is rather a tool for specification work. For the network side the division of functions between physical and higher layers is more important, since there the interface between physical and higher layers is represented by the Iub-interface between the base station and Radio Network Controller (RNC) as described in Chapter 5. In the 3GPP specification the interfacing between physical layer and higher layers is covered in [6].

Two types of transport channels exist: dedicated channels and common channels. The main difference between them is that a common channel is a resource divided between all or a group of users in a cell, whereas a dedicated channel resource, identified by a certain code on a certain frequency, is reserved for a single user only. The transport channels are compared in Section 10.3 for the transmission of packet data.

6.2.1   Dedicated Transport Channel

The only dedicated transport channel is the dedicated channel, for which the term DCH is used in the 25-series of the UTRA specification. The dedicated transport channel carries all the information intended for the given user coming from layers above the physical layer, including data for the actual service as well as higher layer control information. The content of the information carried on the DCH is not visible to the physical layer, thus higher layer control information and user data are treated in the same way. Naturally the physical layer parameters set by UTRAN may vary between control and data.

The familiar GSM channels, the traffic channel (TRCH) or associated control channel (ACCH), do not exist in UTRA physical layer. The dedicated transport channel carries both the service data, such as speech frames, and higher layer control information, such as handover commands or measurement reports from the terminal. In WCDMA a separate transport channel is not needed because of the support of variable bit rate and service multiplexing.

The dedicated transport channel is characterised by features such as fast power control, fast data rate change on a frame-by-frame basis, and the possibility of transmission to a certain part of the cell or sector with varying antenna weights with adaptive antenna systems. The dedicated channel supports soft handover.

6.2.2   Common Transport Channels

There are currently six different common transport channel types defined for UTRA, which are introduced in the following sections. There are a few differences from second generation systems, for example transmission of packet data on the common channels, and a downlink shared channel for transmitting packet data. Common channels do not have soft handover but some of them can have fast power control.

6.2.2.1 Broadcast Channel

The Broadcast Channel (BCH) is a transport channel that is used to transmit information specific to the UTRA network or for a given cell. The most typical data needed in every network is the available random access codes and access slots in the cell, or the types of transmit diversity methods used with other channels for that cell. As the terminal cannot register to the cell without the possibility of decoding the broadcast channel, this channel is needed for transmission with relatively high power in order to reach all the users within the intended coverage area. From a practical viewpoint the information rate on the broadcast channel is limited by the ability of low-end terminals to decode the data rate of the broadcast channel, resulting in a low and fixed data rate for the UTRA broadcast channel.

6.2.2.2 Forward Access Channel

The Forward Access Channel (FACH) is a downlink transport channel that carries control information to terminals known to locate in the given cell. This is so, for example, after a random access message has been received by the base station. It is also possible to transmit packet data on the FACH. There can be more than one FACH in a cell. One of the forward access channels must have such a low bit rate that it can be received by all the terminals in the cell area. When there is more than one FACH, the additional channels can have a higher data rate as well. The FACH does not use fast power control, and the messages transmitted need to include inband identification information to ensure their correct receipt.

6.2.2.3 Paging Channel

The Paging Channel (PCH) is a downlink transport channel that carries data relevant to the paging procedure, that is, when the network wants to initiate communication with the terminal. The simplest example is a speech call to the terminal: the network transmits the paging message to the terminal on the paging channel of those cells belonging to the location area that the terminal is expected to be in. The identical paging message can be transmitted in a single cell or in up to a few hundreds of cells, depending on the system configuration. The terminals must be able to receive the paging information in the whole cell area. The design of the paging channel affects also the terminal's power consumption in the standby mode. The less often the terminal has to tune the receiver in to listen for a possible paging message, the longer will the terminal's battery last in the standby mode.

6.2.2.4 Random Access Channel

The Random Access Channel (RACH) is an uplink transport channel intended to be used to carry control information from the terminal, such as requests to set up a connection. It can also be used to send small amounts of packet data from the terminal to the network. For proper system operation the random access channel must be heard from the whole desired cell coverage area, which also means that practical data rates have to be rather low, at least for the initial system access and other control procedures. The coverage of the random access channel compared to the dedicated channel is presented in Section 11.2.2.

6.2.2.5 Uplink Common Packet Channel

The uplink common packet channel (CPCH) is an extension to the RACH channel that is intended to carry packet-based user data in the uplink direction. The pair providing the data in the downlink direction is the FACH. In the physical layer, the main differences from the RACH are the use of fast power control, a physical layer-based collision detection mechanism and a CPCH status monitoring procedure. The uplink CPCH transmission may last several frames in contrast with one or two frames for the RACH message.

6.2.2.6 Downlink Shared Channel

The downlink shared channel (DSCH) is a transport channel intended to carry dedicated user data and/or control information; it can be shared by several users. In many respects it is similar to the forward access channel, but the shared channel supports the use of fast power control as well as variable bit rate on a frame-by-frame basis. The DSCH does not need to be heard in the whole cell area and can employ the different modes of transmit antenna diversity methods that are used with the associated downlink DCH. The downlink shared channel is always associated with a downlink DCH.

6.2.2.7 Required Transport Channels

The common transport channels needed for the basic network operation are RACH, FACH and PCH, while the use of DSCH and CPCH is optional and can be decided by the network.

6.2.3 Mapping of Transport Channels onto the Physical Channels

The way different transport channels are mapped to different physical channels is shown in Figure 6.2, though some of the transport channels are carried by identical (or even the same) physical channel.

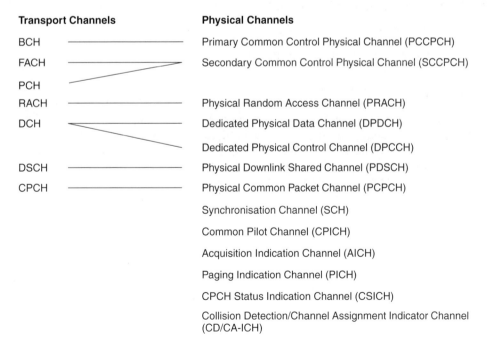

Figure 6.2. Transport-channel to physical-channel mapping

In addition to the transport channels introduced earlier, there exist physical channels to carry only information relevant to physical layer procedures. The Synchronisation Channel (SCH), the Common Pilot Channel (CPICH) and the Acquisition Indication Channel (AICH) are not directly visible to higher layers and are mandatory from the system function point of view, to be transmitted from every base station. The CPCH Status Indication Channel (CSICH) and the Collision Detection/Channel Assignment Indication Channel (CD/CA-ICH) are needed if CPCH is used.

The dedicated channel (DCH) is mapped onto two physical channels. The Dedicated Physical Data Channel (DPDCH) carries higher layer information, including user data, while the Dedicated Physical Control Channel (DPCCH) carries the necessary physical layer control information. These two dedicated physical channels are needed to support efficiently the variable bit rate in the physical layer. The bit rate of DPCCH is constant, while the bit rate of DPDCH can change from frame to frame.

6.2.4 Frame Structure of Transport Channels

The UTRA channels use the 10 ms radio frame structure. The longer period used is the system frame period. The System Frame Number (SFN) is a 12-bit number used by several procedures that span more than a single frame. Physical layer procedures, such as the paging procedure or random access procedure, are examples of procedures that need a longer period than 10 ms for correct definition.

6.3   Spreading and Modulation

6.3.1 Scrambling

The concept of spreading the information in a CDMA system is introduced in Chapter 3. In addition to spreading, part of the process in the transmitter is the scrambling operation. This is needed to separate terminals or base stations from each other. Scrambling is used on top of spreading, so it does not change the signal bandwidth but only makes the signals from different sources separable from each other. With the scrambling, it would not matter if the actual spreading were done with identical code for several transmitters. Figure 6.3 shows the relation of the chip rate in the channel to spreading and scrambling in UTRA. As the chip rate is already achieved in the spreading by the channelisation codes, the symbol rate is not affected by the scrambling. The concept of channelisation codes is covered in the following section.

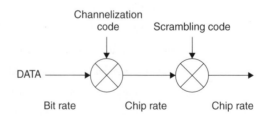

Figure 6.3. Relation between spreading and scrambling

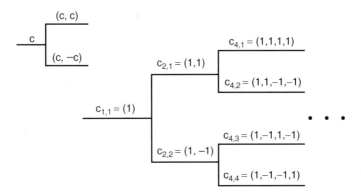

Figure 6.4. Beginning of the channelisation code tree

6.3.2 Channelisation Codes

Transmissions from a single source are separated by channelisation codes, i.e. downlink connections within one sector and the dedicated physical channel in the uplink from one terminal. The spreading/channelisation codes of UTRA are based on the Orthogonal Variable Spreading Factor (OVSF) technique, which was originally proposed in [7].

The use of OVSF codes allows the spreading factor to be changed and orthogonality between different spreading codes of different lengths to be maintained. The codes are picked from the code tree, which is illustrated in Figure 6.4. In case the connection uses a variable spreading factor, the proper use of the code tree also allows despreading according to the smallest spreading factor. This requires only that channelisation codes are used from the branch indicated by the code used for the smallest spreading factor.

There are certain restrictions as to which of the channelisation codes can be used for a transmission from a single source. Another physical channel may use a certain code in the tree if no other physical channel to be transmitted using the same code tree is using a code that is on an underlying branch, i.e. using a higher spreading factor code generated from the intended spreading code to be used. Neither can a smaller spreading factor code on the path to the root of the tree be used. The downlink orthogonal codes within each base station are managed by the radio network controller (RNC) in the network.

The functionality and characteristics of the scrambling and channelisation codes are summarised in Table 6.1. Their usage will be described in more detail in Section 6.3.3.

The definition for the same code tree means that for transmission from a single source, from either a terminal or a base station, one code tree is used with one scrambling code on top of the tree. This means that different terminals and different base stations may operate their code trees totally independently of each other; there is no need to coordinate the code tree resource usage between different base stations or terminals.

6.3.3 Uplink Spreading and Modulation

6.3.3.1 Uplink Modulation

In the uplink direction there are basically two additional terminal-oriented criteria that need to be taken into account in the definition of the modulation and spreading methods. The

Table 6.1. Functionality of the channelisation and scrambling codes

	Channelisation code	Scrambling code
Usage	Uplink: Separation of physical data (DPDCH) and control channels (DPCCH) from same terminal Downlink: Separation of downlink connections to different users within one cell	Uplink: Separation of terminal Downlink: Separation of sectors (cells)
Length	4–256 chips (1.0–66.7 μs) Downlink also 512 chips	Uplink: (1) 10 ms = 38400 chips or (2) 66.7μs = 256 chips Option (2) can be used with advanced base station receivers Downlink: 10 ms = 38400 chips
Number of codes	Number of codes under one scrambling code = spreading factor	Uplink: Several millions Downlink: 512
Code family	Orthogonal Variable Spreading Factor	Long 10 ms code: Gold code Short code: Extended S(2) code family
Spreading	Yes, increases transmission bandwidth	No, does not affect transmission bandwidth

uplink modulation should be designed so that the terminal amplifier efficiency is maximised and/or the audible interference from the terminal transmission is minimised.

Discontinuous uplink transmission can cause audible interference to audio equipment that is very close to the terminal, such as hearing aids. This is a completely separate issue from the interference in the air interface. The audible interference is only a nuisance for the user and does not affect network performance, such as its capacity. With GSM operation we are familiar with the occasional audible interference with audio equipment that is not properly protected. The interference from GSM has a frequency of 217 Hz, which is determined by the GSM frame frequency. This interference falls into the band that can be heard by the human ear. With a CDMA system, the same issues arise when discontinuous uplink transmission is used, for example with a speech service. During the silent periods no information bits need to be transmitted, only the information for link maintenance purposes, such as power control with a 1.5 kHz command rate. With such a rate the transmission of the pilot and the power control symbols with time multiplexing in the uplink direction would cause audible interference in the middle of the telephony voice frequency band. Therefore, in a WCDMA uplink the two dedicated physical channels are not time multiplexed but I-Q/code multiplexing is used.

The continuous transmission achieved with an I-Q/code multiplexed control channel is shown in Figure 6.5. Now, as the pilot and the power control signalling are maintained on a separate continuous channel, no pulsed transmission occurs. The only pulse occurs when the data channel DPDCH is switched on and off, but such switching happens quite seldom.

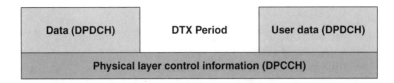

Figure 6.5. Parallel transmission of DPDCH and DPCCH when data is present/absent (DTX)

The average interference to other users and the cellular capacity remain the same as in the time-multiplexed solution. In addition, the link level performance is the same in both schemes if the energy allocated to the pilot and the power control signalling is the same.

For the best possible power amplifier efficiency, the terminal transmission should have as low peak-to-average (PAR) ratio as possible to allow the terminal to operate with a minimal amplifier back-off requirement, mapping directly to the amplifier power conversion efficiency, which in turn is directly proportional to the terminal talk time. With the I-Q/code multiplexing, called also dual-channel QPSK modulation, the power levels of the DPDCH and DPCCH are typically different, especially as data rates increase and would lead in extreme cases to BPSK-type transmission when transmitting the branches independently. This has been avoided by using a complex-valued scrambling operation after the spreading with channelisation codes.

The signal constellation of the I-Q/code multiplexing before complex scrambling is shown in Figure 6.6. The same constellation is obtained after descrambling in the receiver for the data detection.

The transmission of two parallel channels, DPDCH and DPCCH, leads to multicode transmission, which increases the peak-to-average power ratio (crest factor). In Figure 6.6 the peak-to-average ratio changes when G (the relative strengths of the DPDCH and DPCCH) is changed. By using the spreading modulation solution shown in Figure 6.7 the transmitter power amplifier efficiency remains the same as for normal balanced QPSK transmission in general. The complex scrambling codes are formed in such a way that the rotations between consecutive chips within one symbol period are limited to $\pm 90°$. The full $180°$ rotation can happen only between consecutive symbols. This method further reduces the peak-to-average ratio of the transmitted signal from the normal QPSK transmission.

The efficiency of the power amplifier remains constant irrespective of the power difference G between DPDCH and DPCCH. This can be explained with Figure 6.8, which shows

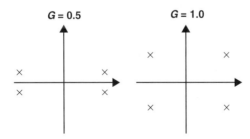

Figure 6.6. Constellation of I-Q/code multiplexing before complex scrambling. G denotes the relative gain factor between DPCCH and DPDCH branches

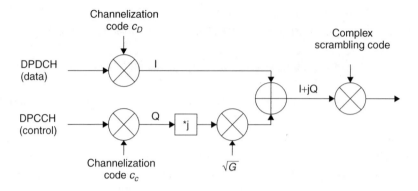

Figure 6.7. I-Q/code multiplexing with complex scrambling

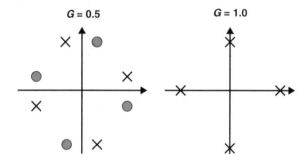

Figure 6.8. Signal constellation for I-Q/code multiplexed control channel with complex scrambling. G denotes the power difference between DPCCH and DPDCH

the signal constellation for the I-Q/code multiplexed control channel with complex spreading. In the middle constellation with $G = 0.5$ the possible constellation points are only circles or only crosses during one symbol period. Their constellation is the same as for the rotated QPSK. Thus the signal envelope variations with complex spreading are very similar to QPSK transmission for all values of G. The I-Q/code multiplexing solution with complex scrambling results in power amplifier output back-off requirements that remain constant as a function of the power difference between DPDCH and DPCCH.

The power difference between DPDCH and DPCCH has been quantified in the UTRA physical layer specifications to 4-bit words, i.e. 16 different values. At a given point in time the gain value for either DPDCH or DPCCH is set to 1 and then for the other channel a value between 0 and 1 is applied to reflect the desired power difference between the channels. Limiting the number of possible values to 4-bit representation is necessary to make the terminal transmitter implementation simple. The power differences can have 15 different values between -23.5 dB and 0.0 dB and one bit combination for no DPDCH when there is no data to be transmitted.

UTRA will face challenges in amplifier efficiency when compared to GSM. The GSM modulation is GMSK (Gaussian Minimum Shift Keying) which has a constant envelope and is thus optimised for amplifier peak-to-average ratio. As a narrowband system, the

GSM signal can be spread relatively more widely in the frequency domain. This allows the use of a less linear amplifier with better power conversion efficiency. The narrowband amplifier is also easier to linearise if necessary. In practice, the efficiency of the WCDMA power amplifier is slightly lower than that of the GSM power amplifier. On the other hand, WCDMA uses fast power control in uplink, which reduces the average required uplink transmission power.

Instead of applying combined I-Q and code multiplexing with complex scrambling, it would be possible to use pure code multiplexing. With code multiplexing, multicode transmission occurs with parallel control and data channels. This approach increases transmitted signal envelope variations and sets higher requirements for power amplifier linearity. Especially for low bit rates, as for speech, the control channel can have an amplitude more than 50% that of the data channel, which causes more envelope variations than the combined I-Q/code multiplexing solution.

6.3.3.2 Uplink Spreading

For the uplink DPCCH spreading code, there is an additional restriction. The same code cannot be used by any another code channel even on a different I or Q branch. The reason for this restriction is that physical channels transmitted with the same channelisation codes on I and Q branches with the dual channel QPSK principle cannot be separated before the DPCCH has been detected and channel phase estimates are available.

In the uplink direction the spreading factor on the DPDCH may vary on a frame-by-frame basis. The spreading codes are always taken from the earlier described code tree. When the channelisation code used for spreading is always taken from the same branch of the code tree, the despreading operation can take advantage of the code tree structure and avoid chip-level buffering. The terminal provides data rate information, or more precisely the Transport Format Combination Indicator (TFCI), on the DPCCH, to allow data detection with a variable spreading factor on the DPDCH.

6.3.3.3 Uplink Scrambling Codes

The transmissions from different sources are separated by the scrambling codes. In the uplink direction there are two alternatives: short and long scrambling codes. The long codes with 25 degree generator polynomials are truncated to the 10 ms frame length, resulting in 38400 chips with 3.84 Mcps. The short scrambling code length is 256 chips. The long scrambling codes are used if the base station uses a Rake receiver. The Rake receiver is described in Section 3.4. If advanced multiuser detectors or interference cancellation receivers are used in the base station, short scrambling codes can be used to make the implementation of the advanced receiver structures easier. The base station multiuser detection algorithms are introduced in Section 11.5.2. Both of the two scrambling code families contain millions of scrambling codes, thus in the uplink direction the code planning is not needed.

The short scrambling codes have been chosen from the extended S(2) code family. The long codes are Gold codes. The complex-valued scrambling sequence is formed in the case of short codes by combining two codes, and in the case of long codes from a single sequence where the other sequence is the delayed version of the first one.

The complex-valued scrambling code can be formed from two real-valued codes c_1 and c_2 with the decimation principle as:

$$c_{\text{scrambling}} = c_1(w_0 + jc_2(2k)w_1), \quad k = 0, 1, 2 \ldots \tag{6.1}$$

with sequences w_0 and w_1 given as chip rate sequences:

$$w_0 = \{1 \quad 1\}, w_1 = \{1 \quad -1\} \tag{6.2}$$

The decimation factor with the second code is 2. This way of creating the scrambling codes will reduce the zero crossings in the constellation and will further reduce the amplitude variations in the modulation process.

6.3.3.4 Spreading and Modulation on Uplink Common Channels

The Random Access Channel (RACH) contains preambles that are sent using the same scrambling code sequence as with the uplink transmission, the difference being that only 4096 chips from the beginning of the code period are needed and the modulation state transitions are limited in a different way. The spreading and scrambling process on the RACH is BPSK-valued, thus only one sequence is used to spread and scramble both the in-phase and quadrature branches. This has been chosen to reduce the complexity of the required matched filter in the base station receivers for the RACH reception.

The RACH message part spreading and modulation, including scrambling, is identical to that for the dedicated channel. The codes available for RACH scrambling use are transmitted on the BCH of each cell.

For the peak-to-average reduction, an additional rotation function is used on the RACH preamble, given as:

$$b(k) = a(k)e^{j\left(\frac{\pi}{4}+\frac{\pi}{2}\right)}, k = 0, 1, 2, \ldots, 4095 \tag{6.3}$$

where $a(k)$ is the binary preamble and $b(k)$ is the resulting complex-valued preamble with limited 90° phase transition between chips. The autocorrelation properties are not affected by this operation.

The RACH preambles have a modulation pattern on top of them, called signature sequences. These have been defined by taking the higher Doppler frequencies as well as frequency errors into account. The sequences have been generated from 16 symbols, which have additionally been interleaved over the preamble duration to avoid large inter-sequence cross-correlations in case of large frequency errors that could otherwise severely degrade the cross-correlation properties between the signature sequences. The 16 signature sequences have been specified for RACH use, but there can be multiple scrambling codes each using the same set of signatures.

The CPCH spreading and modulation are identical to those of the RACH in order to maximise the commonality for both terminal and base station implementation when supporting CPCH. RACH and CPCH processes will be described in more detail in connection with the physical layer procedures.

6.3.4 Downlink Spreading and Modulation

6.3.4.1 Downlink Modulation

In the downlink direction normal QPSK modulation has been chosen with time-multiplexed control and data streams. The time-multiplexed solution is not used in the uplink because it would generate audible interference during discontinuous transmission. The audible interference generated with DTX is not a relevant issue in the downlink since the common channels have continuous transmission in any case. Also, as there exist several parallel code transmissions in the downlink, similar optimisation for peak-to-average (PAR) ratio as with single code (pair) transmission is not relevant. Also, reserving a channelisation code just for DPCCH purposes results in slightly worse code resource utilisation when sending several transmissions from a single source.

Since the I and Q branches have equal power, the scrambling operation does not provide a similar difference to the envelope variations as in the uplink. The discontinuous transmission is implemented by gating the transmission on and off.

6.3.4.2 Downlink Spreading

The spreading in the downlink is based on the channelisation codes, as in the uplink. The code tree under a single scrambling code is shared by several users; typically only one scrambling code and thus only one code tree is used per sector in the base station. The common channels and dedicated channels share the same code tree resource. There is one exception for the physical channels: the synchronisation channel (SCH), which is not under a downlink scrambling code. The SCH spreading codes are covered in a later section.

In the downlink, the dedicated channel spreading factor does not vary on a frame-by-frame basis; the data rate variation is taken care of with either a rate matching operation or with discontinuous transmission, where the transmission is off during part of the slot.

In the case of multicode transmission for a single user, the parallel code channels have different channelisation codes and are under the same scrambling code as normally are all the code channels transmitted from the base station. The spreading factor is the same for all the codes with multicode transmission. Each coded composite transport channel (CCTrCh) may have different spreading factor even if received by the same terminal.

The special case in the downlink direction is the downlink shared channel (DSCH) which may use a variable spreading factor on a frame-by-frame basis. In this case the channelisation codes taking care of the spreading are allocated from the same branch of the code tree to ease the terminal implementation. The restriction specified is illustrated in Figure 6.9 which shows the spreading factor for maximum data rate and the part of the code tree that may be used by the network to allocate codes when the lower data rate is needed. In such a frame-by-frame operation the DPCCH of the dedicated channel contains the TFCI information,

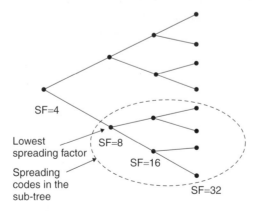

Figure 6.9. DSCH code tree example

which informs the receiver of the spreading code used, as well as other transport format parameters for the DSCH.

6.3.4.3   Downlink Scrambling

The downlink scrambling uses long codes, the same Gold codes as in the uplink. The complex-valued scrambling code is formed from a single code by simply having a delay between the I and Q branches. The code period is truncated to 10 ms; no short codes are used in the downlink direction. The downlink set of the (primary) scrambling codes is limited to 512 codes, otherwise the cell search procedure described in the physical layer procedures section would become too excessive. The scrambling codes must be allocated to the sectors in the network planning. Because the number of scrambling codes is so high, the scrambling code planning is a trivial task and can be done automatically by the network planning tool. The 512 primary scrambling codes are expected to be enough from the cell planning perspective, especially as the secondary scrambling codes can be used in the case of beam steering as used on dedicated channels. This allows the capacity to evolve with adaptive antenna techniques without consuming extra primary scrambling codes and causing problems for downlink code planning.

The actual code period is very long with the 18-degree code generator, but only 38400 chips are used from the beginning. Limiting the code period was necessary from the system perspective; the terminals would have difficulty in finding the correct code phase with a code period spanning several frames and 512 different codes to choose from.

The secondary downlink scrambling codes can be applied with the exception of those common channels that need to be heard in the whole cell and/or prior to the initial registration. Only one scrambling code should be used per cell or sector to maintain the orthogonality between different downlink code channels. With adaptive antennas the beams provide additional spatial isolation and the orthogonality between different code channels is less important. However, in all cases the best strategy is still to keep as many users as possible under a single scrambling code to minimise downlink interference. If a secondary scrambling code needs to be introduced in the cell, then only those users not fitting under the primary scrambling code should use the secondary code. The orthogonality is degraded most if the users are shared evenly between two different scrambling codes.

6.3.4.4   Synchronisation Channel Spreading and Modulation

The downlink synchronisation channel (SCH) is a special type of physical channel that is not visible above the physical layer. It contains two channels, primary and secondary SCHs. These channels are utilised by the terminal to find the cells, and are not under the cell-specific primary scrambling code. The terminal must be able to synchronise to the cell before knowing the downlink scrambling code.

The primary SCH contains a code word with 256 chips, with an identical code word in every cell. The primary SCH code word is sent without modulation on top. The code word is constructed from shorter 16-chip sequences in order to optimise the required hardware at the terminal. When detecting this sequence there is normally no prior timing information available and typically a matched filter is needed for detection. Therefore, for terminal complexity and power consumption reasons, it was important to optimise this synchronisation sequence for low-complexity matched filter implementation.

The secondary SCH code words are similar sequences but vary from one base station to another, with a total of 16 sequences in use. These 16 sequences are used to generate a total of 64 different code words which identify to which of the 64 code groups a base station belongs. Like the primary SCH, the secondary SCH is not under the base station-specific scrambling code, but the code sequences are sent without scrambling on top. The SCH code words contain modulation to indicate the use of open loop transmit diversity on the BCH. The SCH itself can use time-switched transmit antenna diversity (TSTD) and is the only channel in UTRA FDD that uses TSTD.

6.3.5 Transmitter Characteristics

The pulse shaping method applied to the transmitted symbols is root-raised cosine filtering with a roll-off factor of 0.22. The same roll-off is valid for both the terminals and the base stations. There are a few other key RF parameters that are introduced here and that have an essential impact on the implementation as well as on system behaviour.

The nominal carrier spacing in WCDMA is 5 MHz but the carrier frequency in WCDMA can be adjusted with a 200 kHz raster. The central frequency of each WCDMA carrier is indicated with an accuracy of 200 kHz. The target of this adjustment is to provide more flexibility for channel spacing within the operator's band.

The Adjacent Channel Leakage Ratio (ACLR) determines how much of the transmitted power is allowed to leak into the first or second neighbouring carrier. The concept of ACLR is illustrated in Figure 6.10, where $ACLR_1$ and $ACLR_2$ correspond to the power level integrated over the first and second adjacent carriers with 5 MHz and 10 MHz carrier separation respectively. No separate values are specified for other values of carrier spacing.

On the terminal side the ACLR values for the power classes of 21 dBm and 24 dBm have been set to 33 dB and 43 dB for $ACLR_1$ and $ACLR_2$ respectively. On the base station side the corresponding values are 45 dB and 50 dB. In the first phase of network deployment it is also likely that most terminals will belong to the 21 dBm power class and the network needs to be planned accordingly.

The higher the ACLR requirement, the more linearity is required from the power amplifier and the lower is the efficiency of the amplifier. The terminal needs to have a value that allows a power-efficient amplifier. The impact of the ACLR on system performance is studied in Section 8.5.

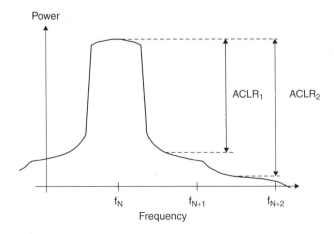

Figure 6.10. Adjacent Channel Leakage Ratio for the first and second adjacent carriers

The frequency accuracy requirements are also directly related to the implementation cost, especially on the terminal side. The terminal frequency accuracy has been defined to be ±0.1 ppm when compared to the received carrier frequency. On the base station side the requirement is tighter: ±0.05 ppm. The baseband timing is tight to the same timing reference as RF. The base station value needs to be tighter than the terminal value, since the base station carrier frequency is the reference for the terminal accuracy. The terminal needs also to be able to search the total frequency uncertainty area caused by the base station frequency error tolerance on top of the terminal tolerances and the error caused by terminal movement. With the 200 kHz carrier raster the looser base station frequency accuracy would start to cause problems. In 3GPP the RF parameters for terminals are specified in [8] and for base stations in [9].

6.4 User Data Transmission

For user data transmission in second generation systems, such as the first versions of GSM, typically only one service has been active at a time, either voice or low-rate data. From the beginning, the technology base has required that the physical layer implementation be defined to the last detail without real flexibility. For example, puncturing patterns in GSM have been defined bit by bit, whereas such a definition for all possible service combinations and data rates is simply not possible for UTRA. Instead, algorithms for generating such patterns are defined. Signal processing technology has also evolved greatly, thus there is no longer a need to have items like puncturing on hardware as in the early days of GSM hardware development.

6.4.1 Uplink Dedicated Channel

As described earlier, the uplink direction uses I-Q/code multiplexing for user data and physical layer control information. The physical layer control information is carried by the Dedicated Physical Control Channel (DPCCH) with a fixed spreading factor of 256. The higher layer information, including user data, is carried on one or more Dedicated Physical Data Channels (DPDCHs), with a possible spreading factor ranging from 256 down to 4. The uplink transmission may consist of one or more Dedicated Physical Data Channels (DPDCH) with a variable spreading factor, and a single Dedicated Physical Control Channel (DPCCH) with a fixed spreading factor.

The DPDCH data rate may vary on a frame-by-frame basis. Typically with a variable rate service the DPDCH data rate is informed on the DPCCH. The DPCCH is transmitted continuously and rate information is sent with Transport Format Combination Indicator (TFCI), the DPCCH information on the data rate on the current DPDCH frame. If the TFCI is not decoded correctly, the whole data frame is lost. Because the TFCI indicates the transport format of the same frame, the loss of the TFCI does not affect any other frames. The reliability of the TFCI is higher than the reliability of the user data detection on the DPDCH. Therefore, the loss of the TFCI is a rare event. Figure 6.11 illustrates the uplink dedicated channel structure in more detail.

The uplink DPCCH uses a slot structure with 15 slots over the 10 ms radio frame. This results in a slot duration of 2560 chips or about 666 µs. This is actually rather close to the GSM burst duration of 577 µs. Each slot has four fields to be used for pilot bits, TFCI,

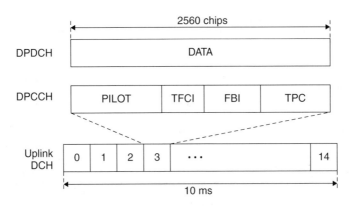

Figure 6.11. Uplink dedicated channel structure

Transmission Power Control (TPC) bits and Feedback Information (FBI) bits. The pilot bits are used for the channel estimation in the receiver, and the TPC bits carry the power control commands for the downlink power control. The FBI bits are used when closed loop transmission diversity is used in the downlink. The use of FBI bits is covered in the physical layer procedures section. There exist a total of six slot structures for uplink DPCCH. The different options are 0,1 or 2 bits for FBI bits and these same alternatives with and without TFCI bits. The TPC and pilot bits are always present and their number varies in such a way that the DPCCH slot is always fully used.

It is beneficial to transmit with a single DPDCH for as long as possible, for reasons of terminal amplifier efficiency, because multicode transmission increases the peak-to-average ratio of the transmission, which reduces the efficiency of the terminal power amplifier. The maximum user data rate on a single code is derived from the maximum channel bit rate, which is 960 kbps without channel coding with spreading factor 4. With channel coding the practical maximum user data rate for the single code case is in the order of 400–500 kbps.

When higher data rates are needed, parallel code channels are used. This allows up to six parallel codes to be used, raising the channel bit rate for data transmission up to 5740 kbps, which can accommodate 2 Mbps user data or an even higher data rate if the coding rate is $\frac{1}{2}$. Therefore, it is possible to offer a user data rate of 2 Mbps even after retransmission. The achievable data rates with different spreading factors are presented in Table 6.2. The rates given assume $\frac{1}{2}$-rate coding and do not include bits taken for coder tail bits or the Cyclic Redundancy Check (CRC). The relative overhead due to tail bits and CRC bits has significance only with low data rates.

The uplink receiver in the base station needs to perform typically the following tasks when receiving the transmission from a terminal:

- The receiver starts receiving the frame and despreading the DPCCH and buffering the DPDCH according to the maximum bit rate, corresponding to the smallest spreading factor.
- For every slot
  - obtain the channel estimates from the pilot bits on the DPCCH
  - estimate the SIR from the pilot bits for each slot

Table 6.2. Uplink DPDCH data rates

DPDCH spreading factor	DPDCH channel bit rate (kbps)	Maximum user data rate with $\frac{1}{2}$-rate coding (approx.)
256	15	7.5 kbps
128	30	15 kbps
64	60	30 kbps
32	120	60 kbps
16	240	120 kbps
8	480	240 kbps
4	960	480 kbps
4,with 6 parallel codes	5740	2.3 Mbps

— send the TPC command in the downlink direction to the terminal to control its uplink transmission power
— decode the TPC bit in each slot and adjust the downlink power of that connection accordingly.

- For every second or fourth slot
  — decode the FBI bits, if present, over two or four slots and adjust the diversity antenna phases, or phases and amplitudes, depending on the transmission diversity mode.
- For every 10 ms frame
  — decode the TFCI information from the DPCCH frame to obtain the bit rate and channel decoding parameters for DPDCH.
- For Transmission Time Interval (TTI, interleaving period) of 10, 20, 40 or 80 ms
  — decode the DPDCH data.

The same functions are valid for the downlink as well, with the following exceptions:

- In the downlink the dedicated channel spreading factor is constant, as well as with the common channels. The only exception is the Downlink Shared Channel (DSCH) which also has a varying spreading factor.
- The FBI bits are not in use in the downlink direction.
- There is a common pilot channel available in addition to the pilot bits on DPCCH. The common pilot can be used to aid the channel estimation.
- In the downlink transmission may occur from two antennas in the case of transmission diversity. The receiver does the channel estimation from the pilot patterns sent from two antennas and consequently accommodates the despread data sent from two different antennas. The overall impact on the complexity is small, however.

6.4.2   Uplink Multiplexing

In the uplink direction the services are multiplexed dynamically so that the data stream is continuous with the exception of zero rate. The symbols on the DPDCH are sent with equal power level for all services. This means in practice that the service coding and channel multiplexing needs in some cases to adjust the relative symbol rates for different services in order to balance the power level requirements for the channel symbols. The rate matching function in the multiplexing chain in Figure 6.12 can be used for such quality balancing

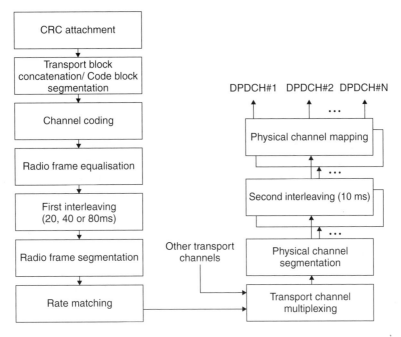

Figure 6.12. Uplink multiplexing and channel coding chain

operations between services on a single DPDCH. For the uplink DPDCH there do not exist fixed positions for different services, but the frame is filled according to the outcome of the rate matching and interleaving operation(s). The uplink multiplexing is done in 11 steps, as illustrated in Figure 6.12.

After receiving a transport block from higher layers, the first operation is CRC attachment. The CRC (Cyclic Redundancy Check) is used for error checking of the transport blocks at the receiving end. The CRC length that can be inserted has four different values: 0,8, 12 16 and 24 bits. The more bits the CRC contains, the lower is the probability of an undetected error in the transport block in the receiver. The physical layer provides the transport block to higher layers together with the error indication from the CRC check.

After the CRC attachment, the transport blocks are either concatenated together or segmented to different coding blocks. This depends on whether the transport block fits the available code block size as defined for the channel coding method. The benefit of the concatenation is better performance in terms of lower overhead due to encoder tail bits and in some cases due to better channel coding performance because of the larger block size. On the other hand, code block segmentation allows the avoidance of excessively large code blocks that could also be a complexity issue. If the transport block with CRC attached does not fit into the maximum available code block, it will be divided into several code blocks.

The channel encoding is performed on the coding blocks after the concatenation or segmentation operation. For some service or bit classes no channel coding is applied. This is so, for example, with AMR class c bits that are sent without channel coding. In that case there is no limitation on the coding block size, as there is no actual coding performed at the physical layer.

The function of radio frame equalisation is to ensure that data can be divided into equal-sized blocks when transmitted over more than a single 10 ms radio frame. This is done by padding the necessary number of bits until the data can be in equal-sized blocks per frame.

The first interleaving or inter-frame interleaving is used when the delay budget allows more than 10 ms of interleaving. The interlayer length of the first interleaving has been defined to be 20, 40 and 80 ms. The interleaving period is directly related to the Transmission Time Interval (TTI), which indicates how often data arrives from higher layers to the physical layer. The start positions of the TTIs for different transport channels multiplexed together for a single connection are time aligned. The TTIs have a common starting point, i.e. a 40 ms TTI goes in twice, even for an 80 ms TTI on the same connection. This is necessary to limit the possible transport format combinations from the signalling perspective. The timing relation with different TTIs is illustrated in Figure 6.13. If the first interleaving is used, the frame segmentation will distribute the data coming from the first interleaving over 2, 4 or 8 consecutive frames in line with the interleaving length.

Rate matching is used to match the number of bits to be transmitted to the number available on a single frame. This is achieved either by puncturing or by repetition. In the uplink direction, repetition is preferred, and basically the only reason why puncturing is used is when facing the limitations of the terminal transmitter or base station receiver. Another reason for puncturing is to avoid multi-code transmission. The rate matching operation in Figure 6.12 needs to take into account the number of bits coming from the other transport channels that are active in that frame. The uplink rate matching is a dynamic operation that may vary on a frame-by-frame basis. When the data rate of the service with lowest TTI varies as in Figure 6.13, the dynamic rate matching adjusts the rate matching parameters for other transport channels as well, so that all the symbols in the radio frame are used. For example, if with two transport channels the other has momentarily zero rate, rate matching increases the symbol rate for the other service sufficiently so that all uplink channel symbols are used, assuming that the spreading factor would stay the same.

The higher layers provide a semi-static parameter, the rate matching attribute, to control the relative rate matching between different transport channels. This is used to calculate the rate matching value when multiplexing several transport channels for the same

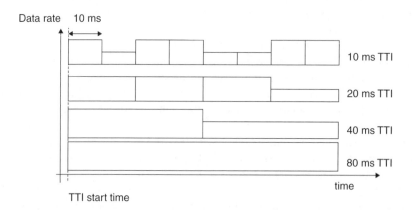

Figure 6.13. TTI start time relation with different TTIs on a single connection

frame. When this rule is applied as specified, with the aid of the rate matching attribute and TFCI the receiver can calculate backwards the rate matching parameters used and perform the inverse operation. By adjusting the rate matching attribute, the quality of different services can be fine-tuned to reach an equal or near-equal symbol power level requirement.

The different transport channels are multiplexed together by the transport channel multiplexing operation. This is a simple serial multiplexing on a frame-by-frame basis. Each transport channel provides data in 10 ms blocks for this multiplexing. In case more than one physical channel (spreading code) is used, physical channel segmentation is used. This operation simply divides the data evenly on the available spreading codes, as currently no cases have been specified where the spreading factors would be different in multicode transmissions. The use of serial multiplexing means also that with multicode transmission the lower rates can be implemented by sending fewer codes than with the full rate.

The second interleaving performs 10 ms radio frame interleaving, sometimes called intra-frame interleaving. This is a block interleaver with intercolumn permutations applied to the 30 columns of the interleaver. It is worth noting that the second interleaving is applied separately for each physical channel, in case more than a single code channel is used. From the output of the second interleaver the bits are mapped on the physical channels. The number of bits given for a physical channel at this stage is exactly the number that the spreading factor of that frame can transmit. Alternatively, the number of bits to transmit is zero and the physical channel is not transmitted at all.

6.4.3   User Data Transmission with the Random Access Channel

In addition to the uplink dedicated channel, user data can be sent on the Random Access Channel (RACH), mapped on the Physical Random Access Channel (PRACH). This is intended for low data rate operation with packet data where continuous connection is not maintained. In the RACH message it will be possible to transmit with a limited set of data rates based on prior negotiations with the UTRA network. The RACH operation does not include power control; thus the validity of the power level obtained with the PRACH power ramping procedure will be valid only for a short period, over one or two frames at most, depending on the environment.

The PRACH has as a specific feature preambles that are sent prior to data transmission. These use a spreading factor of 256 and contain a signature sequence of 16 symbols, resulting in a total length of 4096 chips for the preamble. Once the preamble has been detected and acknowledged with the Acquisition Indicator Channel (AICH), the 10 ms (or 20 ms) message part is transmitted. The spreading factor for the message part may vary from 256 up to 32 depending on the transmission needs, but is subject to prior agreement with the UTRA network. Additionally, the 20 ms message length has been defined for range improvement reasons; this is studied in detail in Section 11.2.2. The AICH structure is covered in the signalling part, while the RACH procedure is covered in detail in the physical layer procedures section.

6.4.4   Uplink Common Packet Channel

As well as the previously covered user data transmission methods, an extension for RACH has been defined. The main differences in the uplink from RACH data transmission are the reservation of the channel for several frames and the use of fast power control, which is not needed with RACH when sending only one or two frames. The uplink Common Packet Channel (CPCH) has as a pair the DPCCH in the downlink direction, providing fast power control information. Also the network has an option to tell the terminals to send an 8-slot power control preamble before the actual message transmission. This is beneficial in some cases as it allows the power control to converge before the actual data transmission starts.

The higher layer downlink signalling to a terminal using uplink CPCH is provided by the Forward Access Channel (FACH). The main reason for not using the DPDCH of the dedicated channel carrying the DPCCH for that is that the CPCH is a fast setup and fast release channel, handled similarly to RACH reception by the physical layer at the base station site. The DPDCH content is taken care of by the higher layer signalling protocols, which are located in a Radio Network Controller (RNC). In case the RNC wants to send a signalling message for a terminal as a response to CPCH activity, an ARQ message for example, the CPCH connection might have already been terminated by the base station. The differences in uplink CPCH operation from the RACH procedure are covered in the physical layer procedures section in more detail.

6.4.5 Downlink Dedicated Channel

The downlink dedicated channel is transmitted on the Downlink Dedicated Physical Channel (Downlink DPCH). The Downlink DPCH applies time multiplexing for physical control information and user data transmission, as shown in Figure 6.14. As in the uplink, the terms Dedicated Physical Data Channel (DPDCH) and Dedicated Physical Control Channel (DPCCH) are used in the 3GPP specification for the downlink dedicated channels.

The spreading factor for the highest transmission rate determines the channelisation code to be reserved from the given code tree. The variable data rate transmission may be implemented in two ways:

- In case TFCI is not present, the positions for the DPDCH bits in the frame are fixed. As the spreading factor is also always fixed in the Downlink DPCH, the lower rates are implemented with Discontinuous Transmission (DTX) by gating the transmission on/off. Since this is done on the slot interval, the resulting gating rate is 1500 Hz. As in the uplink, there are 15 slots per 10 ms radio frame; this determines the gating rate. The data rate, in case of more than one alternative, is determined with Blind Transport Format Detection (BTFD) which is based on the use of a guiding transport channel or channels that have different CRC positions for different Transport Format Combinations (TFCs). For a terminal it is mandatory to have BTFD capability with relatively low rates only, such as with AMR speech service. With higher data rates also the benefits from avoiding the TFCI overhead are insignificant and the complexity of BTFD rates starts to increase.

- With TFCI available it is also possible to use flexible positions, and it is up to the network to select which mode of operation is used. With flexible positions it is possible to keep continuous transmission and implement the DTX with repetition of the bits. In such a case the frame is always filled as in the uplink direction.

The downlink multiplexing chain in Figure 6.16 (Section 6.4.6) is also impacted by the DTX, the DTX indication having been inserted before the first interleaving.

In the downlink the spreading factors range from 4 to 512, with some restrictions on the use of spreading factor 512 in the case of soft handover. The restrictions are due to the timing adjustment step of 256 chips in soft handover operation, but in any case the use of a spreading factor of 512 for soft handover is not expected to occur very often. Typically, such a spreading factor is used to provide information on power control, etc., when providing services with minimal downlink activity, as with file uploading and so on. This is also the case with the CPCH where power control information for the limited duration uplink transmission is provided with a DPCCH with spreading factor 512. In such a case soft handover is not needed either.

Modulation causes some differences between the uplink and downlink data rates. While the uplink DPDCH consists of BPSK symbols, the downlink DPDCH consists of QPSK symbols. Although from the downlink DPDCH part of the time is reserved for DPCCH, especially with high data rates, the bit rate that can be accommodated in a single code in the downlink DPDCH is almost double that in the uplink DPDCH with the same spreading factor. These downlink data rates are given in Table 6.3 with raw bit rates calculated from the QPSK-valued symbols in the downlink reserved for data use.

The Downlink DPCH can use either open loop or closed loop transmit diversity to improve performance. The use of such enhancements is not required from the network side

Table 6.3. Downlink Dedicated Channel symbol and bit rates

Spreading factor	Channel symbol rate (kbps)	Channel bit rate (kbps)	DPDCH channel bit rate range (kbps)	Maximum user data rate with $\frac{1}{2}$-rate coding (approx.)
512	7.5	15	3–6	1–3 kbps
256	15	30	12–24	6–12 kbps
128	30	60	42–51	20–24 kbps
64	60	120	90	45 kbps
32	120	240	210	105 kbps
16	240	480	432	215 kbps
8	480	960	912	456 kbps
4	960	1920	1872	936 kbps
4,with 3 parallel codes	2880	5760	5616	2.3 Mbps

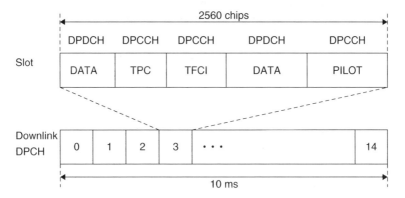

Figure 6.14. Downlink Dedicated Physical Channel (Downlink DPCH) control/data multiplexing

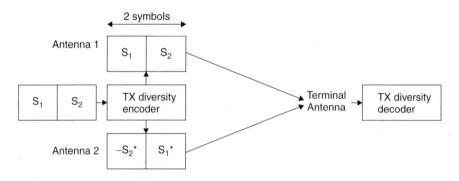

Figure 6.15. Open loop transmit diversity encoding

but is mandatory in terminals. It was made mandatory as it was felt that this kind of feature has a strong relation to such issues as network planning and system capacity, so it was made a baseline implementation capability. The open loop transmit diversity coding principle is shown in Figure 6.15, where the information is coded to be sent from two antennas. The method is also denoted in the 3GPP specification as space time block coding based transmit diversity (STTD). Another possibility is to use feedback mode transmit diversity, where the signal is sent from two antennas based on the feedback information from the terminal. The feedback mode uses phase, and in some cases also amplitude, offsets between the antennas. The feedback mode of transmit diversity is covered in the physical layer procedures section.

6.4.6 Downlink Multiplexing

The multiplexing chain in the downlink is mainly similar to that in the uplink but there are also some functions that are done differently.

As in the uplink, the interleaving is implemented in two parts, covering both intra-frame and inter-frame interleaving. Also the rate matching allows one to balance the required channel symbol energy for different service qualities. The services can be mapped to more than one code as well, which is necessary if the single code capability in either the terminal or base station is exceeded.

There are differences in the order in which rate matching and segmentation functions are performed. Whether fixed or flexible bit positions are used determines the DTX indication insertion point. The DTX indication bits are not transmitted over the air; they are just inserted to inform the transmitter at which bit positions the transmission should be turned off. They were not needed in the uplink where the rate matching was done in a more dynamic way, always filling the frame when there was something to transmit on the DPDCH.

The use of fixed positions means that for a given transport channel, the same symbols are always used. If the transmission rate is below the maximum, then DTX indication bits are used for those symbols. The different transport channels do not have a dynamic impact on the rate matching values applied for another channel, and all transport channels can use the maximum rate simultaneously as well. The use of fixed positions is partly related to the possible use of blind rate detection. When a transport channel always has the same position regardless of the data rate, the channel decoding can be done with a single decoding 'run' and

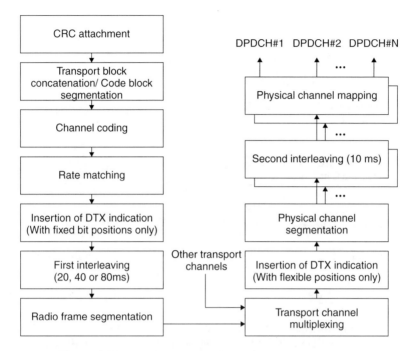

Figure 6.16. Downlink multiplexing and channel coding chain

the only thing that needs to be tested is which position of the output block is matched with the CRC check results. This naturally requires that different rates have different numbers of symbols.

With flexible positions the situation is different since now the channel bits unused by one service may be utilised by another service. This is useful when it is possible to have such a transport channel combination that they do not all need to be able to reach the full data rate simultaneously, but can alternate with the need for full rate transmission. This allows the necessary spreading code occupancy in the downlink to be reduced. The concept of flexible versus fixed positions in the downlink is illustrated in Figure 6.17. The use of blind rate detection is also possible in principle with flexible positions, but is not required by the specifications. If the data rate is not too high and number of possible data rates is not very high, the terminal can run channel decoding for all the combinations and check which of the cases comes out with the correct CRC result.

6.4.7 Downlink Shared Channel

Transmitting data with high peak rate and low activity cycle in the downlink quickly causes the channelisation codes under a single scrambling code to start to run out. To avoid this problem, basically two alternatives exist: use of either additional scrambling codes or common channels. The additional scrambling code approach loses the advantage of the transmissions being orthogonal from a single source, and thus should be avoided. Using a shared channel resource maintains this advantage and at the same time reduces the downlink code resource consumption. As such resource sharing cannot provide a 100% guarantee of

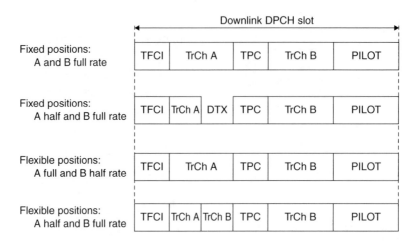

Figure 6.17. Flexible and fixed transport channel slot positions in the downlink

available physical channel resource at all times, its applicability in practice is limited to packet-based services.

As in a CDMA system one has to ensure the availability of power control and other information continuously, the Downlink Shared Channel (DSCH) has been defined to be always associated with a Downlink Dedicated Channel (Downlink DCH). The DCH provides, in addition to the power control information, an indication to the terminal when it has to decode the DSCH and which spreading code from the DSCH it has to despread. For this indication two alternatives have been specified: either TFCI based on a frame-by-frame basis or higher layer signalling based on a longer allocation period. Thus the DSCH data rate without coding is directly the channel bit rate indicated in Table 6.3 for the Downlink DCH. The small difference from the downlink DCH spreading codes is that spreading factor 512 is not supported by DSCH. The DSCH also allows mixing terminals with different data rate capabilities under a single branch from the code resource, making the configuration manageable with evolving terminal capabilities. The DSCH code tree was illustrated in Figure 6.9 in connection with the downlink spreading section.

With DSCH the user may be allocated different data rates, for example 384 kbps with spreading factor 8 and then 192 kbps with spreading factor 16. The DSCH code tree definition allows sharing the DSCH capacity on a frame-by-frame basis, for example with either a single user active with a high data rate or with several lower-rate users active in parallel. The DSCH may be mapped to a multicode case as well; for example, three channelisation codes with spreading factor 4 provide a DSCH with 2 Mbps capability.

In the uplink direction, such concerns for code resource usage do not exist, but there is the question how to manage the total interference level and in some cases the resource usage on the receiver side. Thus an operation similar to DSCH is not specified in the uplink in UTRA FDD.

The physical channel carrying the DSCH is the Physical Downlink Shared Channel (PDSCH). The timing relation of the PDSCH to the associated downlink Dedicated Physical Channel (DPCH) is shown in Figure 6.18. The PDSCH frame may not start before three slots after the end of the associated dedicated channel frame. This ensures that buffering

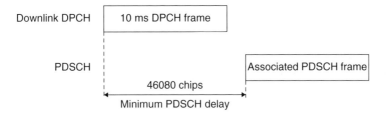

Figure 6.18. PDSCH timing relation to DPCH

requirements for DSCH reception do not increase compared to the other buffering needs in the receiver.

6.4.8 Forward Access Channel for User Data Transmission

The Forward Access Channel (FACH) can be used for transmission of user (packet) data. The channel is typically multiplexed with the paging channel to the same physical channel, but can exist as a standalone channel as well. The main difference with the dedicated and shared channels is that FACH does not allow the use of fast power control and applies either slow power control or no power control at all. Slow power control is possible if a lot of data is transmitted between the base station and the terminal and the latter provides feedback on the quality of the received packets. This type of power control cannot combat the effect of the fading channel but more the longer-term changes in the propagation environment. For less frequent transmission FACH needs to use more or less the full power level. The power control for FACH is also typically very slow, since the FACH data transmission is controlled by RNC, which means rather a large delay for any feedback information from the base station.

Whether the FACH contains pilot symbols or not depends on whether it applies beam forming techniques. Normally FACH does not contain pilot symbols and the receiver uses the common pilot channel as phase reference.

As FACH needs to be received by all terminals, the primary FACH cannot use high data rates. If higher data rates were desired of FACH, this would require a separate physical channel where only the capabilities in terms of maximum data rates of those terminals allocated to that channel need to be taken into account. The necessary configuration would become rather complicated when terminals with different capabilities are included. The FACH has a fixed spreading factor, and reserving FACH for very high data rates is not optimised from the code resource point of view, especially if not all the terminals can decode the high data rate FACH.

Messages on FACH normally need in-band signalling to tell for which user the data was intended. In order to read such information, the terminal must decode FACH messages first. Running such decoding continuously is not desirable due to power consumption, especially with higher FACH rates.

6.4.9 Channel Coding for User Data

In UTRA two channel coding methods have been defined. Half-rate and $\frac{1}{3}$-rate convolutional coding are intended to be used with relatively low data rates, equivalent to the data rates led by second generation cellular networks today, though an upper limit has not been specified. For higher data rates $\frac{1}{3}$-rate turbo coding can be applied and typically brings performance benefits when large enough block sizes are achieved. It has been estimated that roughly 300 bits should be available per TTI in order to give turbo coding some gain over convolutional coding. This also depends on the required quality level and operational environment.

The convolutional coding is based on constraint length 9 coding with the use of tail bits. The selected turbo encoding/decoding method is 8-state PCCC (parallel concatenated convolutional code). The main motivation for turbo coding for higher bit rates has been performance, while for low rates the main reason not to use it has been both low rate or low block length performance as well as the desire to allow the use of simple blind rate detection with low rate services such as speech. Blind rate detection with turbo coding typically requires detection of all transmission rates, while with convolutional coding trial methods can allow only a single Viterbi pass for determining which transmission rate was used. This is performed together with the help of CRC and applying a proper interleaving technique.

Turbo coding has specific interleaving which has been designed with a large variety of data rates in mind. The maximum turbo coding block size has been limited to 5114 information bits, since after that block size only memory requirements increase but no significant effect on the performance side can be observed. For the higher amount of data per interleaving period, several blocks are used, with a block size as equal as possible at or below 5114 bits. The actual block size for data is a little smaller, since the tail bits as well as CRC bits are to be accommodated in the block size.

The minimum block size for turbo coding was initially defined to be 320 bits, which corresponds to 32 kbps with 10 ms interleaving or down to 4 kbps with 80 ms interleaving. The possible range of block sizes was, however, extended down to 40 bits, since with variable rate connection it is not desirable to change the codec 'on the fly' when coming down from the maximum rate. Nor may a transport channel change the channel coding method on a frame-by-frame basis. Data rates below 40 bits can be transmitted with turbo coding as well, but in such a case padding with dummy bits is used to fill the 40 bits minimum size interleaver.

With speech service, AMR coding uses an unequal error protection scheme. This means that the three different classes of bits have different protection. Class A bits—those that contribute the most to voice quality—have the strongest protection, while class C bits are sent without channel coding. This gives around 1 dB gain in Eb/No compared to the equal error protection scheme. The coding methods usable by different channels are summarised in Table 6.4. Although the FACH has two options given, the cell access use of FACH is based on convolutional coding, as not all terminals support turbo coding.

Table 6.4. Channel coding options with different channels

DCH	Turbo coding or convolutional coding
CPCH	Turbo coding or convolutional coding
DSCH	Turbo coding or convolutional coding
FACH	Turbo coding or convolutional coding
Other common channels	$\frac{1}{2}$-rate convolutional coding

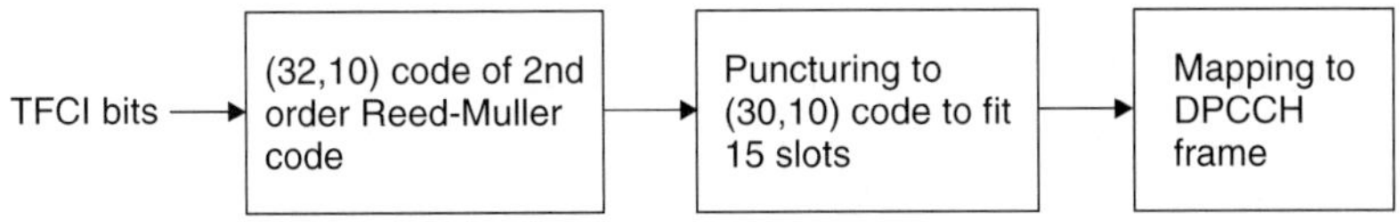

Figure 6.19. TFCI information coding

6.4.10   Coding for TFCI information

The Transport Format Combination Indicator (TFCI) may carry from 1 to 10 bits of transport format information. As well as the normal mode of operation, there is also 'split' mode where the TFCI code word is sent with two different code words and not every cell necessarily sends both code words. In this case both code words are capable of carrying 5 bits. The typical split mode operation would be that an RNC for a downlink dedicated channel would be different from an RNC for controlling a DSCH. The split mode is valid for the downlink direction only.

The coding in the normal mode is second-order Reed Muller code punctured from 32 bits to 30 bits, carrying up to 10 bits of information. The TFCI coding is illustrated in Figure 6.19. The coding with split mode is biorthogonal (16,5) block code.

6.5   Signalling

For signalling purposes a lot of information needs to be transmitted between the network and the terminals. The following chapters describe the methods used for transmitting signalling messages generated above the physical layer, as well as the required physical layer control channels needed for system operation but not necessarily visible for higher layer functionality.

6.5.1   Common Pilot Channel (CPICH)

The common pilot channel is an unmodulated code channel, which is scrambled with the cell-specific primary scrambling code. The function of the CPICH is to aid the channel estimation at the terminal for the dedicated channel and to provide the channel estimation reference for the common channels when they are not associated with the dedicated channels or not involved in the adaptive antenna techniques.

UTRA has two types of common pilot channels, primary and secondary. The difference is that the Primary CPICH is always under the primary scrambling code with a fixed channelisation code allocation and there is only one such channel for a cell or sector. The Secondary CPICH may have any channelisation code of length 256 and may be under a secondary scrambling code as well. The typical area of Secondary CPICH usage would be operations with narrow antenna beams intended for service provision at specific 'hot spots' or places with high traffic density.

An important area for the primary common pilot channel is the measurements for the handover and cell selection/reselection. The use of CPICH reception level at the terminal for handover measurements has the consequence that by adjusting the CPICH power level the cell load can be balanced between different cells. Reducing the CPICH power causes part of the terminals to hand over to other cells, while increasing it invites more terminals to hand over to the cell as well as to make their initial access to the network in that cell.

The CPICH does not carry any higher layer information, neither is there any transport channel mapped to it. The CPICH uses the spreading factor of 256. It may be sent from two antennas in case transmission diversity methods are used in the base station. In this case, the transmissions from the two antennas are separated by a simple modulation pattern on the CPICH transmitted from the diversity antenna, called diversity CPICH. The diversity pilot is used with both open loop and closed loop transmission diversity schemes.

6.5.2 Synchronisation Channel (SCH)

The Synchronisation Channel (SCH) is needed for the cell search. It consists of two channels, the primary and secondary synchronisation channels.

The Primary SCH uses a 256-chip spreading sequence identical in every cell. The system-wide sequence has been optimised for matched filter implementations, as described in connection with SCH spreading and modulation in Section 6.3.4.4.

The Secondary SCH uses sequences with different code word combination possibilities representing different code groups. Once the terminal has identified the secondary synchronisation channel, it has obtained frame and slot synchronisation as well as information on the group the cell belongs to. There are 64 different code groups in use, pointed out by the 256 chip sequences sent on the secondary SCHs. Such a full cell search process with a need to search for all groups is needed naturally only at the initial search upon terminal power-on or when entering a coverage area, otherwise a terminal has more information available on the neighbouring cells and not all the steps are always necessary.

As with the CPICH, no transport channel is mapped on the SCH, as the code words are transmitted for cell search purposes only. The SCH is time multiplexed with the Primary Common Control Physical Channel. For the SCH there are always 256 chips out of 2560 chips from each slot. The Primary and Secondary SCH are sent in parallel, as illustrated in Figure 6.20. Further details on the cell search procedure are covered in Section 6.6.

6.5.3 Primary Common Control Physical Channel (Primary CCPCH)

The Primary Common Control Physical Channel (Primary CCPCH) is the physical channel carrying the Broadcast Channel (BCH). It needs to be demodulated by all the terminals in

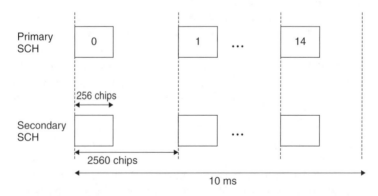

Figure 6.20. Primary and secondary synchronisation channel principle

the system. As a result, the parameters with respect to, for example, the channel coding and spreading code contain no flexibility, as they need to be known by all terminals made since the publication of the Release-99 specifications. The contents of the signalling messages have room for flexibility as long as the new message structures are such that they do not cause unwanted or unpredictable behaviour in the terminals deployed in the network.

The Primary CCPCH contains no Layer 1 control information as it is fixed rate and does not carry power control information for any of the terminals. The pilot symbols are not used, since the Primary CCPCH needs to be available over the whole cell area and does not use specific antenna techniques but is sent with the same antenna radiation pattern as the common pilot channel. This allows the common pilot channel to be used for channel estimation with coherent detection in connection with the Primary CCPCH.

The channel bit rate is 30 kbps with spreading ratio of the permanently allocated channelisation code of 256. The total bit rate is reduced further as the Primary CCPCH alternates with the Synchronisation Channel (SCH), reducing the bit rate without coding available for system information to 27 kbps. This is illustrated in Figure 6.21, where the 256-chip idle period on the Primary CCPCH is shown.

The channel coding with the Primary CCPCH is $\frac{1}{2}$-rate convolutional coding with 20 ms interleaving over two consecutive frames. It is important to keep the data rate with the Primary CCPCH low, as in practice it will be transmitted with very high power from the base station to reach all terminals, having a direct impact on system capacity. If Primary CCPCH decoding fails, the terminals cannot access the system if they are unable to obtain the critical system parameters such as random access codes or code channels used for other common channels.

As a performance improvement method, the Primary CCPCH may apply open loop transmission diversity. In such a case the use of open transmission diversity on the Primary CCPCH is indicated in the modulation of the Secondary SCH. This allows the terminals to have the information before attempting to decode the BCH with the initial cell search.

6.5.4 Secondary Common Control Physical Channel (Secondary CCPCH)

The Secondary Common Control Physical Channel (Secondary CCPCH) carries two different common transport channels: the Forward Access Channel (FACH) and the Paging Channel

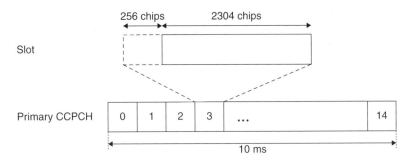

Figure 6.21. Primary CCPCH frame structure

(PCH). The two channels can share a single Secondary CCPCH or can use different physical channels. This means that in the minimum configuration each cell has at least one Secondary CCPCH. In case of a single Secondary CCPCH fewer degrees of freedom exist in terms of data rates, and so on, since again all the terminals in the network need to be able to detect the FACH and PCH. Since there can be more than one FACH or PCH, however, for the additional Secondary CCPCHs the data rates can vary more, as long as the terminals not capable of demodulating higher data rates using another, lower data rate Secondary CCPCH.

The spreading factor used in a Secondary CCPCH is fixed and determined according to the maximum data rate. The data rate may vary with DTX or rate matching parameters, but the channelisation code is always reserved according to the maximum data rate. The maximum data rate usable is naturally dependent on the terminal capabilities. As with the Primary CCPCH, the channel coding method is $\frac{1}{2}$-rate convolutional coding when carrying the channels used for cell access, FACH or PCH. When used to carry PCH, the interleaving period is always 10 ms. For data transmission with FACH, turbo coding or $\frac{1}{3}$-rate convolutional coding may also be applied.

The Secondary CCPCH does not contain power control information, and for other layer 1 control information the following combinations can be used:

- Neither pilot symbols nor rate information (TFCI). Used with PCH and FACH when no adaptive antennas are in use and a channel needs to be detected by all terminals.

- No pilot symbols, but rate information with TFCI. Used typically with FACH when it is desired to use FACH for data transmission with variable transport format and data rate. In such a case variable transmission rates are implemented by DTX or repetition.

- Pilot symbol with or without rate information (TFCI). Typical for the case when an uplink channel is used to derive information for adaptive antenna processing purposes and user-specific antenna radiation patterns or beams are used.

The FACH and PCH can be multiplexed to a single Secondary CCPCH, as the paging indicators used together with the PCH are multiplexed to a different physical channel, called the Paging Indicator Channel (PICH). The motivation for multiplexing the channels together is base station power budget. Since both of the channels need to be transmitted at full power for all the terminals to receive, avoiding the need to send them simultaneously obviously reduces base station power level variations. In order to enable this multiplexing, it has been necessary to terminate both FACH and PCH at RNC.

As a performance improvement method, open loop transmission diversity can be used with a Secondary CCPCH as well. The performance improvement of such a method is higher for common channels in general, as neither Primary nor Secondary CCPCH can use fast power control. Also, since they are often sent with full power to reach the cell edge, reducing the required transmission power level improves downlink system capacity.

6.5.5 Random Access Channel (RACH) for Signalling Transmission

The Random Access Channel (RACH) is typically used for signalling purposes, to register the terminal after power-on to the network or to perform location update after moving from one location area to another or to initiate a call. The structure of the physical RACH for signalling purposes is the same as when using the RACH for user data transmission, as described in connection with the user data transmission. With signalling use the major difference is that the data rate needs to be kept relatively low, otherwise the range achievable with RACH signalling starts to limit the system coverage. This is more critical, the lower the data rates used as a basis for network coverage planning. RACH range issues are studied in detail in Chapter 11. The detailed RACH procedure will be covered in connection with the physical layer procedures.

The RACH that can be used for initial access has a relatively low payload size, since it needs to be usable by all terminals. The ability to support 16 kbps data rate on RACH is a mandatory requirement for all terminals regardless of what kind of services they provide.

6.5.6 Acquisition Indicator Channel (AICH)

In connection with the Random Access Channel, the Acquisition Indicator Channel (AICH) is used to indicate from the base station the reception of the random access channel signature sequence. The AICH uses an identical signature sequence as the RACH on one of the downlink channelisation codes of the base station to which the RACH belongs. Once the base station has detected the preamble with the random access attempt, then the same signature sequence that has been used on the preamble will be echoed back on AICH. As the structure of AICH is the same as with the RACH preamble, it also uses a spreading factor of 256 and 16 symbols as the signature sequence. There can be up to 16 signatures, acknowledged on the AICH at the same time. Both signature sets can be used with AICH. The procedure with AICH and RACH is described in the physical layer procedures section.

For the detection of AICH the terminal needs to obtain the phase reference from the common pilot channel. The AICH also needs to be heard by all terminals and needs to be sent typically at high power level without power control.

The AICH is not visible to higher layers but is controlled directly by the physical layer in the base station, as operation via a radio network controller would make the response time too slow for a RACH preamble. There are only a few timeslots to detect the RACH preamble and to transmit the response to the terminal on AICH.

6.5.7 Paging Indicator Channel (PICH)

The Paging Channel (PCH) is operated together with the Paging Indicator Channel (PICH) to provide terminals with efficient sleep mode operation. The paging indicators use a

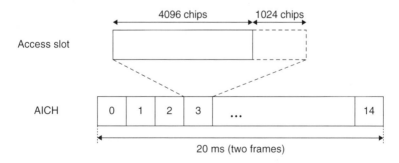

Figure 6.22. AICH access slot structure

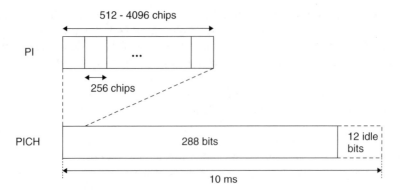

Figure 6.23. PICH structure with different PI repetition rates

channelisation code of length 256. The paging indicators occur once per slot on the corresponding physical channel, the Paging Indicator Channel (PICH). Each PICH frame carries 288 bits to be used by the paging indicator bit, and 12 bits are left idle. Depending on the paging indicator repetition ratio, there can be 18, 36, 72 or 144 paging indicators per PICH frame. How often a terminal needs to listen to the PICH is parameterised, and the exact moment depends on running the system frame number (SFN).

For detection of the PICH the terminal needs to obtain the phase reference from the CPICH, and as with the AICH, the PICH needs to be heard by all terminals in the cell and thus needs to be sent at high power level without power control. The PICH frame structure with different PI repetition factors is illustrated in Figure 6.23.

6.5.8 Physical Channels for CPCH Access Procedure

For the CPCH access procedure, a set of CPCH specific physical channels has been specified. These channels carry no transport channels, but only information needed in the CPCH access procedure. The channels are:

- CPCH Status Indication Channel (CSICH)
- CPCH Collision Detection Indicator Channel (CD-ICH)
- CPCH Channel Assignment Indicator Channel (CA-ICH)
- CPCH Access Preamble Acquisition Channel (AP-AICH)

The CSICH uses the part of the AICH channel that is defined as unused, as shown in Figure 6.22. The CSICH bits indicate the availability of each physical CPCH channel and are used to tell the terminal to initiate access only on a free channel but, on the other hand, to accept a channel assignment command to an unused channel. The CSICH shares the downlink channelisation code resource with the AP-AICH.

The CD-ICH carries the collision detection information to the terminal. When the CA-ICH channel is used, the CD-ICH and CA-ICH are sent in parallel to the terminal. Both have 16 different bit patterns specified.

The AP-AICH is identical to the AICH used with RACH and may share the same channelisation code when sharing access resources with RACH. In this case CSICH uses also the same channelisation code as the CPCH and RACH AICH channels.

6.6 Physical Layer Procedures

In the physical layer of a CDMA system there are many procedures essential for system operation. Examples include the fast power control and random access procedures. Other important physical layer procedures are paging, handover measurements and operation with transmit diversity. These procedures have been naturally shaped by the CDMA-specific properties of the UTRA FDD physical layer.

6.6.1 Fast Closed Loop Power Control Procedure

The fast closed loop power control procedure is denoted in the UTRA specifications as inner loop power control. It is known to be essential in a CDMA-based system due to the uplink near–far problem illustrated in Chapter 3. The fast power control operation operates on a basis of one command per slot, resulting in a 1500 Hz command rate. The basic step size is 1 dB. Additionally, multiples of that step size can be used and smaller step sizes can be emulated. The emulated step size means that the 1 dB step is used, for example, only every second slot, thus emulating the 0.5 dB step size. 'True' step sizes below 1 dB are difficult to implement with reasonable complexity, as the achievable accuracy over the large dynamic range is difficult to ensure. The specifications define the relative accuracy for a 1 dB power control step to be ± 0.5 dB. The other 'true' step size specified is 2 dB.

Fast power control operation has two special cases: operation with soft handover and with compressed mode in connection with handover measurements. Soft handover needs special concern as there are several base stations sending commands to a single terminal, while with compressed mode operation breaks in the command stream are periodically provided to the terminal.

In soft handover the main issue for terminals is how to react to multiple power control commands from several sources. This has been solved by specifying the operation such that the terminal combines the commands but also takes the reliability of each individual command decision into account in deciding whether to increase or decrease the power.

In the compressed mode case, the fast power control uses a larger step size for a short period after a compressed frame. This allows the power level to converge more quickly to the correct value after a break in the control stream. The need for this method depends heavily on the environment and it is not relevant for the lower terminal or very short transmission gap lengths.

The SIR target for closed loop power control is set by the outer loop power control. The latter power control is introduced in Section 3.5 and described in detail in Section 9.2.2.

On the terminal side it is specified rather strictly what is expected to be done inside a terminal in terms of (fast) power control operation. On the network side there is much greater freedom to decide how a base station should behave upon reception of a power control command, as well as the basis on which the base station should tell a terminal to increase or decrease the power.

6.6.2 Open Loop Power Control

In UTRA FDD there is also open loop power control, which is applied only prior to initiating the transmission on the RACH or CPCH. Open loop power control is not very accurate, since it is difficult to measure large power dynamics accurately in the terminal equipment. The mapping of the actual received absolute power to the absolute power to be transmitted shows large deviations, due to variation in the component properties as well as to the impact of environmental conditions, mainly temperature. Also, the transmission and reception occur at different frequencies, but the internal accuracy inside the terminal is the main source of uncertainty. The requirement for open loop power control accuracy is specified to be within ± 9 dB in normal conditions.

Open loop power control was used in earlier CDMA systems, such as IS-95, being active in parallel with closed loop power control. The motivation for such usage was to allow corner effects or other sudden environmental changes to be covered. As the UTRA fast power control has almost double the command rate, it was concluded that a 15 dB adjustment range does not need open loop power control to be operated simultaneously. Additionally, the fast power control step size can be increased from 1 dB to 2 dB, which would allow a 30 dB correction range during a 10 ms frame.

The use of open loop power control while in active mode also has some impact on link quality. The large inaccuracy of open loop power control can cause it to make adjustments to the transmitted power level even when they are not needed. As such behaviour depends on terminal unit tolerances and on various environmental variables, running open loop power control makes it more difficult from the network side to predict how a terminal will behave in different conditions.

6.6.3 Paging Procedure

The Paging Channel (PCH) operation is organised as follows. A terminal, once registered to a network, has been allocated a paging group. For the paging group there are Paging Indicators (PI) which appear periodically on the Paging Indicator Channel (PICH) when there are paging messages for any of the terminals belonging to that paging group.

Once a PI has been detected, the terminal decodes the next PCH frame transmitted on the Secondary CCPCH to see whether there was a paging message intended for it. The terminal may also need to decode the PCH in case the PI reception indicates low reliability of the decision. The paging interval is illustrated in Figure 6.24.

The less often the PIs appear, the less often the terminal needs to wake up from the sleep mode and the longer the battery life becomes. The trade-off is obviously the response time to the network-originated call. An infinite paging indicator interval does not lead to infinite battery duration, as there are other tasks the terminal needs to perform during idle mode as well.

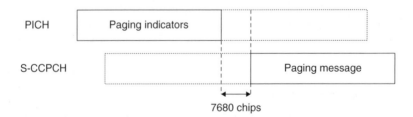

Figure 6.24. PICH relation to PCH

6.6.4   RACH Procedure

The Random Access procedure in a CDMA system has to cope with the near–far problem, as when initiating the transmission there is no exact knowledge of the required transmission power. The open loop power control has a large uncertainty in terms of absolute power values from the received power measurement to the transmitter power level setting value, as stated in connection with the open loop description. In UTRA the RACH procedure has the following phases:

- The terminal decodes the BCH to find out the available RACH sub-channels and their scrambling codes and signatures.
- The terminal selects randomly one of the RACH sub-channels from the group its access class allows it to use. Furthermore, the signature is also selected randomly from among the available signatures.
- The downlink power level is measured and the initial RACH power level is set with the proper margin due to the open loop inaccuracy.
- A 1 ms RACH preamble is sent with the selected signature.
- The terminal decodes AICH to see whether the base station has detected the preamble.
- In case no AICH is detected, the terminal increases the preamble transmission power by a step given by the base station, as multiples of 1 dB. The preamble is retransmitted in the next available access slot.
- When an AICH transmission is detected from the base station, the terminal transmits the 10 ms or 20 ms message part of the RACH transmission.

The RACH procedure is illustrated in Figure 6.25, where the terminal transmits the preamble until acknowledgement is received on AICH, and then the message part follows.

In the case of data transmission on RACH, the spreading factor and thus the data rate may vary; this is indicated with the TFCI on the DPCCH on PRACH. Spreading factors from 256 to 32 have been defined to be possible, thus a single frame on RACH may contain up to 1200 channel symbols which, depending on the channel coding, maps to around 600 or 400 bits. For the maximum number of bits the achievable range is naturally less than what can be achieved with the lowest rates, especially as RACH messages do not use methods such as macro-diversity as in the dedicated channel.

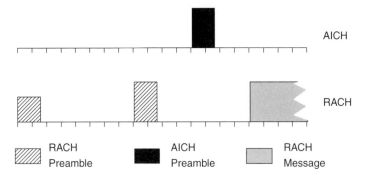

Figure 6.25. PRACH ramping and message transmission

6.6.5   CPCH Operation

Uplink Common Packet Channel (CPCH) operation is rather similar to RACH operation. The main difference is the Layer 1 Collision Detection (CD) based on a signal structure similar to that of the RACH preamble. The operation follows the RACH procedure until the terminal detects AICH, as illustrated in Figure 6.26. After that a CD preamble with the same power level is still sent back with another signature, randomly chosen from a given set. Then the base station is expected to echo this signature back to the terminal on the CD Indication Channel (CD-ICH) and in this way to create a method of reducing the collision probability on layer 1. After the correct preamble has been sent by the base station on the collision detection procedure, the terminal starts the transmission, which may last over several frames. The longer duration of the transmission highlights the need for the physical layer-based collision detection mechanism. In RACH operation only one RACH message may end up lost due to collision, whereas with CPCH operation an undetected collision may cause several frames to be sent and cause only extra interference.

The fast power control on CPCH helps to reduce the interference due to the data transmission while it also highlights the importance of the added collision detection to RACH. A terminal transmitting data over several frames and following a power control command stream intended for another terminal would create a severe interference problem in the cell, especially when high data rates are involved. At the beginning of the CPCH transmission, an optional power control preamble can be sent before actual data transmission is initiated. This is to allow power control to converge, as there is a longer delay with CPCH than with RACH between the acknowledged preamble and actual data frame transmission. The 8-slot power control preamble also uses a 2 dB step size for faster power control convergence.

A CPCH transmission needs to have a restriction on maximum duration, since CPCH supports neither soft handover nor compressed mode to allow inter-frequency or inter-system measurements. UTRAN sets the maximum CPCH transmission during service negotiations.

The latest addition to CPCH operation is the status monitoring and channel assignment functionality. The CPCH Status Indication Channel (CSICH) is a separate physical channel, sent from the base station, that has indicator bits to indicate the status of different CPCH channels. This avoids unnecessary access attempts when all CPCH channels are busy, so it will also improve CPCH throughput. The Channel Assignment functionality is a system option, in the form of a CA message that may direct the terminal to a CPCH channel other than the one used for the access procedure. The CA message is sent in parallel with the collision detection message.

6.6.6   Cell Search Procedure

The cell search procedure or synchronisation procedure in an asynchronous CDMA system differs greatly from the procedure in a synchronous system like IS-95. Since the cells in an asynchronous UTRA CDMA system use different scrambling codes and not just different code phase shifts, terminals with today's technology cannot search for 512 codes of 10 ms duration without any prior knowledge. There would be too many comparisons to make and users would experience too long an interval from power-on to the service availability indication in the terminal.

The cell search procedure using the synchronisation channel has basically three steps, though from the standards point of view there will be no requirements as to which steps

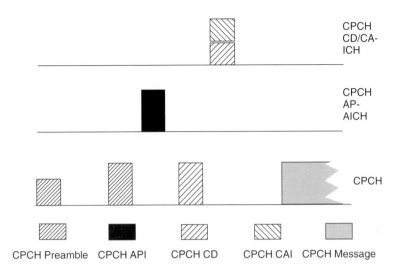

Figure 6.26. CPCH access procedure operation

to perform and when. Rather the standard will set requirements for performance in terms of maximum search duration in reference test conditions. The basic steps for the initial cell search are typically as follows:

1. The terminal searches the 256-chip primary synchronisation code, being identical for all cells. As the primary synchronisation code is the same in every slot, the peak detected corresponds to the slot boundary.

2. Based on the peaks detected for the primary synchronisation code, the terminal seeks the largest peak from the Secondary SCH code word. There are 64 possibilities for the secondary synchronisation code word. The terminal needs to check all 15 positions, as the frame boundary is not available before Secondary SCH code word detection.

3. Once the Secondary SCH code word has been detected, the frame timing is known. The terminal then seeks the primary scrambling codes that belong to that particular code group. Each group consists of eight primary scrambling codes. These need to be tested for a single position only, as the starting point is known already.

When setting the network parameters, the properties of the synchronisation scheme need to be taken into account for optimum performance. For the initial cell search there is no practical impact, but the target cell search in connection with handover can be optimised. Basically, since there are rather a large number of code groups, in a practical planning situation one can, in most cases, implement the neighbouring cell list so that all the cells in the list for one cell belong to a different code group. Thus the terminal can search for the target cell and skip step 3 totally, just confirming detection without needing to compare the different primary scrambling codes for that step.

Further ways of improving cell search performance include the possibility of providing information on the relative timing between cells. This kind of information, which is being measured by the terminals for soft handover purposes in any case, can be used to improve especially the step 2 performance. The more accurate the relative timing information, the fewer slot positions need to be tested for the Secondary SCH code word, and the better is the probability of correct detection.

6.6.7 Transmit Diversity Procedure

As was mentioned in connection with the downlink channels, UTRA uses two types of transmit diversity transmission for user data performance improvement, as studied in Chapter 11. These methods are classified as open loop and closed loop methods. In this section the feedback procedure for closed loop transmit diversity is described. The open loop method was covered in connection with the downlink dedicated channel description.

In the case of closed loop transmit diversity, the base station uses two antennas to transmit the user information. The use of these two antennas is based on the feedback from the terminal, transmitted in the Feedback (FB) bits in the uplink DPCCH. The closed loop transmit diversity itself has two modes of operation.

In mode 1, the terminal feedback commands control the phase adjustments that are expected to maximise the power received by the terminal. The base station thus maintains the phase with antenna 1 and then adjusts the phase of antenna 2 based on the sliding averaging over two consecutive feedback commands. With this method, thus four different phase settings are applied to antenna 2.

In mode 2, the amplitude is adjusted in addition to the phase adjustment. The same signalling rate is used, but now the command is spread over four bits in four uplink DPCCH slots, with a single bit for amplitude and three bits for phase adjustment. This gives a total of eight different phase and two different amplitude combinations, thus a total of 16 combinations for signal transmission from the base station. The amplitude values have been defined to be 0.2 and 0.8, while the phase values are naturally distributed evenly for the antenna phase offsets, from $-135°$ to $+180°$ phase offset. In this mode the last three slots of the frame contain only phase information, while amplitude information is taken from the previous four slots. This allows the command period to go even with 15 slots as with mode 1, where the average at the frame boundary is slightly modified by averaging the commands from slot 13 and slot 0 to avoid discontinuities in the adjustment process.

The closed loop method may be applied only on the dedicated channels or with a DSCH together with a dedicated channel. The open loop method may be used on both the common and dedicated channels.

6.6.8 Handover Measurements Procedure

Within the UTRA FDD the possible handovers are as follows:

- Intra-mode handover, which can be soft handover, softer handover or hard handover. Hard handover may take place as intra- or inter-frequency handover.
- Inter-mode handover as handover to the UTRA TDD mode.
- Inter-system handover, which in Release-99 means only GSM handover. The GSM handover may take place to a GSM system operating at 900 MHz, 1800 MHz and 1900 MHz. Release-2000 is expected to cover additional details needed for hard handover to the Multi-Carrier CDMA, described in Chapter 13.

The main relevance of the handover to the physical layer is what to measure for handover criteria and how to obtain the measurements.

6.6.8.1 Intra-Mode Handover

The UTRA FDD intra-mode handover relies on the Ec/No measurement performed from the common pilot channel (CPICH). The quantities defined that can be measured by the terminal from the CPICH are as follows:

- Received Signal Code Power (RSCP), which is the received power on one code after despreading, defined on the pilot symbols.
- Received Signal Strength Indicator (RSSI), which is the wideband received power within the channel bandwidth.
- Ec/No, representing the received signal code power divided by the total received power in the channel bandwidth, which is defined as RSCP/RSSI.

There are also other items that can be used as a basis for handover decisions in UTRAN, as the actual handover algorithm decisions are left as an implementation issue. One such parameter mentioned in the standardisation discussions has been the dedicated channel SIR, giving information on the cell orthogonality and being measured in any case for power control purposes.

Additional essential information for soft handover purposes is the relative timing information between the cells. As in an asynchronous network, there is a need to adjust the transmission timing in soft handover to allow coherent combining in the Rake receiver, otherwise the transmissions from the different base stations would be difficult to combine, and especially the power control operation in soft handover would suffer additional delay. The timing measurement in connection with the soft handover operation is illustrated in Figure 6.27. The new base station adjusts the downlink timing in steps of 256 chips based on the information it receives from the RNC.

When the cells are within the 10 ms window, the relative timing can be found from the primary scrambling code phase, since the code period used is 10 ms. If the timing uncertainty is larger, the terminal needs to decode the System Frame Number (SFN) from the Primary CCPCH. This always takes time and may suffer from errors, which requires

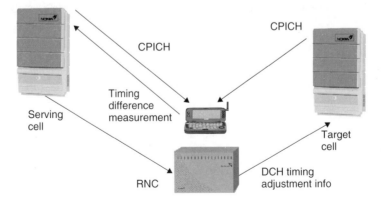

Figure 6.27. Timing measurement for soft handover

also a CRC check to be made on the SFN. The 10 ms window has no relevance when the timing information is provided in the neighbouring cell list. In such a case only the phase difference of the scrambling codes needs to be considered, unless the base stations are synchronised to chip level.

For the hard handover between frequencies such accurate timing information on chip level is not needed. Obtaining the other measurements is slightly more challenging as the terminal must make the measurements on a different frequency. This is typically done with the aid of compressed mode, which is described later in this chapter.

6.6.8.2   Inter-Mode Handover

On request from UTRAN, the dual-mode FDD–TDD terminals operating in FDD measure the power level from the TDD cells available in the area. The TDD CCPCH bursts sent twice during the 10 ms TDD frame can be used for measurement, since they are guaranteed to always exist in the downlink. The TDD cells in the same coverage area are synchronised, thus finding one slot with the reference midamble means that other TDD cells have roughly the same timing for their burst with reference power. UTRA TDD is covered in further detail in Chapter 12.

6.6.8.3   Inter-System Handover

For UTRA–GSM handover, basically similar requirements are valid as for GSM–GSM handover. Normally the terminal receives the GSM Synchronisation Channel (GSM SCH) during compressed frames in UTRA FDD to allow measurements from other frequencies. GSM 1800 set special requirements for compressed mode and required that compressed mode was specified for the uplink also. This was also needed for TDD measurements.

Other systems will be covered in Release-2000, with the focus on Multi-carrier CDMA (MC mode). The main concern for the FDD mode is to measure the pilot channel reception level from the MC mode downlink. The handover between UTRA FDD (also called DS mode) and the MC mode is always hard handover, such as handover to GSM. The need for the compressed mode depends on the terminal capability as well as on the location of the frequency band used by the MC mode. In general the same principles are valid from the measurements point of view as from the FDD–FDD inter-frequency point of view, as long as sufficient information on the MC mode system parameters is provided to the terminal via UTRAN. The Release-99 measurement procedures, like the compressed mode technique, are expected to be usable to provide measurements from the MC mode as well.

6.6.9   Compressed Mode Measurement Procedure

The compressed mode, often referred to as the slotted mode, is needed when making measurements from another frequency in a CDMA system without a full dual receiver terminal. The compressed mode means that transmission and reception are halted for a short time, in the order of a few milliseconds, in order to perform measurements on the other frequencies. The intention is not to lose data but to compress the data transmission in the time domain. Frame compression can be achieved with three different methods:

- Lowering the data rate from higher layers, as higher layers have knowledge of the compressed mode schedule for the terminal.

- Increasing the data rate by changing the spreading factor. For example, using spreading factor 64 instead of spreading factor 128 doubles the number of available symbols and makes it very straightforward to achieve the desired compression ratio for the frame.

- Reducing the symbol rate by puncturing at the physical layer multiplexing chain. In practice, this is limited to the rather short Transmission Gap Lengths (TGL), since puncturing has some practical limits. The benefit is obviously in keeping the existing spreading factor and not causing new requirements for channelisation code usage.

The compressed frames are provided normally in the downlink and in some cases in the uplink as well. If they appear in the uplink, they need to be simultaneous with the downlink frames, as illustrated in Figure 6.28.

The specified TGL lengths are 3, 4, 7, 10 and 14 slots. TGL lengths of 3, 4 and 7 can be obtained with both single- and double-frame methods. For TGL lengths of 10 or 14 only the double-frame method can be used. An example of the double-frame method is illustrated in Figure 6.29, where the idle slots are divided between two frames. This allows minimising the impact during a single frame and keeping, for example, the required increment in the transmission power lower than with the single frame method.

The case when uplink compressed frames are always needed with UTRA is the GSM 1800 measurements, where the close proximity of the GSM 1800 downlink frequency band

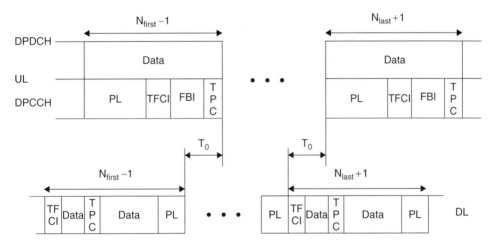

Figure 6.28. Compressed frames in the uplink and downlink

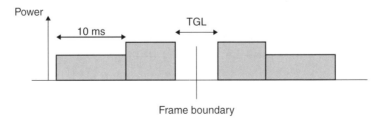

Figure 6.29. Compressed mode with the double frame method

to the core UTRA FDD uplink frequency band at 1920 MHz and upwards is too close to allow simultaneous transmission and reception.

Use of the compressed mode in the uplink with GSM 900 measurements or UTRA inter-frequency handover depends on terminal capability. For maintaining the continuous uplink, the terminal needs to have a means of generating the additional frequency parallel while maintaining the existing frequency. In practice, this means additional oscillators for frequency generation as well as some other duplicated components, which add to terminal power consumption.

The use of compressed mode has an inevitable impact on link performance, as studied in [10] for the uplink compressed mode and in [11] for the downlink. Link performance does not deteriorate very much if the terminal is not at the cell edge, since there is room to compensate the momentary performance loss with fast power control. The impact is largest at the cell edge; the difference in uplink performance between compressed mode and non-compressed mode is very slight until headroom is less than 4 dB. At 0 dB headroom the difference from normal transmission is between 2 and 4 dB, depending on the transmission gap duration with compressed frames. The 0 dB headroom corresponds to terminal operation at full power at the cell edge with no possibility of (soft) handover and with no room to run fast power control any more. The use of soft handover (or handover in general) will improve the situation, since low headroom values are less likely to occur, as with typical planning there is some overlap in the cell coverage area and the 0 dB headroom case should occur only when leaving the coverage area. The compressed mode performance is analysed in Section 9.3.2.

The actual time available for sampling on another frequency is reduced from the above values, due to the time taken by the hardware to switch the frequency; thus very short values of 1 or 2 slots have been excluded, since there is no really practical time available for measurements. The smallest value used in the specifications is 3, which itself allows only a very short measurement time window and should be considered for use only in specific cases.

6.6.10 Other Measurements

In the base station other measurements are needed to give RNC sufficient information on uplink status and base station transmission power resource usage. The following have been specified for the base station, to be supported by signalling between base station and RNC:

- RSSI, to give information on the uplink load.
- Uplink SIR on the DPCCH.
- Total transmission power on a single carrier at a base station transmitter, giving information on the available power resources at the base station.
- The transmission code on a single code for one terminal. This is used, for example, in balancing power between radio links in soft handover.
- Block Error Rate (BLER) and Bit Error Rate (BER) estimates for different physical channels.

The BLER measurement is to be supported by the terminals as well. The main function of terminal BLER measurement is to provide feedback for outer loop power control operation in setting the SIR target for fast power control operation.

Support of position location functionality needs measurements from the physical layer. For that purpose a second type of timing measurements has been specified that gives the timing difference between the primary scrambling codes of different cells with $\frac{1}{4}$-chip resolution for improved position location accuracy. The achievable position accuracy in theory can thus be estimated from the fact that a single chip corresponds to roughly 70 m in distance. In a cellular environment there are obviously further factors contributing to the achievable accuracy. To alleviate the impact of the near–far problem for a terminal that is very close to a base station, the specifications contain also a method of introducing idle periods in base station transmission. This enables timing measurements from base stations that would otherwise be too weak due to close proximity of the serving base station.

6.6.11   Operation with Adaptive Antennas

UTRA has been designed to allow the use of adaptive antennas, also known as beamforming, both in the uplink and downlink direction. Basically there are two types of beamforming one may use. Either a beam may use the secondary common pilot channel (S-CPICH) or then any may use only the dedicated pilot symbols. From the physical layer point of view the use of adaptive antennas is fully covered with Release-99 but the exact performance requirements for the terminals in different operation scenarios shall be covered for the later Release only, starting with Release 4.

Table 6.5. Application of beamforming concepts on downlink physical channel types

Physical channel type	Beamforming with S-CPICH	Beamforming without S-CPICH
P-CCPCH	No	No
SCH	No	No
S-CCPCH	No	No
DPCH	Yes	Yes
PICH	No	No
PDSCH (with associated DPCH)	Yes	Yes
AICH	No	No
CSICH	No	No

Whether beamforming may be applied to particular channels depends on certain factors. For example, does the channel contain dedicated pilot symbols?, or does the base station know, in the case of common channels, for which terminal is the data in the downlink direction intended? Table 6.5 shows the application of the beamforming concepts on different downlink physical channel types.

If it is desired to use beamforming together with any of the transmit diversity modes, then S-CPICH needs to be transmitted, including the diversity pilot, in the same antenna beam.

6.7   Terminal radio access capabilities

As explained in Chapter 2, the class mark approach of GSM in not applied in the same way with UMTS. Instead a terminal upon connection establishment informs a network of a large set of capability parameters and not only one or more class mark values. The reason for this approach has been the large variety of capabilities and data rates with UMTS terminals, which would have resulted to very high number of different class marks. For practical guidance reference classes were specified anyway.

Table 6.6. Terminal radio access capability parameter combinations for downlink decoding

Reference combination	32 kbps class	64 kbps class	128 kbps class	384 kbps class	768 kbps class	2048 kbps class
Transport channel parameters						
Maximum sum of number of bits of all transport blocks being received at an arbitrary time instant	640	3840	3840	6400	10240	20480
Maximum sum of number of bits of all turbo coded transport blocks being received at an arbitrary time instant	Not Supported	3840	3840	6400	10240	20480
Maximum number of simultaneous Coded Composite Transport Channels (CCTrCHs), higher value with PDSCH support	1	2/1	2/1	2/1	2	2
Maximum total number of transport blocks received within TTIs that end at the same time	8	8	16	32	64	96
Maximum number of Transport Format Combinations (TFC) in the TFC Set (TFCS)	32	48	96	128	256	1024
Maximum number of Transport Formats	32	64	64	64	128	256
Physical channel parameters						
Maximum number of DPCH/PDSCH codes simultaneously received, higher value with DSCH support	1	2/1	2/1	3	3	3
Maximum number of physical channel bits received in any 10 ms interval (DPCH, PDSCH, S-CCPCH), higher value with DSCH support.	1200	3600/2400	7200/4800	19200	28800	57600
Support of Physical DSCH	No	Yes/No	Yes/No	Yes/No	Yes	Yes

The reference classes in [12] have a few of common values as well, which are not covered here. For example the support for spreading factor 512 is not expected to be covered by any of the classes by default. For the channel coding methods the turbo coding is supported with classes above 32 kbps class and with higher classes the higher data rates above 64 kbps are

Table 6.7. Terminal radio access capability parameter combinations for uplink encoding

Reference combination	32 kbps class	64 kbps class	128 kbps class	384 kbps class	768 kbps class
Transport channel parameters					
Maximum sum of number of bits of all transport blocks being transmitted at an arbitrary time instant	640	3840	3840	6400	10240
Maximum sum of number of bits of all turbo coded transport blocks being transmitted at an arbitrary time instant	Not Supported	3840	3840	6400	10240
Maximum total number of transport blocks transmitted within TTIs that start at the same time	4	8	8	16	32
Maximum number of Transport Format Combinations (TFC) in the TFC Set (TFCS)	16	32	48	64	128
Maximum number of Transport Formats	32	32	32	32	64
Physical channel parameters					
Maximum number of DPDCH bits transmitted per 10 ms	1200	2400	4800	9600	19200

supported with turbo coding only as can be seen in tables 6.6 and 6.7. For the convolutional coding all the classes have the value of 640 bits at an arbitrary time instant for both encoding and decoding. This is needed in any case for decoding of broadcast channels. All the classes, except 32 kbps uplink, support atleast 8 parallel transport channels.

The value given for the number of bits received at an arbitrary time instant needs to be converted to the maximum data rate supported by considering at the same time the interleaving length (or TTI length with 3GPP terminology). For example the value 6400 bits for 384 kbps class can be converted to the maximum data rate with particular TTI as follows: The data rate of the application is 256 kbps, thus the number of bits per 10 ms is 2560 bits. With 10 ms or 20 ms TTI lengths the number of bits per interleaving period stays below 6400 bits but with 40 ms TTI the 6400 limit would be exceeded and the terminal would not have enough memory to operate with such a configuration. Respectively 384 kbps data rate with a terminal of a same class could be maintained with 10 ms TTI but 20 ms TTI would exceed the limit.

The value ranges given in [12] range from beyond what the classes contain, for example it is possible for a terminal to indicate values allowing 2 Mbps with 80 ms TTI. The minimum values haven been determined by the necessary capabilities needed to access the system, e.g. to listen the BCH or to access the RACH.

The key physical channel parameter is the maximum number of physical channel bits received/transmitted per 10 ms interval. This determines which spreading factors are supported. For example value 1200 bits for the 32 kbps class indicated that in the downlink the spreading factors supported are 256,128 and 64 while in the uplink the smallest value supported would be 64. The difference is coming from the from the use of QPSK modulation in the downlink and BPSK modulation in the uplink as explained earlier in this chapter in the section on modulation.

There are also parameters that are not dependant on a particular reference combination. Such a parameters indicate, for example, support for a particular terminal position location method. In the RF side the class independent parameters allow to indicate, for example, supported frequency bands or the terminal power class.

The parameters in table 6.7 and 6.8 cover The UTRA FDD while for UTRA TDD there are few additional TDD specific parameters in the complete tables [12] such as number of slots to be received etc.

References

[1]   3GPP Technical Specification 25.211, Physical Channels and Mapping of Transport Channels onto Physical Channels (FDD).

[2]   3GPP Technical Specification 25.212, Multiplexing and Channel Coding (FDD).

[3]   3GPP Technical Specification 25.213, Spreading and Modulation (FDD).

[4]   3GPP Technical Specification 25.214, Physical Layer Procedures (FDD).

[5]   3GPP Technical Specification 25.215, Physical Layer—Measurements (FDD).

[6]   3GPP Technical Specification 25.302, Services Provided by the Physical Layer.

[7]   Adachi, F., Sawahashi, M., and Okawa, K., 'Tree-structured Generation of Orthogonal Spreading Codes with Different Lengths for Forward Link of DS-CDMA Mobile', *Electronics Letters*, 1997, Vol. 33, No. 1, pp. 27–28.

[8]   3GPP Technical Specification 25.101, UE Radio Transmission and Reception (FDD).

[9]   3GPP Technical Specification 25.104, UTRA (BS) FDD; Radio Transmission and Reception.

[10] Toskala, A., Lehtinen, O. and Kinnunen, P., 'UTRA GSM Handover from Physical Layer Perspective', *Proc. ACTS Summit 1999*, Sorrento, Italy, June 1999.

[11] Gustafsson, M., Jamal, K., and Dahlman, E., 'Compressed Mode Techniques for Inter-Frequency Measurements in a Wide-band DS-CDMA System', *Proc. IEEE Int. Conf. on Personal Indoor and Mobile Radio Communications*, PIMRC'97, Helsinki, Finland, 1–4 September 1997, Vol. 1, pp. 231–235.

[12] 3GPP Technical Specification 25.306, UE Radio Access Capabilities, version 3.0.0., December 2000.

7

Radio Interface Protocols

Jukka Vialén

7.1 Introduction

The radio interface protocols are needed to set up, reconfigure and release the Radio Bearer services (including the UTRA FDD/TDD service), which were discussed in Chapter 2.

The protocol layers above the physical layer are called the data link layer (layer 2) and the network layer (layer 3). In the UTRA FDD radio interface, layer 2 is split into sublayers. In the control plane, layer 2 contains two sublayers—Medium Access Control (MAC) protocol and Radio Link Control (RLC) protocol. In the user plane, in addition to MAC and RLC, two additional service-dependent protocols exist: Packet Data Convergence Protocol (PDCP) and Broadcast/Multicast Control Protocol (BMC). Layer 3 consists of one protocol, called Radio Resource Control (RRC), which belongs to the control plane. The other network layer protocols, such as Call Control, Mobility Management, Short Message Service, and so on, are transparent to UTRAN and are not described in this book.

In this chapter the general radio interface protocol architecture is first described before going into deeper details of each protocol. For each protocol, the logical architecture and main functions are described. In the MAC section also the logical channels (services offered by MAC) and mapping between logical channels and transport channels are explained. For MAC and RLC, an example layer model is defined to describe what happens to a data packet passing through these protocols. In the RRC section, the RRC service states are described together with the main (RRC) functions and signalling procedures.

7.2 Protocol Architecture

The overall radio interface protocol architecture [1] is shown in Figure 7.1. This figure contains only the protocols that are visible in UTRAN.

The physical layer offers services to the MAC layer via transport channels [2] that were characterised by *how and with what characteristics* data is transferred (transport channels were discussed in Chapter 6).

WCDMA for UMTS, edited by Harri Holma and Antti Toskala
© 2001 John Wiley & Sons, Ltd

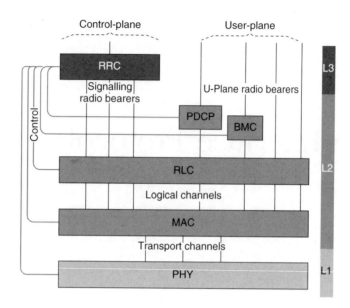

Figure 7.1. UTRA FDD Radio Interface protocol architecture

The MAC layer, in turn, offers services to the RLC layer by means of logical channels. The logical channels are characterised by *what type of data* is transmitted. Logical channels are described in detail in Section 7.3.

The RLC layer offers services to higher layers via service access points (SAPs), which describe how the RLC layer handles the data packets and if, for example, the automatic repeat request (ARQ) function is used. On the control plane, the RLC services are used by the RRC layer for signalling transport. On the user plane, the RLC services are used either by the service- specific protocol layers PDCP or BMC or by other higher-layer u-plane functions (e.g. speech codec). The RLC services are called Signalling Radio Bearers in the control plane and Radio Bearers in the user plane for services not utilising the PDCP or BMC protocols. The RLC protocol can operate in three modes—transparent, unacknowledged and acknowledged mode. These are further discussed in Section 7.4.

The Packet Data Convergence Protocol (PDCP) exists only for the PS domain services. Its main function is header compression. Services offered by PDCP are called Radio Bearers.

The Broadcast Multicast Control protocol (BMC) is used to convey over the radio interface messages originating from Cell Broadcast Center. In Release'99 of the 3GPP specifications, the only specified broadcasting service is the SMS Cell Broadcast service, which is derived from GSM. The service offered by BMC protocol is also called a Radio Bearer.

The RRC layer offers services to higher layers (to the Non Access Stratum) via service access points, which are used by the higher layer protocols in the UE side and by the Iu RANAP protocol in the UTRAN side. All higher layer signalling (mobility management, call control, session management, and so on) is encapsulated into RRC messages for transmission over the radio interface.

The control interfaces between the RRC and all the lower layer protocols are used by the RRC layer to configure characteristics of the lower layer protocol entities, including parameters for the physical, transport and logical channels. The same control interfaces are used by the RRC layer, for example to command the lower layers to perform certain types of measurements and by the lower layers to report measurement results and errors to the RRC.

7.3 The Medium Access Control Protocol

In the Medium Access Control (MAC) layer [5] the logical channels are mapped to the transport channels. The MAC layer is also responsible for selecting an appropriate transport format (TF) for each transport channel depending on the instantaneous source rate(s) of the logical channels. The transport format is selected with respect to the transport format combination set (TFCS) which is defined by the admission control for each connection.

7.3.1 MAC Layer Architecture

The MAC layer logical architecture is shown in Figure 7.2.

The MAC layer consists of three *logical entities*:

— **MAC-b** handles the broadcast channel (BCH). There is one MAC-b entity in each UE and one MAC-b in the UTRAN (located in Node B) for each cell.

— **MAC-c/sh** handles the common channels and shared channels—paging channel (PCH), forward link access channel (FACH), random access channel (RACH), uplink Common Packet Channel (CPCH) and Downlink Shared Channel (DSCH). There is one MAC-c/sh entity in each UE that is using shared channel(s) and one MAC-c/sh in the UTRAN (located in the controlling RNC) for each cell. Note that the BCCH logical channel can be mapped to either the BCH or FACH transport channel. Since the MAC header format for the BCCH depends on the transport channel used, two BCCH instances are shown in the figure. For PCCH, there is no MAC header, thus the only function of the MAC layer is to forward the data received from PCCH to the PCH at the time instant determined by RRC.

— **MAC-d** is responsible for handling dedicated channels (DCH) allocated to a UE in connected mode. There is one MAC-d entity in the UE and one MAC-d entity in the UTRAN (in the serving RNC) for each UE.

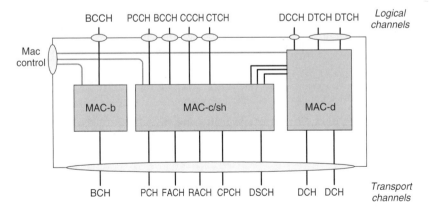

Figure 7.2. MAC layer architecture

7.3.2   MAC Functions

The functions of the MAC layer include:

— **Mapping between logical channels and transport channels**.

— **Selection of appropriate Transport Format (from the Transport Format Combination Set) for each Transport Channel, depending on the instantaneous source rate**.

— **Priority handling between data flows of one UE**. This is achieved by selecting 'high bit rate' and 'low bit rate' Transport Formats for different data flows.

— **Priority handling between UEs by means of dynamic scheduling**. A dynamic scheduling function may be applied for common and shared downlink transport channels FACH and DSCH.

— **Identification of UEs on common transport channels**. When a common transport channel (RACH, FACH or CPCH) carries data from dedicated-type logical channels (DCCH, DTCH), the identification of the UE (Cell Radio Network Temporary Identity (C-RNTI) or UTRAN Radio Network Temporary Identity (U-RNTI)) is included in the MAC header.

— **Multiplexing/demultiplexing of higher layer PDUs into/from transport blocks delivered to/from the physical layer on common transport channels**. MAC handles service multiplexing for common transport channels (RACH/FACH/CPCH). This is necessary, since it cannot be done in the physical layer.

— **Multiplexing/demultiplexing of higher layer PDUs into/from transport block sets delivered to/from the physical layer on dedicated transport channels**. MAC allows service multiplexing also for dedicated transport channels. While the physical layer multiplexing makes it possible to multiplex any type of service, including services with different quality of service parameters, MAC multiplexing is possible only for services with the same QoS parameters. Physical layer multiplexing is described in Chapter 6.

— **Traffic volume monitoring**. MAC receives RLC PDUs together with status information on the amount of data in the RLC transmission buffer. MAC compares the amount of data corresponding to a transport channel with the thresholds set by RRC. If the amount of data is too high or too low, MAC sends a measurement report on traffic volume status to RRC. The RRC can also request MAC to send these measurements periodically. The RRC use these reports for triggering reconfiguration of Radio Bearers and/or Transport Channels.

— **Dynamic Transport Channel type switching**. Execution of the switching between common and dedicated transport channels is based on a switching decision derived by RRC.

— **Ciphering**. If a radio bearer is using transparent RLC mode, ciphering is performed in the MAC sub-layer (MAC-d entity). Ciphering is a XOR operation (as in GSM and GPRS) where data is XORed with a ciphering mask produced by a ciphering algorithm. In MAC ciphering, the time-varying input parameter (COUNT-C) for the ciphering algorithm is incremented at each transmission time interval (TTI), that is, once every 10,20,40 or 80 ms depending on the transport channel configuration. Each radio bearer is ciphered separately. The ciphering details are described in 3GPP specification TS 33.102 [10].

— **Access Service Class (ASC) selection for RACH transmission**. The PRACH resources (i.e. access slots and preamble signatures for FDD) may be divided between different Access Service Classes in order to provide different priorities of RACH usage. Maximum number of ASCs is 8. MAC indicates the ASC associated with a PDU to the physical layer.

7.3.3 Logical Channels

The data transfer services of the MAC layer are provided on logical channels. A set of logical channel types is defined for the different kinds of data transfer services offered by MAC. A general classification of logical channels is into two groups: Control Channels and Traffic Channels. Control Channels are used to transfer control plane information, and Traffic Channels for user plane information.

The Control Channels are:

— Broadcast Control Channel (BCCH). A downlink channel for broadcasting system control information.

— Paging Control Channel (PCCH). A downlink channel that transfers paging information.

— Dedicated Control Channel (DCCH). A point-to-point bidirectional channel that transmits dedicated control information between a UE and the RNC. This channel is established during the RRC connection establishment procedure.

— Common Control Channel (CCCH). A bidirectional channel for transmitting control information between the network and UEs. This logical channel is always mapped onto RACH/FACH transport channels. A long UTRAN UE identity is required (U-RNTI, which includes SRNC address), so that the uplink messages can be routed to the correct serving RNC even if the RNC receiving the message is not the serving RNC of this UE.

The Traffic Channels are:

— Dedicated Traffic Channel (DTCH). A Dedicated Traffic Channel (DTCH) is a point-to-point channel, dedicated to one UE, for the transfer of user information. A DTCH can exist in both uplink and downlink.

— Common Traffic Channel (CTCH). A point-to-multipoint downlink channel for transfer of dedicated user information for all or a group of specified UEs.

7.3.4 Mapping Between Logical Channels And Transport Channels

The mapping between logical channels and transport channels is shown in Figure 7.3. The following connections between logical channels and transport channels exist:

— PCCH is connected to PCH.

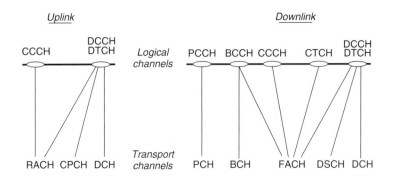

Figure 7.3. Mapping between logical channels and transport channels, uplink and downlink directions

— BCCH is connected to BCH and may also be connected to FACH.
— DCCH and DTCH can be connected to either RACH and FACH, to CPCH and FACH, to RACH and DSCH, to DCH and DSCH, or to a DCH and DCH.
— CCCH is connected to RACH and FACH.
— CTCH is connected to FACH.

7.3.5   Example Data Flow Through The MAC Layer

To illustrate the operation of the MAC layer, a block diagram in Figure 7.4 shows the MAC functions when data is processed through the layer. To keep the figure readable, the viewpoint is selected to be a network side transmitting entity, and uplink transport channels RACH and CPCH are omitted. The right-hand side of the figure describes the building of a MAC PDU when a packet received from DCCH or DTCH logical channel is processed by the MAC functions, which are shown in the left-hand side of the figure. In this example the MAC PDU is forwarded to the FACH transport channel.

A data packet arriving from the DCCH/DTCH logical channel triggers first the transport channel type selection in the MAC layer. In this example, the FACH transport channel is selected. In the next phase, the multiplexing unit adds a C/T field indicating the logical channel instance where the data originates. For common transport channels, such as FACH, this field is always needed. For dedicated transport channels (DCH) it is needed only if several logical channel instances are configured to use the same transport channel. The C/T field is 4 bits, allowing up to 15 simultaneous logical channels per transport channel (the value '1111' for the C/T field is reserved for future use). The priority tag (not part of the MAC PDU) for FACH and DSCH is set in MAC-d and used by MAC-c/sh when scheduling data onto transport channels. Priority for FACH can be set per UE; for DSCH it can be set per PDU. A flow control function in Iur interface (chapter 5) is needed to limit buffering between MAC-d and MAC-c/sh (which can be located in different RNCs). After receiving the data from MAC-d, the MAC-c/sh entity first adds the UE identification type (2 bits), the actual UE identification (C-RNTI 16 bits, or U-RNTI 32 bits), and the Target Channel Type Field (TCTF, in this example 2 bits) which is needed to separate the logical channel type using the transport channel (for FACH, the possible logical channel types could be BCCH, CCCH, CTCH or DCCH/DTCH). Now the MAC PDU is ready and the task for the scheduling/priority

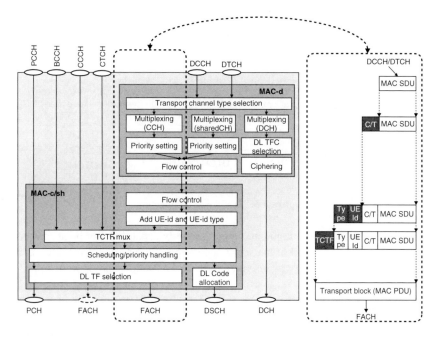

Figure 7.4. UTRAN side MAC entity (left side of figure) and building of MAC PDU when data received from DTCH or DCCH is mapped to FACH (right side of figure)

handling function is to decide the exact timing when the PDU is passed to layer 1 via the FACH transport channel (with an indication of the transport format to be used).

7.4 The Radio Link Control Protocol

The radio link control protocol [6] provides segmentation and retransmission services for both user and control data. Each RLC instance is configured by RRC to operate in one of three modes: transparent mode (Tr), unacknowledged mode (UM) or acknowledged mode (AM). The service the RLC layer provides in the control plane is called Signalling Radio Bearer (SRB). In the user plane, the service provided by the RLC layer is called a Radio Bearer (RB) only if the PDCP and BMC protocols are not used by that service, otherwise the RB service is provided by the PDCP or BMC.

7.4.1 RLC Layer Architecture

The RLC layer architecture is shown in Figure 7.5. All three RLC entity types and their connection to RLC-SAPs and to logical channels (MAC-SAPs) are shown. Note that the transparent and unacknowledged mode RLC entities are defined to be unidirectional, whereas the acknowledged mode entities are described as bidirectional.

For all RLC modes, the CRC error detection is performed on physical layer and the result of the CRC check is delivered to RLC together with the actual data.

In **transparent mode** no protocol overhead is added to higher layer data. Erroneous protocol data units (PDUs) can be discarded or marked erroneous. Transmission can be of the streaming type in which higher layer data is not segmented, though in special

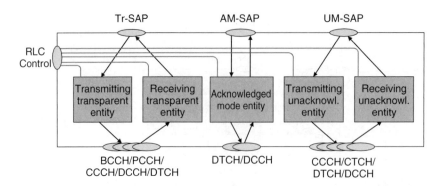

Figure 7.5. RLC layer architecture

cases transmission with limited segmentation/reassembly capability can be accomplished. If segmentation/reassembly is used, it has to be negotiated in the radio bearer setup procedure. The UMTS Quality of Service classes, including the streaming class, were introduced in Chapter 2.

In **unacknowledged mode** no retransmission protocol is in use and data delivery is not guaranteed. Received erroneous data is either marked or discarded depending on the configuration. On the sender side, a timer based discard without explicit signalling function is applied, thus RLC SDUs which are not transmitted within a specified time are simply removed from the transmission buffer. The PDU structure includes sequence numbers so that the integrity of higher layer PDUs can be observed. Segmentation and concatenation are provided by means of header fields added to the data. An RLC entity in unacknowledged mode is defined as unidirectional, because no association between uplink and downlink is needed. The unacknowledged mode is used, for example, for certain RRC signalling procedures, where the acknowledgement and retransmissions are part of the RRC procedure. Examples of user services that could utilize unacknowledged mode RLC are the cell broadcast service (see Section 7.6) and voice over IP (VoIP).

In the **acknowledged mode** an automatic repeat request (ARQ) mechanism is used for error correction. The quality vs. delay performance of the RLC can be controlled by RRC through configuration of the number of retransmissions provided by RLC. In case RLC is unable to deliver the data correctly (max number of retransmissions reached or the transmission time exceeded), the upper layer is notified and the RLC SDU is discarded. Also the peer entity is informed of a SDU discard operation by sending a Move Receiving Window command (in a STATUS message), so that also the receiver removes all AMD PDUs belonging to the discarded RLC SDU. An acknowledged mode RLC entity is bidirectional and capable of 'piggybacking' an indication of the status of the link in the opposite direction into user data. RLC can be configured for both in-sequence and out-of-sequence delivery. With in-sequence delivery the order of higher layer PDUs is maintained, whereas out-of-sequence delivery forwards higher layer PDUs as soon as they are completely received. In addition to data PDU delivery, *status* and *reset* control procedures can be signalled between peer RLC entities. The control procedures can even use a separate logical channel, thus one AM RLC entity can use either one or two logical channels. The acknowledged mode is the normal RLC mode for packet-type services, such as Internet browsing and email downloading, for example.

7.4.2 RLC Functions

The functions of the RLC layer are:

— **Segmentation and reassembly**. This function performs segmentation/reassembly of variable-length higher layer PDUs into/from smaller RLC Payload Units (PUs). One RLC PDU carries one PU. The RLC PDU size is set according to the smallest possible bit rate for the service using the RLC entity. Thus, for variable rate services, several RLC PDUs need to be transmitted during one transmission time interval when any bit rate higher than the lowest one is used.

— **Concatenation**. If the contents of an RLC SDU do not fill an integral number of RLC PUs, the first segment of the next RLC SDU may be put into the RLC PU in concatenation with the last segment of the previous RLC SDU.

— **Padding**. When concatenation is not applicable and the remaining data to be transmitted does not fill an entire RLC PDU of given size, the remainder of the data field is filled with padding bits.

— **Transfer of user data**. RLC supports acknowledged, unacknowledged and transparent data transfer. Transfer of user data is controlled by QoS setting.

— **Error correction**. This function provides error correction by retransmission in the acknowledged data transfer mode.

— **In-sequence delivery of higher layer PDUs**. This function preserves the order of higher layer PDUs that were submitted for transfer by RLC using the acknowledged data transfer service. If this function is not used, out-of-sequence delivery is provided.

— **Duplicate detection**. This function detects duplicated received RLC PDUs and ensures that the resultant higher layer PDU is delivered only once to the upper layer.

— **Flow control**. This function allows an RLC receiver to control the rate at which the peer RLC transmitting entity may send information.

— **Sequence number check (Unacknowledged data transfer mode)**. This function guarantees the integrity of reassembled PDUs and provides a means of detecting corrupted RLC SDUs through checking the sequence number in RLC PDUs when they are reassembled into a RLC SDU. A corrupted RLC SDU is discarded.

— **Protocol error detection and recovery**. This function detects and recovers from errors in the operation of the RLC protocol.

— **Ciphering** is performed in the RLC layer for acknowledged and unacknowledged RLC modes. The same ciphering algorithm is used as for MAC layer ciphering, the only difference being the time-varying input parameter (COUNT-C) for the algorithm, which for RLC is incremented together with the RLC PDU numbers. For retransmission, the same ciphering COUNT-C is used as for the original transmission (resulting in the same ciphering mask); this would not be so if ciphering were on the MAC layer. An identical ciphering mask for retransmission is essential, for example for Hybrid Type II ARQ, which is not part of a 3GPP Release-99 but a working item that may be added later to the standards. The ciphering details are described in 3GPP specification TS 33.102 [10].

— **Suspend/resume function for data transfer**. Suspension is needed during the security mode control procedure so that the same ciphering keys are always used by the peer entities. Suspensions and resumptions are local operations commanded by RRC via the control interface.

7.4.3 Example Data Flow Through The RLC Layer

This section takes a closer look at how data packets pass through the RLC layer. Figure 7.6 shows a simplified block diagram of an AM-RLC entity. The figure shows only how an AMD PDU can be constructed. It does not show how separate control PDUs (*status, reset*) between RLC entities are build.

Data packets (RLC SDUs) received from higher layers via AM-SAP are segmented and/or concatenated to payload units (PU) of fixed length. The PU length is a semi-static value that is decided in the Radio Bearer setup and can only be changed through the (RRC) Radio Bearer reconfiguration procedure. For concatenation or padding purposes, bits carrying information on the length and extension are inserted into the beginning of the last PU where data from an SDU is included. If several SDUs fit into one PU, they are concatenated and the appropriate length indicators are inserted into the beginning of the PU.

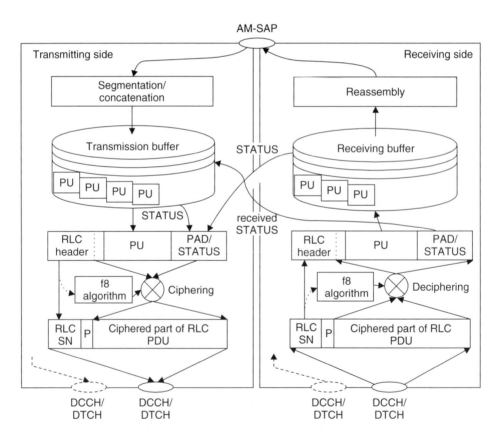

Figure 7.6. A simplified block diagram of an RLC AM entity

The PUs are then placed in the transmission buffer, which, in this example, also takes care of retransmission management.

An RLC AMD PDU is constructed by taking one PU from the transmission buffer, adding a header for it and, if the data in the PU does not fill the whole RLC AMD PDU, a PADding field or piggybacked STATUS message is appended. The piggybacked STATUS message can originate either from the receiving side (if the peer entity has requested a status report) or from the transmitting side to indicate a RLC SDU discard. The header contains the RLC PDU sequence number SN (12 bits for AM-RLC), poll bit P (which is used to request STATUS from the peer entity) and optionally a length indicator (7 or 15 bits), which is used if concatenation of SDUs, padding or a piggybacked STATUS PDU takes place in the RLC PDU.

Next, the AM RLC PDU is ciphered, excluding the two first octets which comprise the PDU sequence number (SN) and the poll bit (P). The PDU sequence number is one input parameter to the ciphering algorithm (forming the least significant bits of a COUNT-C parameter), and it must be readable by the peer entity to be able to perform deciphering. The details of the ciphering process are described in 3GPP specification TS 33.102 [10].

After this the PDU is ready to be forwarded to the MAC layer via a logical channel. In Figure 7.6, extra logical channels are shown by dashed lines, indicating that one RLC entity can be configured to send control PDUs and data PDUs using different logical channels. Note, however, that Figure 7.6 does not describe how the separate control PDUs are constructed.

The receiving side of the AM entity receives RLC AMD PDUs through one of the logical channels from the MAC sublayer. Errors are checked with the (physical layer) CRC, which is calculated over the whole RLC PDU. The actual CRC check is performed in the physical layer and the RLC entity receives result of this CRC check together with the data. After deciphering, the whole header and possible piggybacked status information can be extracted from the RLC PDU. If the received PDU was a control message or if status information was piggybacked to an AMD PDU, the control information (STATUS message) is passed to the transmitting side, which will check its retransmission buffer against the received status information. The PDU number from the RLC header is needed for deciphering and also when storing the deciphered PU into the receiving buffer. Once all PUs belonging to a complete SDU are in the receiving buffer, the SDU is reassembled. After this (not shown in the figure), the checks for in-sequence delivery and duplicate detection are performed before the RLC SDU is delivered to the higher layer.

7.5   The Packet Data Convergence Protocol

The Packet Data Convergence Protocol (PDCP) [7] exists only in the user plane and only for services from the PS domain. The PDCP contains compression methods, which are needed to get better spectral efficiency for services requiring IP packets to be transmitted over the radio. For 3GPP Release-99 standards, a header compression method is defined, for which several header compression algorithms can be used. As an example of why header compression is valuable, the size of the combined RTP/UDP/IP headers is at least 40 bytes for IPv4 and at least 60 bytes for IPv6, while the payload, for example for IP voice service, can be about 20 bytes or less.

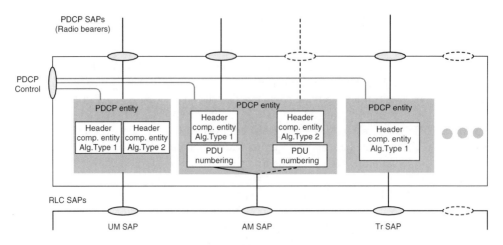

Figure 7.7. The PDCP layer architecture

7.5.1 PDCP Layer Architecture

An example of the PDCP layer architecture is shown in Figure 7.7. Multiplexing of Radio Bearers in the PDCP layer is not part of 3GPP Release-99 but is one possible feature for future releases. The multiplexing possibility is illustrated in Figure 7.7 with the two PDCP SAPs (one with dashed lines) provided by one PDCP entity using AM RLC. Every PDCP entity uses zero, one or several header compression algorithm types with a set of configurable parameters. Several PDCP entities may use the same algorithm types. The algorithm types and their parameters are negotiated during the RRC Radio Bearer establishment or reconfiguration procedures and indicated to the PDCP through the PDCP Control Service Access Point.

7.5.2 PDCP Functions

The main PDCP functions are:

— Compression of redundant protocol control information (e.g. TCP/IP and RTP/UDP/IP headers) at the transmitting entity, and decompression at the receiving entity. The header compression method is specific to the particular network layer, transport layer or upper layer protocol combinations, for example TCP/IP and RTP/UDP/IP. The only compression method that is mentioned in the PDCP Release-99 specification is RFC2507 [13].

— Transfer of user data. This means that the PDCP receives a PDCP SDU from the non-access stratum and forwards it to the appropriate RLC entity and vice versa.

— Support for lossless SRNS relocation. In practice this means that those PDCP entities which are configured to support lossless SRNS relocation have PDU sequence numbers, which, together with unconfirmed PDCP packets are forwarded to the new SRNC during relocation. Only applicable when PDCP is using acknowledged mode RLC with in-sequence delivery.

7.6   The Broadcast/Multicast Control Protocol

The other service-specific layer 2 protocol—the Broadcast/Multicast Control (BMC) protocol [8]—exists also only in the user plane. This protocol is designed to adapt broadcast and multicast services, originating from the Broadcast domain, on the radio interface. In the Release-99 of the standard, the only service utilising this protocol is the SMS Cell Broadcast service. This service is directly taken from GSM. It utilises UM RLC using the CTCH logical channel which is mapped into the FACH transport channel. Each SMS CB message is targeted to a geographical area, and RNC maps this area into cells.

7.6.1   BMC Layer Architecture

The BMC protocol, shown in Figure 7.8, does not have any special logical architecture.

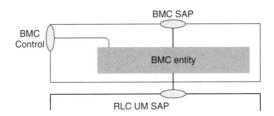

Figure 7.8. The Broadcast/Multicast Control layer architecture

7.6.2   BMC Functions

The main functions of the BMC protocol are:

— **Storage of Cell Broadcast messages**. The BMC in RNC stores the Cell Broadcast messages received over the CBC—RNC interface for scheduled transmission.

— **Traffic volume monitoring and radio resource request for CBS**. On the UTRAN side, the BMC calculates the required transmission rate for the Cell Broadcast Service based on the messages received over the CBC—RNC interface, and requests appropriate CTCH/FACH resources from RRC.

— **Scheduling of BMC messages**. The BMC receives scheduling information together with each Cell Broadcast message over the CBC—RNC interface. Based on this scheduling information, on the UTRAN side the BMC generates schedule messages and schedules BMC message sequences accordingly. On the UE side, the BMC evaluates the schedule messages and indicates scheduling parameters to RRC, which are used by RRC to configure the lower layers for CBS discontinuous reception.

— **Transmission of BMC messages to UE**. This function transmits the BMC messages (Scheduling and Cell Broadcast messages) according to the schedule.

— **Delivery of Cell Broadcast messages to the upper layer**. This UE function delivers the received non-corrupted Cell Broadcast messages to the upper layer.

When sending SMS CB messages to a cell for the first time, appropriate capacity has to be allocated in the cell. The CTCH has to be configured and the transport channel used has to be indicated to all UEs via (RRC) system information broadcast on the BCH. The capacity allocated for SMS CB is cell-specific and may vary over time to allow efficient use of the radio resources.

7.7   The Radio Resource Control Protocol

The major part of the control signalling between UE and UTRAN is Radio Resource Control (RRC) [3][9] messages. RRC messages carry all parameters required to set up, modify and release layer 2 and layer 1 protocol entities. RRC messages carry in their payload also all higher layer signalling (MM, CM, SM, etc.). The mobility of user equipment in the connected mode is controlled by RRC signalling (measurements, handovers, cell updates, etc.).

7.7.1   RRC Layer Logical Architecture

The RRC layer logical architecture is shown in Figure 7.9.

The RRC layer can be described with four *functional entities:*

— The Dedicated Control Function Entity (DCFE) handles all functions and signalling specific to one UE. In the SRNC there is one DCFE entity for each UE having an RRC connection with this RNC. DCFE uses mostly acknowledged mode RLC (AM-SAP), but some messages are sent using unacknowledged mode SAP (e.g. RRC Connection Release) or transparent SAP (e.g. Cell Update). DCFE can utilize services from all Signalling Radio Bearers, which are described in Chapter 7.7.3.4.

— The Paging and Notification control Function Entity (PNFE) handles paging of idle mode UE(s). There is at least one PNFE in the RNC for each cell controlled by this RNC. The PNFE uses the PCCH logical channel normally via transparent SAP of RLC. However, the specification mentions that PNFE could utilize also UM-SAP. In this example architecture the PNFE in RNC, when receiving a paging message from an Iu interface, needs to check with the DCFE whether or not this UE already has

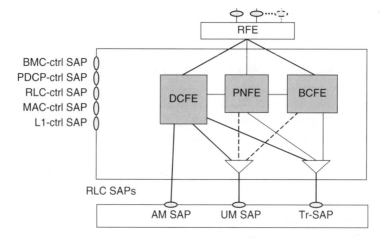

Figure 7.9. RRC layer architecture

an RRC connection (signalling connection with another CN domain); if it does, the paging message is sent (by the DCFE) using the existing RRC connection.

— The broadcast control function entity (BCFE) handles the system information broadcasting. There is at least one BCFE for each cell in the RNC. The BCFE uses either BCCH or FACH logical channels, normally via transparent SAP. The specification mentions that BFCE could utilize also UM-SAP.

— The fourth entity is normally drawn outside of the RRC protocol, but still belonging to access stratum and 'logically' to RRC layer, since the information required by this entity is part of RRC messages. The entity is called Routing Function Entity (RFE) and its task is the routing of higher layer (non access stratum) messages to different MM/CM entities (UE side) or different core network domains (UTRAN side). Every higher layer message is piggybacked into the RRC *Direct Transfer* messages (three types of Direct Transfer messages are specified, *Initial Direct Transfer* (uplink), *Uplink Direct Transfer* and *Downlink Direct Transfer*).

7.7.2   RRC Service States

The two basic operational modes of a UE are *idle mode* and *connected mode*. The connected mode can be further divided into service states, which define what kind of physical channels a UE is using. Figure 7.10 shows the main RRC service states in the connected mode. It also shows the transitions between idle mode and connected mode and the possible transitions within the connected mode.

In the idle mode [4], after the UE is switched on, it selects (either automatically or manually) a PLMN to contact. The UE looks for a suitable cell of the chosen PLMN, chooses that cell to provide available services, and tunes to its control channel. This choosing is known as 'camping on a cell'. The cell search procedure described in Chapter 6 is part of this camping process. After camping on a cell in idle mode, the UE is able to receive system information and cell broadcast messages. The UE stays in idle mode until it transmits a request to establish an RRC connection (Section 7.7.3.4). In idle mode the UE is identified by non-access stratum identities such as IMSI, TMSI and P-TMSI. In addition, the UTRAN has no information of its own about the individual idle mode UEs and can only address, for example, all UEs in a cell or all UEs monitoring a paging occasion.

In the **Cell_DCH** state a dedicated physical channel is allocated to the UE, and the UE is known by its serving RNC on a cell or active set level. The UE performs measurements and sends measurement reports according to measurement control information received from

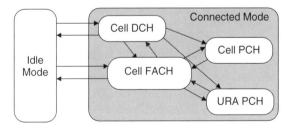

Figure 7.10. UE modes and RRC states in connected mode

RNC. The DSCH can also be used in this state, and UEs with certain capabilities are also able to monitor the FACH channel for system information messages.

In the **Cell_FACH** state no dedicated physical channel is allocated for the UE, but RACH and FACH channels are used instead, for transmitting both signalling messages and small amounts of user plane data. In this state the UE is also capable of listening to the broadcast channel (BCH) to acquire system information. The CPCH channel can also be used when instructed by UTRAN. In this state the UE performs cell reselections, and after a reselection always sends a Cell Update message to the RNC, so that the RNC knows the UE location on a cell level. For identification, a C-RNTI in the MAC PDU header separates UEs from each other in a cell. When the UE performs cell reselection it uses the U-RNTI when sending the Cell Update message, so that UTRAN can route the Cell Update message to the current serving RNC of the UE, even if the first RNC receiving the message is not the current SRNC. The U-RNTI is part of the RRC message, not in the MAC header. If the new cell belongs to another radio access system, such as GPRS, the UE enters idle mode and accesses the other system according to that system's access procedure.

In the **Cell_PCH** state the UE is still known on a cell level in SRNC, but it can be reached only via the paging channel (PCH). In this state the UE battery consumption is less than in the Cell_FACH state, since the monitoring of the paging channel includes a discontinuous reception (DRX) functionality. The UE also listens to system information on BCH. A UE supporting the Cell Broadcast Service (CBS) is also capable of receiving BMC messages in this state. If the UE performs a cell reselection, it moves autonomously to the Cell_FACH state to execute the Cell Update procedure, after which it re-enters the Cell_PCH state if no other activity is triggered during the Cell Update procedure. If a new cell is selected from another radio access system, the UTRAN state is changed to idle mode and access to the other system is performed according to that system's specifications.

The **URA_PCH** state is very similar to the Cell_PCH, except that the UE does not execute Cell Update after each cell reselection, but instead reads UTRAN Registration Area (URA) identities from the broadcast channel, and only if the URA changes (after cell reselection) does UE inform its location to the SRNC. This is achieved with the URA Update procedure, which is similar to the Cell Update procedure (the UE enters the Cell_FACH state to execute the procedure and then reverts to the URA_PCH state). One cell can belong to one or many URAs, and only if the UE cannot find its latest URA identification from the list of URAs in a cell does it need to execute the URA Update procedure. This 'overlapping URA' feature is needed to avoid ping-pong effects in a possible network configuration, where geographically succeeding base stations are controlled by different RNCs.

The UE leaves the connected mode and returns to idle mode when the RRC connection is released or at RRC connection failure.

7.7.2.1 Enhanced State Model For Multimode Terminals

Figure 7.11 presents an overview of the possible state transitions of a multimode terminal, in this example a UTRA FDD—GSM/GPRS terminal. With these terminal types it is possible to perform inter-system handover between UTRA FDD and GSM, and inter-system cell reselection from UTRA FDD to GPRS. The actual signalling procedures that relate to the thick arrows in Figure 7.11 are described in Section 7.7.3.

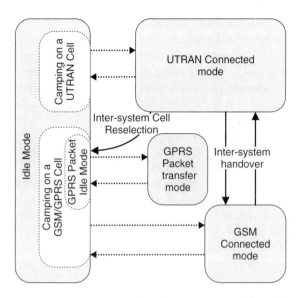

Figure 7.11. UE RRC states for a dual mode UTRA FDD—GSM/GPRS terminal

7.7.3 RRC Functions And Signalling Procedures

Since the RRC layer handles the main part of control signalling between the UEs and UTRAN, it has a long list of functions to perform. Most of these functions are part of the RRM algorithms, which are discussed in Chapters 9 and 10, but since the information is carried in RRC layer messages, the specifications list the functions as part of the RRC protocol. The main RRC functions are:

— Broadcast of system information, related to access stratum and non-access stratum
— Paging
— Initial cell selection and reselection in idle mode
— Establishment, maintenance and release of an RRC connection between the UE and UTRAN
— Control of Radio Bearers, transport channels and physical channels
— Control of security functions (ciphering and integrity protection)
— Integrity protection of signalling messages
— UE measurement reporting and control of the reporting
— RRC connection mobility functions
— Support of SRNS relocation
— Support for downlink outer loop power control in the UE
— Open loop power control
— Cell broadcast service related functions
— Support for UE Positioning functions.

In the following sections, these functions and related signalling procedures are described in more detail.

7.7.3.1 Broadcast Of System Information

The broadcast system information originates from the Core Network, from RNC and from Node Bs. The *System Information* messages are sent on a BCCH logical channel, which can be mapped to the BCH or FACH transport channel. A System Information message carries *system information blocks* (SIBs), which group together system information elements of the same nature. Dynamic (i.e. frequently changing) parameters are grouped into different SIBs from the more static parameters. One System Information message can carry either several SIBs or only part of one SIB, depending on the size of the SIBs to be transmitted. One System Information message will always fit into the size of a BCH or FACH transport block. If padding is required, it is inserted by the RRC layer.

The system information blocks are organised as a tree (Figure 7.12). A *master information block* (MIB) gives references and scheduling information to a number of system information blocks in a cell. It may also include reference and scheduling information to one or two *scheduling blocks*, which give references and scheduling information for all additional system information blocks. The master information block is sent regularly on the BCH and its scheduling is static. In addition to scheduling information of other SIBs and scheduling blocks, the master information block contains only the parameters 'Supported PLMN Types' and depending on which PLMN types are supported, either 'PLMN iden-tity' (GSM MAP) or 'ANSI-41 Core Network Information'. The system information blocks contain all the other actual system information.

The scheduling information (included in the MIB or in scheduling blocks) for SIBs containing frequently changing parameters contains a SIB specific timers (value in frames), which can be used by the UE to trigger re-reading of each block.

For the other SIBs (with more 'static' parameters) the master information block, or the 'parent' SIB, contains, as part of the scheduling information, a 'value tag' that the UE

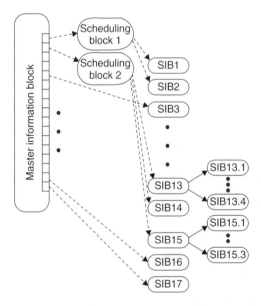

Figure 7.12. The overall structure of system information blocks in 3GPP Release-99. Dotter arrows show an example where scheduling information for each SIB could be included.

compares to the latest read 'value tag' of this system information block. Only if the value tag has changed after the last reading of the SIB in question does the UE re-read it. Thus, by monitoring the master information block and the scheduling blocks, the UE can notice if any of the system information blocks (of the more 'static' nature) has changed. UTRAN can also inform of the change in system information with *Paging* messages sent on the PCH transport channel (see Section 7.7.3.2) or with a *System Information Change Indication* message on the FACH transport channel. With these two messages, all UEs needing information about a change in the system information (all UEs in the Cell_FACH, Cell_PCH and URA_PCH states) can be reached.

The number of system information blocks in 3GPP Release-99 is one master information block, two scheduling blocks and 17 SIBs. Only SIB number 10, containing information needed only in Cell_DCH state, is sent using FACH transport channel, all other SIBs (incl. MIB and scheduling blocks) are sent on BCH. Scheduling information for each SIB can be included only into one place, either in MIB or in one of the scheduling blocks.

7.7.3.2 Paging

The RRC layer can broadcast paging information on the PCCH from the network to selected UEs in a cell. The paging procedure can be used for three purposes:

— At core network-originated call or session setup. In this case the request to start paging comes from the Core Network via the Iu interface

— To change the UE state from Cell_PCH or URA_PCH to Cell_FACH. This can be initiated, for example, by downlink packet data activity.

— To indicate change in the system information. In this case RNC sends a paging message with no paging records but with information describing a new 'value tag' for the master information block. This type of paging is targeted to all UEs in a cell.

7.7.3.3 Initial Cell Selection And Reselection In Idle Mode

The most suitable cell is selected, based on idle mode measurements and cell selection criteria. The cell search procedure described in Chapter 6 is part of the cell selection process.

7.7.3.4 Establishment, Maintenance And Release Of RRC Connection

The establishment of an RRC connection and Signalling Radio Bearers (SRB) between UE and UTRAN (RNC) is initiated by a request from higher layers (non-access stratum) on the UE side. In a network- originated case the establishment is preceded by an RRC *Paging* message. The request from non-access stratum is actually a request to set up a Signalling Connection between UE and CN (Signalling Connection consists of a RRC Connection and a Iu Connection). Only if the UE is in idle mode, thus no RRC connection exists, the UE initiates RRC Connection Establishment procedure. There can always be only zero or one RRC connections between one UE and UTRAN. If more than one signalling connection between UE and CN nodes exist, they all 'share' the same RRC connection.

The 'maintenance' of RRC connection refers to the RRC Connection Re-establishment functionality, which can be used to re-establish a connection after radio link failure. Timers are used to control the allowed time for a UE to return to "in-service-area" and to execute the re-establishment. The re-establishment functionality is included into the Cell Update procedure (7.7.3.9).

The RRC connection establishment procedure is shown in Figure 7.13. There is no need for a contention resolution step such as in GSM [12], since the UE identifier used in the connection equest and setup messages is a unique UE identity (for GSM based core network P-TMSI+RAI, TMSI+LAI or IMSI). In the RRC connection establishment procedure this initial UE identifier is used only for the purpose of uniqueness and can be discarded by UTRAN after the procedure ends. Thus, when these UE identities are later needed for the higher layer (Non Access Stratum) signalling, they must be resent (in the higher layer messages). The RRC *Connection Setup* message may include a dedicated physical channel assignment for the UE (move to Cell_DCH state), or it can command the UE to use common channels (move to Cell_FACH state). In the latter case, a radio network temporary identity (U-RNTI and possibly C-RNTI) to be used as UE identity on common transport channels is allocated to the UE.

The channel names in Figure 7.13 indicate either the logical channel or logical/transport channel used for each message.

The RRC connection establishment procedure creates three (optionally four) Signalling Radio Bearers (SRBs) designated by the RB identities #1 ... #4 (RB identity #0 is reserved for signalling using CCCH). The SRBs can later be created, reconfigured or even deleted with the normal Radio Bearer control procedures. The SRBs are used for RRC signalling according to the following rules:

— RB#1 is used for all messages sent on the DCCH and RLC-UM.

— RB#2 is used for all messages sent on the DCCH and RLC-AM, except for the *Direct Transfer* messages.

— RB#3 is used for the *Direct Transfer* messages (using DCCH and RLC-AM), which carries higher layer signalling. The reason for reserving a dedicated signalling radio bearer for the *Direct Transfer* is to enable prioritisation of UE-UTRAN signalling over the UE-CN signalling by using the RLC services (no need for extra RRC functionality).

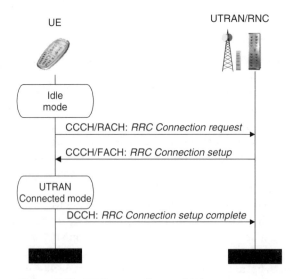

Figure 7.13. RRC connection establishment procedure

— RB#4 is optional and, if it exists, is also used for the *Direct Transfer* messages (using DCCH and RLC-AM). With two SRBs carrying higher layer signalling, UTRAN can handle prioritisation on signalling, RB#4 being used for 'low priority' and RB#3 for 'high priority' NAS signalling. The priority level is indicated to RRC with the actual NAS message to be carried over the radio. An example of low priority signalling could be the SMS.

— For RRC messages utilizing transparent mode RLC and CCCH logical channel (e.g. Cell Update, URA Update), RB identity #0 is used. A special function required in RRC layer for these messages is padding, because RLC in transparent mode does neither impose size requirements nor perform padding but the message size must still equal to a Transport Block size.

7.7.3.5 Control Of Radio Bearers, Transport Channels And Physical Channels

On request from higher layers, RRC performs the establishment, reconfiguration and release of Radio Bearers. At establishment and reconfiguration, UTRAN (RNC) performs admission control and selects parameters describing the Radio Bearer processing in layer 2 and layer 1. The SRBs are normally set up during the RRC Connection Establishment procedure (Section 7.7.3.4) but can also be controlled with the normal Radio Bearer procedures.

The transport channel and physical channel parameters are included in the Radio Bearer procedures but can also be configured separately with transport channel and physical channel dedicated procedures. These are needed, for example, if temporary congestion occurs in the network or when switching the UE between Cell_DCH and Cell_FACH states.

7.7.3.6 Control Of Security Functions

The RRC *Security Mode Control* procedure is used to start ciphering and integrity protection between the UE and UTRAN and to trigger the change of the ciphering and integrity keys during the connection.

The ciphering key is CN domain specific; thus in a typical network configuration (see Chapter 5), two ciphering keys can be used simultaneously for one UE—one for the PS domain services and one for the CS domain services. For the signalling (that uses common Radio Bearer(s) for both CN domains) the newer of these two keys is used. Ciphering is executed on the RLC layer for services using unacknowledged or acknowledged RLC and on the MAC layer for services using transparent RLC.

Integrity protection (see next section) is used only for signalling. In a typical network configuration two integrity keys would be available, but since only one RRC Connection can exist per UE, all signalling is protected with one and the same integrity key, which is always the newer of the keys IK_{CS} and IK_{PS}.

7.7.3.7 Integrity Protection Of Signalling

The RRC layer inserts a 32-bit integrity checksum, called a Message Authentication Code MAC-I, into most RRC PDUs. The integrity checksum is used by the receiving RRC entity to verify the origin and integrity of the messages. The receiving entity also calculates MAC-I and compares it to the one received with the signalling message. Messages received with wrong or missing message authentication code are discarded. Since all higher layer (non access stratum) signalling is carried in *RRC Direct Transfer* messages, all higher layer messages are automatically also integrity protected. The only exception to this is the initial higher layer message carried in *Initial Direct Transfer* message.

The checksum is calculated using UMTS integrity algorithm (UIA) that uses a secret 128-bit integrity key (IK) as one input parameter. The key is generated together with the ciphering key (CK) during the authentication procedure [11]. Figure 7.14 illustrates the calculation of MAC-I using the integrity algorithm f9 [15]. In addition to the IK, other parameters used as input to the algorithm are COUNT-I, which is incremented by one for each integrity protected message, a random number FRESH generated by RNC, DIRECTION bit (uplink/downlink), and the actual signalling message. Also the signalling radio bearer identity should affect the calculation of MAC-I. Since the algorithm f9 was ready when this requirement was identified, no new input parameter could be added to the f9 algorithm. The signalling radio bearer identity is inserted into the MESSAGE before it is given to the integrity algorithm.

Only a few RRC messages cannot be integrity protected; examples are the messages exchanged during RRC Connection Establishment procedure, since the algorithms and parameters are not yet negotiated when these messages are sent.

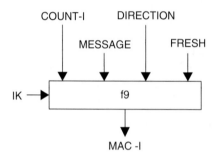

Figure 7.14. Calculation of message authentication code MAC-I

7.7.3.8   UE Measurement Reporting And Control

The measurements performed by the UE are controlled by the RNC using RRC protocol messages, in terms of what to measure, when to measure and how to report, including both UTRA radio interface and other systems. RRC signalling is also used in reporting of the measurements from the UE to the UTRAN (RNC).

Measurement Control

The measurement control (and reporting) procedure is designed to be very flexible. Serving RNC may start, stop or modify a number of parallel measurements in the UE and each of these measurements (including how they are reported) can be controlled independently of each other. The measurement control information is included in *System Information Block Type 12* and *System Information Block Type 11*. For UEs in Cell_DCH state also a dedicated *Measurement Control* message can be used. This is illustrated in Figure 7.15.

The measurement control information includes:

— *Measurement identity number*: A reference number that is used by the UTRAN at modification or release of the measurement and by the UE in the measurement report.

— *Measurement command*: May be either setup, modify or release.

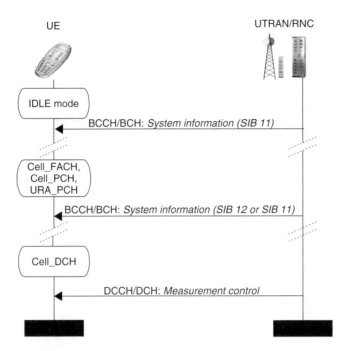

Figure 7.15. Measurement control procedures in different UE states

— *Measurement type*: One of the seven types from a predefined list, where each type describes what the UE measures. The seven types of measurements are defined as:

   — Intra-frequency measurements: measurements on downlink physical channels at the same frequency as the active set
   — Inter-frequency measurements: measurements on downlink physical channels at frequencies that differ from the frequency of the active set
   — Inter-system measurements: measurements on downlink physical channels belonging to a radio access system other than UTRAN, e.g. GSM
   — Traffic volume measurements: measurements on uplink traffic volume, e.g. RLC buffer payload for each Radio Bearer
   — Quality measurements: measurements of quality parameters, e.g. downlink transport channel block error rate
   — Internal measurements: measurements of UE transmission power and UE received signal level.
   — Measurements for Location Services (LCS) [14]. The basic measurement provided by the UE for the network based OTDOA-IPDL positioning method is Observed Time Difference of system frame numbers (SFN) between measured cells.

— *Measurement objects*: The objects the UE shall measure, and corresponding object information. In handover measurements this is the cell information needed by the UE to make measurements on certain intra-frequency, inter-frequency or inter-system cells. In traffic volume measurements this parameter contains transport channel identification.

— *Measurement quantity*: The quantity the UE measures.

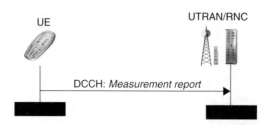

Figure 7.16. Measurement reporting procedure

— *Measurement reporting quantities*: The quantities the UE includes in the report.

— *Measurement reporting criteria*: The criteria that trigger the measurement report, such as periodical or event-triggered reporting.

— *Reporting mode*: This specifies whether the UE transmits the measurement report using acknowledged or unacknowledged data transfer of RLC.

Measurement Reporting

The measurement reporting procedure—shown in Figure 7.16—is initiated from the UE side when the reporting criteria are met. The UE sends a *Measurement Report* message, including the measurement identity number and the measurement results.

The *Measurement Report* message is used in the Cell_DCH and Cell_FACH states. In the Cell_FACH state, it is used only for a traffic volume measurement report. Traffic volume measurements may be triggered also in Cell_PCH and URA_PCH states, but the UE has to first change to Cell_FACH state before being able to send a measurement report.. In order to receive measurement information needed for the immediate establishment of macrodiversity when establishing a dedicated physical channel, the UTRAN may also request the UE to append radio link-related (intra-frequency) measurement reports to the following messages when they are sent on the RACH channel:

— *RRC Connection Request* message sent to establish an RRC connection.

— *Initial Direct Transfer* and *Uplink Direct Transfer* messages.

— *Cell Update* message

— *Measurement Report* message sent to report uplink traffic volume in Cell_FACH state.

7.7.3.9   RRC Connection Mobility Functions

RRC 'connection mobility' means keeping track of a UE's location (on a cell or active set level) while the UE is in UTRAN Connected mode. For this, a number of RRC procedures are defined. When dedicated channels are allocated to a UE, a normal way to perform mobility control is to use *Active Set Update* and *Hard Handover* procedures. When the UE is using only common channels (RACH/FACH/PCH) while in the UTRAN Connected mode, specific procedures are used to keep track of UE location either on cell or on UTRAN Registration Area (URA) level.

The UE mobility-related RRC procedures include:

— *Active Set Update* to update the UE's active set while in the Cell_DCH state.

— *Hard Handover* to make inter-frequency or intra-frequency hard handovers while in the Cell_DCH state.

— *Inter-system handover* between UTRAN and another radio access system (e.g. GSM).

— *Inter-system cell reselection* between UTRAN and another radio access system (e.g. GPRS).

— *Inter-system cell change order* between UTRAN and another radio access system (e.g. GPRS)

— *Cell Update* to report the UE location into RNC while in the Cell_FACH or Cell_PCH state.

— *URA Update* to report the UE location into RNC while in the URA_PCH state.

These procedures are described in the following sections.

Active Set Update

The purpose of the active set update procedure is to update the active set of the connection between the UE and UTRAN while the UE is in the Cell_DCH state. The procedure—shown in Figure 7.17—can have one of the following three functions: radio link addition, radio link removal, or combined radio link addition and removal. The maximum number of simultaneous radio links is 8 and it is possible to remove even all of them with one Active Set Update command. The soft handover algorithm and its performance are discussed in Section 9.3.1.

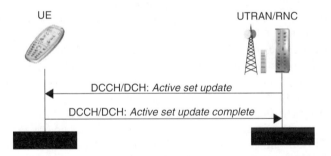

Figure 7.17. Active Set Update procedure

Hard Handover

The Hard Handover procedure can be used to change the radio frequency band of the connection between the UE and UTRAN or to change the cell on the same frequency when no network support of macro diversity exists. It can also be used to change the mode between FDD and TDD. This procedure is used only in the Cell_DCH state. No dedicated signalling messages have been defined for the Hard Handover but the functionality can be performed as part of the following RRC procedures: Physical channel reconfiguration, Radio bearer establishment, Radio bearer reconfiguration, Radio bearer release and Transport channel reconfiguration.

Inter-System Handover From UTRAN

The inter-system handover from UTRAN procedure is shown in Figure 7.18. This procedure is used for handover from UTRAN to another radio access system when the UE has at least

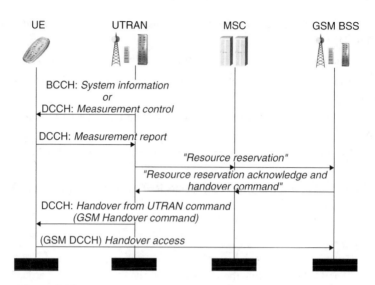

Figure 7.18. Inter-system handover procedure from UTRAN to GSM

one RAB in use for a CS domain service. For Release'99 UE, only support of handover of one RAB is expected, although the specification allows also handover of multiple RABs and even RABs from CS and PS domains simultaneously. In this example the target system is GSM, but the specifications also support handover to PCS 1900 and cdma2000 radio access systems. This procedure may be used in the Cell_DCH and Cell_FACH states. The UE receives the GSM neighbour cell parameters [12] either on *System Information or in a Measurement* Control message. These parameters are required to be able to measure candidate GSM cells. Based on the measurement report from UE, including GSM measurements, RNC makes a handover decision. After resources have been reserved from GSM BSS, the RNC sends an *Handover From UTRAN Command* message that carries a piggybacked *GSM Handover Command*. At this point the GSM RR protocol in UE takes control and sends a *GSM Handover Access* message to GSM BSC. After successful completion of the handover procedure, GSM BSS initiates resource release from UTRAN which will release the radio connection and remove all context information for the UE concerned.

Inter-System Handover To UTRAN

The inter-system handover to UTRAN procedure is shown in Figure 7.19. This procedure is used for handover from a non-UTRAN system to UTRAN. In this example, the other system is again GSM. The dual mode UE receives the UTRAN neighbour cell parameters on *GSM System Information* messages. The parameters required to be able to measure UTRA FDD cells include Downlink Center frequency or UTRA Absolute Radio Frequency Channel Number (UARFCN), Downlink Bandwidth (in 3GPP Release-99, only 5 MHz allowed, but in future other bandwidths may appear), Downlink Scrambling Code or scrambling code group for the Primary Common Pilot Channel (CPICH), and reference time difference for UTRA cell (timing between the current GSM cell and the UMTS cell that is to be measured).

After receiving a measurement report from GSM MS, including UTRA measurements, and after making a handover decision, the GSM BSC initiates resource reservation from UTRAN RNC. In the next phase, GSM BSC sends an *GSM Inter-System Handover Command* [12]

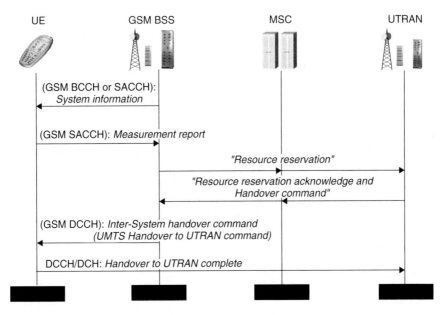

Figure 7.19. Inter-system handover procedure from GSM to UTRAN

including a piggybacked UMTS *Handover To UTRAN Command* message, which contains all the information required to set up connection to a UTRA cell. The GSM handover message (*Inter-System Handover Command*) must fit into one 23-octet data link layer PDU. Since the amount of information that could be included in the *Handover To UTRAN Command* is great, a preconfiguration mechanism is included in the standards. The preconfiguration means that only a reference number to a predefined set of UTRA parameters (Radio Bearer, Transport Channel and Physical Channel parameters) is included in the message. Naturally, the preconfiguration has to be transmitted to the UE beforehand. This can be done by GSM signalling or, if the UE has been earlier in UMTS mode it has been able to read the preconfiguration information from System Information Block type 16. The UE completes the procedure with a *Handover to UTRAN Complete* message to RNC. After successful completion of the handover procedure, RNC initiates resource release from GSM BSS.

Inter-System Cell Reselection From UTRAN
The inter-system cell reselection procedure from UTRAN is used to transfer a connection between the UE and UTRAN to another radio access system, such as GSM/GPRS. This procedure may be initiated in states Cell_FACH, Cell_PCH or URA_PCH. It is controlled mainly by the UE, but to some extent also by UTRAN. When UE has initiated an establishment of a connection to the other radio access system, it shall release all UTRAN specific resources.

Inter-System Cell Reselection To UTRAN
The inter-system cell reselection procedure to UTRAN is used to transfer a connection between the UE and another radio access system, such as GSM/GPRS, to UTRAN. This procedure is controlled mainly by the UE, but to some extent also by the other radio access system. The UE initiates an RRC connection establishment procedure to UTRAN with cause

value 'Inter-system cell reselection', and releases all resources specific to the other radio access system.

Inter-System Cell Change Order From UTRAN

The inter-system cell change order procedure—illustrated in Figure 7.20—can be used by the UTRAN to order UE to another radio access system. This procedure is used for UEs having at least one RAB for PS domain services. This procedure may be used in Cell_DCH and Cell_FACH states. Like in the case of inter-system handover from UTRAN, Release'99 UE is expected to be able to perform inter-system cell change with only one PS domain RAB, but the specification does no restrict this.

The procedure is initiated by UTRAN with the *Cell Change Order from UTRAN* message. The message contains at least required information of the target cell. After successful establishment of connection between UE and the other radio access system (e.g. GSM/GPRS), the other radio access system initiates release of the used UTRAN radio resources and UE context information.

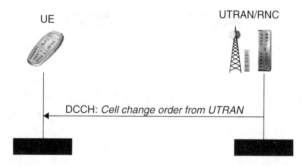

Figure 7.20. Inter-system cell change order from UTRAN

Inter-System Cell Change Order To UTRAN

This procedure is used by the other radio access system (e.g. GSM/GPRS) to command UE to move to UTRAN cell. The "cell change order" message in the other radio access system shall include the identity of the target UTRAN cell. In UTRAN side, the UE shall initiate an RRC connection establishment procedure with "establishment cause" set to "Inter-RAT cell change order".

Cell Update

The Cell Update procedure can be triggered by several reasons, including cell reselection, expiry of periodic cell update timer, initiation of uplink data transmission, UTRAN-originated paging and radio link failure in Cell_DCH state.

The *Cell Update Confirm* may include UTRAN mobility information elements (new U-RNTI and C-RNTI) for the UE. In this case, it responds with an *UTRAN Mobility Information Confirm* message so that the RNC knows that the new identities are taken into use.

The Cell Update Confirm may also include a radio bearer release, radio bearer reconfiguration, transport channel reconfiguration or physical channel reconfiguration. In these cases, the UE responds with suitable 'complete' message, see Figure 7.21.

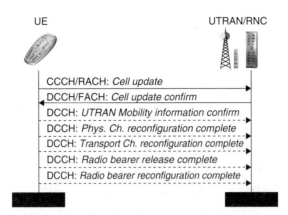

Figure 7.21. Cell Update procedure

URA Update

The UTRAN Registration Area (URA) Update procedure is used in the URA_PCH state. It can be triggered either after cell reselection, if the new cell does not broadcast the URA identifier that the UE is following, or by expiry of periodical URA Update timer. Since no uplink activity is possible in URA_PCH state, the UE has to temporarily switch to Cell_FACH state to execute the signal processing procedure, as shown in Figure 7.22.

UTRAN registration areas may be hierarchical to avoid excessive signalling. This means that several URA identifiers may be broadcast in one cell and that different UEs in one cell may reside in different URAs. A UE in the URA_PCH state always has one and only one valid URA. If a cell broadcasts several URAs, the RNC assigns one URA to a UE in the *URA Update Confirm* message.

The *URA Update Confirm* may assign a new URA Identity that the UE has to follow. It may also assign new RNTIs for the UE. In these cases UE responds with an *UTRAN Mobility Information Confirm* message so that the RNC knows that the new identities are taken into use.

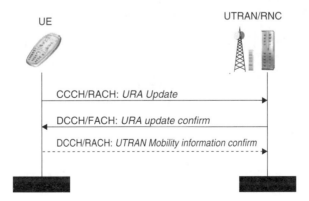

Figure 7.22. URA Update procedure

7.7.3.10 Support Of SRNS Relocation

In the serving RNS relocation procedure (see Chapter 5), the SRNC RRC layer builds a special RRC message—*RRC Information to Target RNC*. The issue that makes this message a 'special' one is it is not targeted for UE but for the new SRNC. Thus, this message is not sent over the air, but carried from the old SRNC to the new one via the Core Network. The initialisation information contains e.g. RRC state information and all the required protocol parameters (RRC, RLC, MAC, PDCP, PHY) that are needed to set up the UE context in the new SRNC. In addition, the expected PDCP sequence numbers (which are normally maintained locally in UE and UTRAN) need to be sent between UE and UTRAN, in any RRC messages which are sent during the SRNS relocation.

7.7.3.11 Support for Downlink Outer Loop Power Control

All RRC messages that can be used to add or reconfigure downlink transport channels (e.g. *Radio Bearer Setup/Reconfiguration/Release, Transport Channel Reconfiguration*) include a parameter 'Quality Target' (BLER quality value) that is used to configure the quality requirement (initial downlink SIR target) for each downlink transport channel separately.

The outer loop power control algorithm and its performance are discussed in Section 9.2.

7.7.3.12 Open Loop Power Control

Prior to PRACH transmission (see Chapter 6), the UE calculates the power for the first preamble as:

$$\text{Preamble\_Initial\_Power} = \text{Primary CPICH DL TX power} - \text{CPICH\_RSCP}$$

$$+ \text{UL interference} + \text{constant value}$$

The value for the CPICH_RSCP is measured by the UE, all other parameters being received on *System Information*.

As long as the physical layer is configured for PRACH transmission, the UE continuously recalculates the Preamble_Initial_Power when any of the broadcast parameters used in the above formula changes. The new Preamble_Initial_Power is then resubmitted to the physical layer.

When establishing the first DPCCH the UE shall start the UL inner loop power control at a power level according to:

$$\text{DPCCH\_Initial\_power} = \text{DPCCH\_Power\_offset} - \text{CPICH\_RSCP}$$

The value for the DPCCH_Power_offset is received from UTRAN on various signalling messages. The value for the CPICH_RSCP shall be measured by the UE.

7.7.3.13 Cell Broadcast Service Related Functions

The CBS-related functions of the RRC layer are as follows:

— Initial configuration of the BMC layer.

— Allocation of radio resources for CBS, in practice allocating the schedule for mapping the CTCH logical channel into the FACH transport channel and further into the S-CCPCH physical channel.

— Configuration of layer 1 and layer 2 for CBS discontinuous reception in the UE.

7.7.3.14 UE Positioning Related Functions

Although the full set of Release'99 UTRAN specifications support only the Cell_ID based positioning method, RRC protocol already is capable of supporting also both UE based and UE assisted OTDOA and GPS methods [14]. This includes capability to transfer positioning related UE measurements to UTRAN and delivery of assistance data for OTDOA and/or GPS from UTRAN to UE, which can be done either with *System Information* or with a dedicated message, called *Assistance Data Delivery*.

References

[1]   3G TS 25.301 Radio Interface Protocol Architecture.
[2]   3G TS 25.302 Services Provided by the Physical Layer.
[3]   3G TS 25.303 UE Functions and Interlayer Procedures in Connected Mode.
[4]   3G TS 25.304 UE Procedures in Idle Mode.
[5]   3G TS 25.321 MAC Protocol Specification.
[6]   3G TS 25.322 RLC Protocol Specification.
[7]   3G TS 25.323 PDCP Protocol Specification.
[8]   3G TS 25.324 Broadcast/Multicast Control Protocol (BMC) Specification.
[9]   3G TS 25.331 RRC Protocol Specification.
[10] 3G TS 33.102 3G Security; Security Architecture.
[11] 3G TS 24.008 Mobile Radio Interface Layer 3 Specification, Core Network Protocols—Stage 3.
[12] GSM 04.18 Digital Cellular Telecommunications System (Phase 2+); Mobile Radio Interface Layer 3 Specification, Radio Resource Control Protocol.
[13] IETF RFC 2507 IP Header Compression.
[14] 3G TS 25.305 Stage 2 Functional Specification of Location Services in UTRAN.
[15] 3G TS 33.105 3G Security; Cryptographic Algorithm Requirements.

8

Radio Network Planning

Harri Holma, Zhi-Chun Honkasalo, Seppo Hämäläinen, Jaana Laiho, Kari Sipilä and Achim Wacker

8.1 Introduction

This chapter presents WCDMA radio network planning, including dimensioning, detailed capacity and coverage planning, and network optimisation. The WCDMA radio network planning process is shown in Figure 8.1. In the dimensioning phase an approximate number of base station sites, base stations and their configurations and other network elements is estimated, based on the operator's requirements and the radio propagation in the area. The dimensioning must fulfil the operator's requirements for coverage, capacity and quality of service. Capacity and coverage are closely related in WCDMA networks, and therefore both must be considered simultaneously in the dimensioning of such networks. The dimensioning of WCDMA networks is introduced in Section 8.2.

In Section 8.3 detailed capacity and coverage planning is presented, together with a WCDMA planning tool. In detailed planning, real propagation maps and operator's traffic estimates in each area are needed. The base station locations and network parameters are selected by the planning tool and/or the planner. Capacity and coverage can be analysed for each cell after the detailed planning. One case study of the detailed planning is presented in Section 8.3 with capacity and coverage analysis. When the network is in operation, its performance can be observed by measurements, and the results of those measurements can be used to visualise and optimise network performance. The planning and the optimisation process can also be automated with intelligent tools and network elements. The optimisation is introduced in Section 8.3.

The adjacent channel interference must be considered in designing any wideband systems where large guard bands are not possible. In Section 8.4 the effect of interference between operators is analysed and network planning solutions are presented.

8.2 Dimensioning

WCDMA radio network dimensioning is a process through which possible configurations and amount of network equipment are estimated, based on the operator's requirements related to the following.

WCDMA for UMTS, edited by Harri Holma and Antti Toskala
© 2001 John Wiley & Sons, Ltd

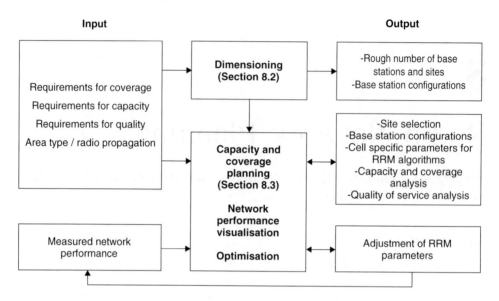

Figure 8.1. WCDMA radio network planning process

Coverage:

−   coverage regions
−   area type information
−   propagation conditions

Capacity:

−   spectrum available
−   subscriber growth forecast
−   traffic density information

Quality of Service:

−   area location probability (coverage probability)
−   blocking probability
−   end user throughput

Dimensioning activities include radio link budget and coverage analysis, capacity estimation, and finally, estimations on the amount of sites and base station hardware, radio
network controllers (RNC), equipment at different interfaces, and core network elements
(i.e. Circuit Switched Domain and Packet Switched Domain Core Networks).

8.2.1   Radio Link Budgets and Coverage Efficiency

The link budget of the WCDMA uplink is presented in this section. There are some
WCDMA-specific parameters in the link budget that are not used in a TDMA-based radio
access system such as GSM. The most important ones are as follows.

– Interference margin:

The interference margin is needed in the link budget because the loading of the cell, the load factor, affects the coverage: see Section 8.2.2. The more loading is allowed in the system, the larger is the interference margin needed in the uplink, and the smaller is the coverage area. For coverage-limited cases a smaller interference margin is suggested, while in capacity-limited cases a larger interference margin should be used. In the coverage-limited cases the cell size is limited by the maximum allowed path loss in the link budget, and the maximum air interface capacity of the base station site is not used. Typical values for the interference margin in the coverage-limited cases are 1.0–3.0 dB, corresponding to 20–50% loading.

– Fast fading margin (= power control headroom):

Some headroom is needed in the mobile station transmission power for maintaining adequate closed loop fast power control. This applies especially to slow-moving pedestrian mobiles where fast power control is able to effectively compensate the fast fading. The power control headroom was studied in [1]. The performance of fast power control is discussed in Section 9.2.1. Typical values for the fast fading margin are 2.0–5.0 dB for slow-moving mobiles.

– Soft handover gain:

Handovers—soft or hard—give a gain against slow fading (= log-normal fading) by reducing the required log-normal fading margin. This is because the slow fading is partly uncorrelated between the base stations, and by making handover the mobile can select a better base station. Soft handover gives an additional macro diversity gain against fast fading by reducing the required E_b/N_0 relative to a single radio link, due to the effect of macro diversity combining. The total soft handover gain is assumed to be between 2.0 and 3.0 dB in the examples below, including the gain against slow and fast fading. The handovers are discussed in Section 9.3 and the macro diversity gain for the coverage in Section 11.2.1.4.

Other parameters in the link budget are discussed in Chapter 7 in [2]. Below, three examples of link budgets are given for typical UMTS services: 12.2 kbps voice service using AMR speech codec, 144 kbps real-time data and 384 kbps non-real-time data, in an urban macro- cellular environment at the planned uplink noise rise of 3 dB. An interference margin of 3 dB is reserved for the uplink noise rise. The assumptions that have been used in the link budgets for the receivers and transmitters are shown in Tables 8.1 and 8.2.

The link budget in Table 8.3 is calculated for 12.2 kbps speech for in-car users, including 8.0 dB in-car loss. No fast fading margin is reserved in this case, since at 120 km/h the fast power control is unable to compensate for the fading. The required E_b/N_0 is assumed to be 5.0 dB. The E_b/N_0 requirement depends on the bit rate, service, multipath profile, mobile speed, receiver algorithms and base station antenna structure. For low mobile speeds the

Table 8.1. Assumptions for the mobile station

	Speech terminal	Data terminal
Maximum transmission power	21 dBm	24 dBm
Antenna gain	0 dBi	2 dBi
Body loss	3 dB	0 dB

Table 8.2. Assumptions for the base station

Noise figure	5.0 dB
Antenna gain	18 dBi (3-sector base station)
E_b/N_0 requirement	Speech: 5.0 dB
	144 kbps real-time data: 1.5 dB
	384 kbps non-real-time data: 1.0 dB
Cable loss	2.0 dB

Table 8.3. Reference link budget of AMR 12.2 kbps voice service (120 km/h, in-car users, Vehicular A type channel, with soft handover)

12.2 kbps voice service (120 km/h, in-car)		
Transmitter (mobile)		
Max. mobile transmission power [W]	0.125	
As above in dBm	21.0	a
Mobile antenna gain [dBi]	0.0	b
Body loss [dB]	3.0	c
Equivalent Isotropic Radiated Power (EIRP) [dBm]	18.0	d = a + b − c
Receiver (base station)		
Thermal noise density [dBm/Hz]	−174.0	e
Base station receiver noise figure [dB]	5.0	f
Receiver noise density [dBm/Hz]	−169.0	g = e + f
Receiver noise power [dBm]	−103.2	h = g + 10 ∗ log(3840000)
Interference margin [dB]	3.0	i
Receiver interference power [dBm]	−103.2	j = 10 ∗ log(10ʹ((h + i)/10) − 10ʹ(h/10))
Total effective noise + interference [dBm]	−100.2	k = 10 ∗ log(10ʹ(h/10) + 10ʹ(j/10))
Processing gain [dB]	25.0	l = 10 ∗ log(3840/12.2)
Required Eb/No [dB]	5.0	m
Receiver sensitivity [dBm]	−120.2	n = m − l + k
Base station antenna gain [dBi]	18.0	o
Cable loss in the base station [dB]	2.0	p
Fast fading margin [dB]	0.0	q
Max. path loss [dB]	154.2	r = d − n + o − p − q
Coverage probability [%]	95	
Standard deviation of log normal fading [dB]	7.0	
Propagation model exponent	3.52	
Log normal fading margin [dB]	7.3	s
Soft handover gain [dB], multi-cell	3.0	t
In-car loss [dB]	8.0	u
Allowed propagation loss for cell range [dB]	**141.9**	v = r − s + t − u

E_b/N_0 requirement is low but, on the other hand, a fast fading margin is required. Typically, the low mobile speeds are the limiting factor in the coverage dimensioning because of the required fast fading margin. Table 8.4 shows the link budget for 144 kbps real-time data service when an indoor location probability of 80% is provided by the outdoor base stations. The main differences between Tables 8.3 and 8.4 are the different processing gain, a higher mobile transmission power and a lower E_b/N_0 requirement. Additionally, a headroom of 4.0 dB is reserved for the fast power control to be able to compensate for the fading at 3 km/h. An average building penetration loss of 15 dB is assumed here.

Table 8.4. Reference link budget of 144 kbps real-time data service (3 km/h, indoor user covered by outdoor base station, Vehicular A type channel, with soft handover)

144 kbps real time data		
Transmitter (mobile)		
Max. mobile transmission power[W]	0.25	
As above in dBm	24.0	a
Mobile antenna gain [dBi]	2.0	b
Body loss [dB]	0.0	c
Equivalent Isotropic Radiated Power (EIRP) [dBm]	26.0	$d = a + b - c$
Receiver (base station)		
Thermal noise density [dBm/Hz]	−174.0	e
Base station receiver noise figure [dB]	5.0	f
Receiver noise density [dBm/Hz]	−169.0	$g = e + f$
Receiver noise power [dBm]	−103.2	$h = g + 10 * \log(3840000)$
Interference margin [dB]	3.0	i
Receiver interference power [dBm]	−103.2	$j = 10 * \log(10^{((h + i)/10)} - 10^{(h/10)})$
Total effective noise + interference [dBm]	−100.2	$k = 10 * \log(10^{(h/10)} + 10^{(j/10)})$
Processing gain [dB]	14.3	$l = 10 * \log(3840/144)$
Required Eb/No [dB]	1.5	m
Receiver sensitivity [dBm]	−113.0	$n = m - l + k$
Base station antenna gain [dBi]	18.0	o
Cable loss in the base station [dB]	2.0	p
Fast fading margin [dB]	4.0	q
Max. path loss [dB]	**151.0**	$r = d - n + o - p - q$
Coverage probability [%]	80	
Standard deviation of log normal fading [dB]	12.0	
Propagation model exponent	3.52	
Log normal fading margin [dB]	4.2	s
Soft handover gain [dB], multi-cell	2.0	t
Indoor loss [dB]	15.0	u
Allowed propagation loss for cell range [dB]	**133.8**	$v = r - s + t - u$

Table 8.5 presents a link budget for 384 kbps non-real-time data service for outdoors. The processing gain is lower than in the previous tables because of the higher bit rate. Also, the E_b/N_0 requirement is lower than that of the lower bit rates. The effect of the bit rate to the E_b/N_0 requirement is discussed in Section 11.2.1.1. This link budget is calculated assuming no soft handover.

The coverage efficiency of WCDMA is defined by the average coverage area per site, in km²/site, for a predefined reference propagation environment and supported traffic density.

From the link budgets above, the cell range R can be readily calculated for a known propagation model, for example the Okumura–Hata model or the Walfish–Ikegami model. The propagation model describes the average signal propagation in that environment, and it converts the maximum allowed propagation loss in dB to the maximum cell range in kilometres. As an example we can take the Okumura–Hata propagation model for an urban macro cell with base station antenna height of 30 m, mobile antenna height of 1.5 m and carrier frequency of 1950 MHz [6]:

$$L = 137.4 + 35.2 \log_{10}(R) \tag{8.1}$$

Table 8.5. Reference link budget of non-real-time 384 kbps data service (3 km/h, outdoor user, Vehicular A type channel, no soft handover)

384 kbps non-real time data, no soft handover		
Transmitter (mobile)		
Max. mobile transmission power [W]	0.25	
As above in dBm	24.0	a
Mobile antenna gain [dBi]	2.0	b
Body loss [dB]	0.0	c
Equivalent Isotropic Radiated Power (EIRP) [dBm]	26.0	d = a + b − c
Receiver (base station)		
Thermal noise density [dBm/Hz]	−174.0	e
Base station receiver noise figure [dB]	5.0	f
Receiver noise density [dBm/Hz]	−169.0	g = e + f
Receiver noise power [dBm]	−103.2	h = g + 10 ∗ log(3840000)
Interference margin [dB]	3.0	i
Receiver interference power [dBm]	−103.2	j = 10 ∗ log(10^((h + i)/10) − 10^(h/10))
Total effective noise + interference [dBm]	−100.2	k = 10 ∗ log(10^(h/10) + 10^(j/10))
Processing gain [dB]	10.0	l = 10 ∗ log(3840/384)
Required Eb/No [dB]	1.0	m
Receiver sensitivity [dBm]	−109.2	n = m − l + k
Base station antenna gain [dBi]	18.0	o
Cable loss in the base station [dB]	2.0	p
Fast fading margin [dB]	4.0	q
Max. path loss [dB]	**147.2**	r = d − n + o − p − q
Coverage probability [%]	95	
Standard deviation of log normal fading [dB]	7.0	
Propagation model exponent	3.52	
Log normal fading margin [dB]	7.3	s
Soft handover gain [dB], multi-cell	0.0	t
Indoor loss [dB]	0.0	u
Allowed propagation loss for cell range [dB]	**139.9**	v = r − s + t − u

where L is the path loss in dB and R is the range in km. For suburban areas we assume an additional area correction factor of 8 dB and obtain the path loss as:

$$L = 129.4 + 35.2 \log_{10}(R) \qquad (8.2)$$

According to Equation (8.2), the cell range of 12.2 kbps speech service with 141.9 dB path loss in Table 8.3 in a suburban area would be 2.3 km. The range of 144 kbps indoors would be 1.4 km. Once the cell range R is determined, the site area, which is also a function of the base station sectorisation configuration, can then be derived. For a cell of hexagonal shape covered by an omnidirectional antenna, the coverage area can be approximated as $2.6R^2$.

8.2.2 Load Factors and Spectral Efficiency

The second phase of dimensioning is estimating the amount of supported traffic per base station site. When the frequency reuse of a WCDMA system is 1, the system is typically interference-limited by the air interface and the amount of interference and delivered cell capacity must thus be estimated.

8.2.2.1   Uplink Load Factor

The theoretical spectral efficiency of a WCDMA cell can be calculated from the load equation whose derivation is shown below. We first define the E_b/N_0, energy per user bit divided by the noise spectral density:

$$(E_b/N_0)_j = \text{Processing gain of user } j \cdot \frac{\text{Signal of user } j}{\text{Total received power (excl. own signal)}} \quad (8.3)$$

This can be written:

$$(E_b/N_0)_j = \frac{W}{v_j R_j} \cdot \frac{P_j}{I_\text{total} - P_j} \quad (8.4)$$

where W is the chip rate, P_j is the received signal power from user j, v_j is the activity factor of user j, R_j is the bit rate of user j, and I_total is the total received wideband power including thermal noise power in the base station. Solving for P_j gives

$$P_j = \frac{1}{1 + \dfrac{W}{(E_b/N_0)_j \cdot R_j \cdot v_j}} I_\text{total} \quad (8.5)$$

We define $P_j = L_j \cdot I_\text{total}$ and obtain the load factor L_j of one connection

$$L_j = \frac{1}{1 + \dfrac{W}{(E_b/N_0)_j \cdot R_j \cdot v_j}} \quad (8.6)$$

The total received interference, excluding the thermal noise P_N, can be written as the sum of the received powers from all N users in the same cell

$$I_\text{total} - P_N = \sum_{j=1}^{N} P_j = \sum_{j=1}^{N} L_j \cdot I_\text{total} \quad (8.7)$$

The noise rise is defined as the ratio of the total received wideband power to the noise power

$$\text{Noise rise} = \frac{I_\text{total}}{P_N} \quad (8.8)$$

and using Equation (8.7) we can obtain

$$\text{Noise rise} = \frac{I_\text{total}}{P_N} = \frac{1}{1 - \displaystyle\sum_{j=1}^{N} L_j} = \frac{1}{1 - \eta_{UL}} \quad (8.9)$$

where we have defined the load factor η_{UL} as

$$\eta_{UL} = \sum_{j=1}^{N} L_j \quad (8.10)$$

When η_{UL} becomes close to 1, the corresponding noise rise approaches to infinity and the system has reached its pole capacity.

Additionally, in the load factor the interference from the other cells must be taken into account by the ratio of other cell to own cell interference, i:

$$i = \frac{\text{other cell interference}}{\text{own cell interference}} \tag{8.11}$$

The uplink load factor can then be written as

$$\eta_{UL} = (1+i) \cdot \sum_{j=1}^{N} L_j = (1+i) \cdot \sum_{j=1}^{N} \frac{1}{1 + \dfrac{W}{(E_b/N_0)_j \cdot R_j \cdot \upsilon_j}}. \tag{8.12}$$

The load equation predicts the amount of noise rise over thermal noise due to interference. The noise rise is equal to $-10 \cdot \log_{10}(1 - \eta_{UL})$. The interference margin in the link budget must be equal to the maximum planned noise rise.

The required E_b/N_0 can be derived from link level simulations and from measurements. It includes the effect of the closed loop power control and soft handover. The effect of soft handover is measured as the macro diversity combining gain relative to the single link E_b/N_0 result. The other cell to own (serving) cell interference ratio i is a function of cell environment or cell isolation (e.g. macro/micro, urban/suburban) and antenna pattern (e.g. omni, 3-sector or 6-sector [4]). The parameters are further explained in Table 8.6.

The load equation is commonly used to make a semi-analytical prediction of the average capacity of a WCDMA cell, without going into system-level capacity simulations. This load equation can be used for the purpose of predicting cell capacity and planning noise rise in the dimensioning process.

For a classical all-voice-service network, where all N users in the cell have a low bit rate of R , we can note that

$$\frac{W}{E_b/N_0 \cdot R \cdot \upsilon} \gg 1 \tag{8.13}$$

and the above uplink load equation can be approximated and simplified to

$$\eta_{UL} = \frac{E_b/N_0}{W/R} \cdot N \cdot \upsilon \cdot (1+i). \tag{8.14}$$

Table 8.6. Parameters used in uplink load factor calculation

	Definitions	Recommended values
N	Number of users per cell	
υ_j	Activity factor of user j at physical layer	**0.67** for speech, assumed 50% voice activity and DPCCH overhead during DTX **1.0** for data
E_b/N_0	Signal energy per bit divided by noise spectral density that is required to meet a predefined Quality of Service (e.g. bit error rate). Noise includes both thermal noise and interference	Dependent on service, bit rate, multipath fading channel, receive antenna diversity, mobile speed, etc.
W	WCDMA chip rate	**3.84 Mcps**
R_j	Bit rate of user j	Dependent on service
i	Other cell to own cell interference ratio seen by the base station receiver	Macro cell with omnidirectional antennas: **55%**

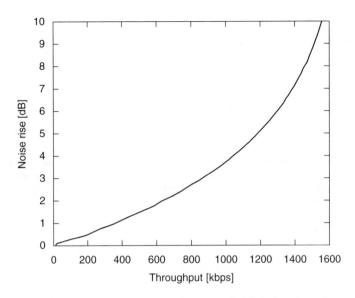

Figure 8.2. Uplink noise rise as a function of uplink data throughput

An example uplink noise rise is shown in Figure 8.2 for data service, assuming an E_b/N_0 requirement of 1.5 dB and $i = 0.65$. The noise rise of 3.0 dB corresponds to a 50% load factor, and the noise rise of 6.0 dB to a 75% load factor. Instead of showing the number of users N, we show the total data throughput per cell of all simultaneous users. In this example, a throughput of 860 kbps can be supported with 3.0 dB noise rise, and 1300 kbps with 6.0 dB noise rise.

8.2.2.2 Downlink Load Factor

The downlink load factor, η_{DL}, can be defined based on a similar principle as for the uplink, although the parameters are slightly different [5]:

$$\eta_{DL} = \sum_{j=1}^{N} v_j \cdot \frac{(E_b/N_0)_j}{W/R_j} \cdot [(1 - \alpha_j) + i_j] \qquad (8.15)$$

where $-10 \cdot \log_{10}(1 - \eta_{DL})$ is equal to the noise rise over thermal noise due to multiple access interference. The parameters are further explained in Table 8.7. Compared to the uplink load equation, the most important new parameter is α_j, which represents the orthogonality factor in the downlink. WCDMA employs orthogonal codes in the downlink to separate users, and without any multipath propagation the orthogonality remains when the base station signal is received by the mobile. However, if there is sufficient delay spread in the radio channel, the mobile will see part of the base station signal as multiple access interference. The orthogonality of 1 corresponds to perfectly orthogonal users. Typically, the orthogonality is between 0.4 and 0.9 in multipath channels.

In the downlink, the ratio of other cell to own cell interference, i, depends on the user location and is therefore different for each user j.

In downlink interference modelling, the effect of soft handover transmission can be modelled as having additional connections in the cell. The soft handover overhead is defined

Table 8.7. Parameters used in downlink load factor calculation

	Definitions	Recommended values for dimensioning
N	Number of connections per cell = number of users per cell * (1 + soft handover overhead)	
υ_j	Activity factor of user j at physical layer	**0.67** for speech, assumed 50% voice activity and DPCCH overhead during DTX **1.0** for data
E_b/N_0	Signal energy per bit divided by noise spectral density, required to meet a predefined Quality of Service (e.g. bit error rate). Noise includes both thermal noise and interference	Dependent on service, bit rate, multipath fading channel, transmit antenna diversity, mobile speed, etc.
W	WCDMA chip rate	**3.84 Mcps**
R_j	Bit rate of user j	Dependent on service
α_j	Orthogonality of channel of user j	Dependent on the multipath propagation 1: fully orthogonal 1-path channel 0: no orthogonality
i_j	Ratio of other cell to own cell base station power, received by user j	Each user sees a different i_j, depending on its location in the cell and log-normal shadowing
$\overline{\alpha}$	Average orthogonality factor in the cell	ITU Vehicular A channel: ~**60%** ITU Pedestrian A channel: ~**90%**
$\overline{i}$	Average ratio of other cell to own cell base station power received by user. Own cell interference is here wideband	Macro cell with omnidirectional antennas: **55%**

Note: The own cell is defined as the best serving cell. If a user is in soft handover, all the other base stations in the active set are counted as part of the 'other cell'.

as the total number of connections divided by the total number of users minus one. At the same time the soft handover gain relative to the single link E_b/N_0 is taken into account. This gain, called the macro diversity combining gain, can be derived from link/system level simulation analysis and is measured as the reduction in the required E_b/N_0 for each user.

The effect of transmit antenna diversity should be included into the required E_b/N_0. The gain of transmit diversity is discussed in Section 11.3.2.

The downlink load factor η_{DL} exhibits very similar behaviour to the uplink load factor η_{UL}, in the sense that when approaching unity, the system reaches its pole capacity and the noise rise over thermal goes to infinity.

For downlink dimensioning, it is important to estimate the total amount of base station transmission power required. This should be based on the *average* transmission power for the user, not the *maximum* transmission power for the cell edge shown by the link budget. The reason is that the wideband technology gives trunking gain in the power amplifier dimensioning: while some users are at the cell edge are requiring high power, other users close to the base station need much less power at the same time. The difference between the maximum and the average path loss is typically 6 dB in macro cells. This effect can be considered as power trunking gain of the wideband technology allowing to use a smaller base station power amplifier than in narrowband technologies.

The minimum required transmission power for each user is determined by the average attenuation between base station transmitter and mobile receiver, that is $\overline{L}$, and the mobile receiver sensitivity, in the absence of multiple access interference (intra- or inter-cell). Then the effect of noise rise due to interference is added to this minimum power and the total represents the transmission power required for a user at an 'average' location in the cell. Mathematically, the total base station transmission power can be expressed by the following equation:

$$BS\_TxP = \frac{N_{rf} \cdot W \cdot \overline{L} \cdot \sum\limits_{j=1}^{N} v_j \dfrac{(E_b/N_0)_j}{W/R_j}}{1 - \overline{\eta_{DL}}} \tag{8.16}$$

where N_{rf} is the noise spectral density of the mobile receiver front-end. The value of N_{rf} can be obtained from

$$N_{rf} = k \cdot T + NF$$
$$= -174.0 \text{ dBm} + NF \text{(assuming } T = 290 \text{ K)} \tag{8.17}$$

where k is Boltzmann constant of $1.381 \cdot 10^{-23}$ J/K, T is temperature in Kelvin and NF is the mobile station receiver noise figure with typical values of 5–9 dB.

The load factor can be approximated by its average value across the cell, that is

$$\overline{\eta_{DL}} = \sum\limits_{j=1}^{N} v_j \cdot \frac{(E_b/N_0)_j}{W/R_j} \cdot \left[(1 - \overline{\alpha}) + \overline{i} \right] \tag{8.18}$$

Downlink Common Channels

Part of the downlink power has to be allocated for the common channels: common pilot channel CPICH, primary and secondary synchronisation channels SCH, and primary and secondary common control physical channels CCPCH. Those channels are introduced in Section 6.5. SCH and CCPCH are time multiplexed with each other. The amount of power of the common channels affects synchronisation time, channel estimation accuracy, and the reception quality of the broadcast channel. On the other hand, the common channels eat up the capacity of the cell that could otherwise be allocated for the traffic channels. Typical power allocations for the common channels are shown in Table 8.8.

8.2.3 Example Load Factor Calculation

In both uplink and downlink the air interface load affects the coverage but the effect is not exactly the same. The difference between the uplink and downlink load curves is described below. The maximum path loss, i.e. coverage, as a function of the load is shown in Figure 8.3 for both uplink and downlink. A 3-sector site is assumed, and the throughputs are shown per sector per 5 MHz carrier. The uplink is calculated for 144 kbps data and the link budget is shown in Table 8.17. An other-cell to own cell interference ratio i of 0.65 is assumed. In the downlink an orthogonality of 0.6 and E_b/N_0 of 5.5 dB are assumed, giving a pole capacity of 1030 kbps/cell. No transmit diversity is assumed in this E_b/N_0. 20% of the total

Table 8.8. Typical powers for the downlink common channels [6]

Downlink common channel	Activity	Percentage of the maximum base station power	Power allocation with 20 W maximum power
Common pilot channel CPICH	100%	10%	2 W
Primary synchronisation channel SCH	10%	6%	1.2 W
Secondary synchronisation channel SCH	10%	4%	0.8 W
Primary common control physical channel CCPCH	90%	5%	1 W
Total common channel powers		~15%	~3 W

power is allocated for the common channels leaving 820 kbps/cell for the traffic channels. The following assumptions were used in Figure 8.3: base station transmission power 10 W, cable loss 2 dB and maximum-to-average path loss ratio 6 dB.

The uplink pole capacity in this example is 1730 kbps/cell. In the downlink the coverage depends more on the load than in the uplink, according to Figure 8.3. The reason is that in the downlink the maximum transmission power is the same 10 W regardless of the number of users and is shared between the downlink users, while in the uplink each additional user has its own power amplifier. Therefore, even with low load in the downlink, the coverage decreases as a function of the number of users.

We note that with the above assumptions the coverage is clearly limited by the uplink for a load below 650 kbps, while the capacity is downlink limited. Therefore, in Chapter 11 the coverage discussion concentrates on the uplink, while the capacity discussion concentrates on downlink.

The capacities presented above depend on the environment and represent only examples. The dependency of capacity and coverage will remain regardless of the assumptions. The effect of the environment on capacities is presented in Section 11.3.1.2.

We need to remember that in third generation networks the traffic can be asymmetric between uplink and downlink, and the load can be different in uplink and in downlink.

In Figure 8.3 a base station maximum power of 10 W is assumed. How much could we improve the downlink coverage and capacity by using more power, such as 20 W? The difference in downlink coverage and capacity between 10 W and 20 W base station output powers is shown in Figure 8.4. If we increase the downlink power by 3.0 dB, we can allow 3.0 dB higher maximum path loss regardless of the load. The capacity improvement is smaller than the coverage improvement because of the load curve. If we now keep the downlink path loss fixed at 153 dB, which is the maximum uplink path loss with 3 dB interference margin, the downlink capacity can be increased by only 10% (0.4 dB) from 680 kbps to 750 kbps. Increasing downlink transmission power is an inefficient approach to increase downlink capacity, since the available power does not affect the pole capacity.

Assume we had 20 W downlink transmission power available. Splitting the downlink power between two frequencies would increase downlink capacity from 750 kbps to 2 × 680 kbps = 1360 kbps, i.e. by 80%. The splitting of the downlink power between two carriers is an efficient approach to increase the downlink capacity without any extra investment

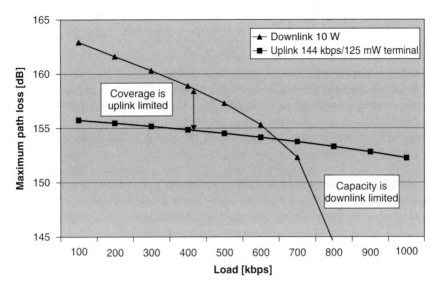

Figure 8.3. Example coverage vs. capacity relation in downlink and uplink in macro cells

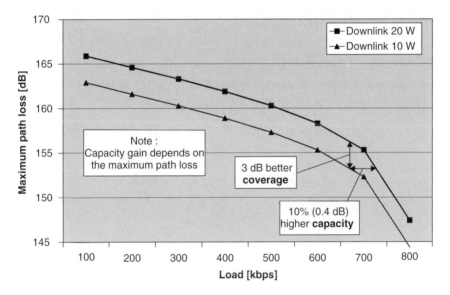

Figure 8.4. Effect of base station output power to downlink capacity and coverage

in power amplifiers. The power splitting approach requires that the operator's frequency allocation allows the use of two carriers in the base station.

The advantages of the wideband technology in WCDMA can be seen in the example above. It is possible to make a trade-off between the downlink capacity and coverage: if there are fewer users, more power can be allocated for one user allowing a higher path loss. The wideband power amplifier also allows the addition of a second carrier without adding a power amplifier. Similar advantages have not been possible in GSM without multi-carrier power amplifiers. On the other hand, WCDMA requires very linear base station power amplifiers which is challenging for the implementation.

Power Splitting Between Sectors

In the initial deployment phase the sectorised uplink is needed to improve the coverage, but the sectorisation may bring more capacity than the what is required by the initial traffic density. In WCDMA it is possible to use sectorised uplink reception while having only one common wideband power amplifier for all the sectors. This solution is illustrated in Figure 8.5 and the performance is described in Table 8.9.

This solution is a low-cost one compared to the real sectorisation where one wideband power amplifier is needed for each sector. The performance of this low-cost solution can be estimated from Figure 8.4. Let's take an example where the uplink uses three sectors while only one 20-W power amplifier is used in downlink: the available power per sector is $20/3 = 6.7$ W. That power gives 630 kbps capacity in Figure 8.4 with 153 dB maximum path loss. The curve for 6.7 W can be obtained by moving the 10-W curve down by $10 * \log 10(6.7/10) = -1.7$ dB. Since the same power amplifier is shared between all sectors, the capacity of 630 kbps is the total capacity of the site. This low-cost solution provides $630/(3 * 750) = 28\%$ of the capacity of the real sectorised solution, and from RNC point of view it is equal to a single sector solution. RNC is not involved in softer handovers between the sectors in this solution, but it is the base station baseband that decides which uplink sectors are used in receiving the user signal.

8.2.4 Capacity Upgrade Paths

When the amount of traffic increases, the downlink capacity can be upgraded in a number of different ways. The most typical upgrade options are:

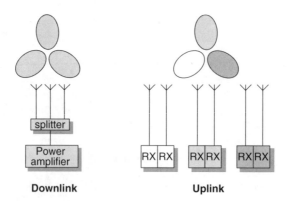

Figure 8.5. 3-sector uplink with receive diversity, single power amplifier in downlink

Table 8.9. 3-sector uplink, omni downlink

Uplink coverage	Equal to normal 3-sector configuration
Downlink capacity	28% of the 3-sector downlink
Number of logical sectors (RNC)	One

— more power amplifiers if initially the power amplifier split between sectors
— two or more carriers if the operator's frequency allocation permits
— transmit diversity with 2nd power amplifier per sector

The availability of these capacity upgrade solutions depends on the base station manufacturer. All these capacity upgrade may not be available in all base station types.

An example capacity upgrade path is shown in Figure 8.6. The initial solution with one power amplifier and one carrier gives 630 kbps capacity while the 2-carrier 3-sector transmit diversity solution gives $2 * 3 * 750 * 1.4 = 6.3$ Mbps assuming 40% capacity increase from the transmit diversity. The transmit diversity procedure is described in Section 6.6.7 and its performance is discussed in Section 11.3.2.

These capacity upgrade solutions do not require any changes to the antenna configurations, only upgrades within the base station cabinet are needed on the site. The uplink coverage is not affected by these upgrades.

The capacity can be improved also by increasing the number of antenna sectors, for example, starting with omni-directional antennas and upgrading to 3-sector and finally to 6-sector antennas. The sectorisation is analysed in more detail in [4]. The drawback of increasing the number of sectors is that the antennas must be replaced. The increased number of sectors also brings improved coverage through a higher antenna gain. The challenge in here is that the uplink coverage would be needed already in the initial deployment phase.

8.2.5 Capacity per km2

Providing high capacity will be challenging in urban area where the offered amount of traffic per km^2 can be very high. In this section we evaluate the maximum capacity that can be provided per km^2 using macro and micro sites. The results are shown in Table 8.10 and illustrated in Figure 8.7.

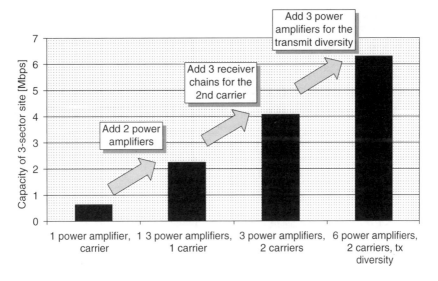

Figure 8.6. An example capacity upgrade path for 3-sector macro site

Table 8.10. Typical capacities per km^2 with macro and micro layers in urban area

	Macro cell layer	Micro cell layer
Capacity per site per carrier with one power amplifier	630 kbps	—
Maximum capacity per site per carrier	3 Mbps with 3 sectors	2 Mbps
Capacity per site with 3 UMTS frequencies	9 Mbps	6 Mbps
Initial sparse site density	0.5 sites / km^2	—
Maximum dense site density	5 sites / km^2	30 sites / km^2
Maximum capacity	45 Mbps / km^2	180 Mbps / km^2

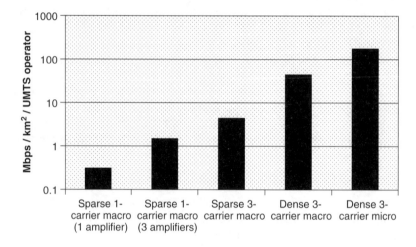

Figure 8.7. Capacity per km^2 for one UMTS operator with macro and micro layers

We assume that the maximum capacity per sector per carrier is 1 Mbps in macro cells and 2 Mbps in micro. These capacities assume that it is possible to obtain an approximate 1.5 dB, i.e. 40%, transmit diversity gains in capacity in addition to the capacity of Table 11.13. The macro and micro layer capacities are evaluated in Section 11.3.1.2.

If the UMTS operation is started using 1-carrier macro cells with one power amplifier per site with 2 km^2 site area, the available capacity will be 630 kbps/carrier/2 km^2 = 315 kbps/km^2. If more capacity is needed, the operator can install more power amplifiers, and further deploy several frequencies per site his frequency license permitting. With three frequencies the capacity will be 3 carriers $*$ 3.0 Mbps/carrier/2 km^2 = 4.5 Mbps/km^2.

When more capacity is needed, an operator can add more macro sites. If we assume a maximum macro site density of 5 per km^2, the capacity of macro cell layer is 3 Mbps/carrier $*$ 3 carriers/site $*$ 5 sites/km^2 = 45 Mbps/km^2. If clearly more capacity is needed, micro cell layer need to be deployed.

For the micro cell layer we assume a maximum site density of 30 sites per km^2. Having an even higher site density is challenging because the other-to-own cell interference tends to increase and the capacity per site decreases. Also, the site acquisition may be difficult if more sites are needed. The maximum micro site density depends heavily on the operator and on the city, and it can be higher than 30 sites per km^2.

Using three carriers the micro cell layer would offer 2 Mbps/carrier $*$ 3 carriers/site $*$ 30 site/km^2 = 180 Mbps/km^2. The total capacity of the 12-carrier UMTS band could be up to 700 Mbps/km^2 if all operators deploy dense micro cell networks. Using indoor pico cells with FDD or with TDD can further enhance the capacity.

8.2.6 *Soft Capacity*

8.2.6.1 **Erlang Capacity**

In the dimensioning in Section 8.2 the number of channels per cell was calculated. Based on those figures, we can calculate the maximum traffic density that can be supported with a given blocking probability. The traffic density can be measured in Erlang and is defined ([7], p. 270) as:

$$\text{Traffic density [Erlang]} = \frac{\text{Call arrival rate [calls/s]}}{\text{Call departure rate [calls/s]}} \qquad (8.19)$$

If the capacity is hard blocked, i.e. limited by the amount of hardware, the Erlang capacity can be obtained from the Erlang B model [7]. If the maximum capacity is limited by the amount of interference in the air interface, it is by definition a soft capacity, since there is no single fixed value for the maximum capacity. For a soft capacity limited system, the Erlang capacity cannot be calculated from the Erlang B formula, since it would give too pessimistic results. The total channel pool is larger than just the average number of channels per cell, since the adjacent cells share part of the same interference, and therefore more traffic can be served with the same blocking probability. The soft capacity can be explained as follows. The less interference is coming from the neighbouring cells, the more channels are available in the middle cell, as shown in Figure 8.8. With a low number of channels per cell, i.e. for high bit rate real-time data users, the average loading must be quite low to guarantee low blocking probability. Since the average loading is low, there is typically extra capacity available in the neighbouring cells. This capacity can be borrowed from the adjacent cells, therefore the interference sharing gives soft capacity. Soft capacity is important for high bit rate real-time data users, e.g. for video connections. It can also be obtained in GSM if the air interface capacity is limited by the amount of interference instead of the number of time slots; this assumes low frequency reuse factors in GSM with fractional loading.

In the soft capacity calculations below it is assumed that the number of subscribers is the same in all cells but the connections start and end independently. In addition, the call arrival interval follows a Poisson distribution. This approach can be used in dimensioning

Equally loaded cells   Less interference in the neighboring cells
                             → higher capacity in the middle cell

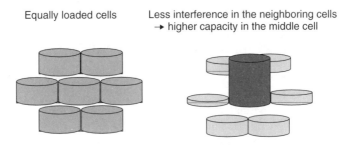

Figure 8.8. Interference sharing between cells in WCDMA

when calculating Erlang capacities. There is an additional soft capacity in WCDMA if also the number of users in the neighbouring cells is smaller.

The difference between hard blocking and soft blocking is shown with a few examples in the uplink below. WCDMA soft capacity is defined in this section as the increase of Erlang capacity with soft blocking compared to that with hard blocking with the same maximum number of channels per cell on average with both soft and hard blocking:

$$\text{Soft capacity} = \frac{\text{Erlang capacity with soft blocking}}{\text{Erlang capacity with hard blocking}} - 1 \qquad (8.20)$$

The wideband power-based admission control strategy is presented in Section 9.5. Such an admission control strategy gives soft blocking and soft capacity.

Uplink soft capacity can be approximated based on the total interference at the base station. This total interference includes both own cell and other cell interference. Therefore, the total channel pool can be obtained by multiplying the number of channels per cell in the equally loaded case by $1 + i$, which gives the single isolated cell capacity, since

$$i + 1 = \frac{\text{other cell interference}}{\text{own cell interference}} + 1 = \frac{\text{other cell interference} + \text{own cell interference}}{\text{own cell interference}}$$

$$= \frac{\text{isolated cell capacity}}{\text{multicell capacity}} \qquad (8.21)$$

The basic Erlang B formula is then applied to this larger channel pool (= interference pool). The Erlang capacity obtained is then shared equally between the cells. The procedure for estimating the soft capacity is summarised below.

1. Calculate the number of channels per cell, N, in the equally loaded case, based on the uplink load factor, Equation (8.12).
2. Multiply that number of channels by $1 + i$ to obtain the total channel pool in the soft blocking case.
3. Calculate the maximum offered traffic from the Erlang B formula.
4. Divide the Erlang capacity by $1 + i$.

8.2.6.2   Uplink Soft Capacity Examples

A few numerical examples of soft capacity calculations are given, with the assumptions shown in Table 8.11.

The capacities obtained, in terms of both channels based on Equation (8.12) and Erlang per cell, are shown in Table 8.12. The trunking efficiency shown in Table 8.12 is defined as the hard blocked capacity divided by the number of channels. The lower the trunking efficiency, the lower is the average loading, the more capacity can be borrowed from the neighbouring cells, and the more soft capacity is available.

We note that there is more soft capacity for higher bit rates than for lower bit rates. This relationship is shown in Figure 8.9.

It should be noted that the amount of soft capacity depends also on the propagation environment and on the network planning which affect the value of i. The soft capacity can be obtained only if the radio resource management algorithms can utilise a higher capacity in one cell if the adjacent cells have lower loading. This can be achieved if the radio resource

Table 8.11. Assumptions in soft capacity calculations

Bit rates	Speech: 12.2 kbps Real-time data: 16–144 kbps
Voice activity	Speech 67% Data 100%
E_b/N_0	Speech: 4 dB Data 16–32 kbps: 3 dB Data 64 kbps: 2 dB Data 144 kbps: 1.5 dB
i	0.55
Noise rise	3 dB (= 50% load factor)
Blocking probability	2%

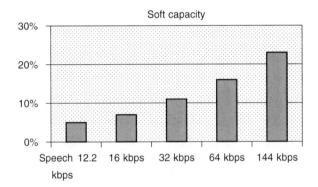

Figure 8.9. Soft capacity as a function of bit rate for real-time connections

Table 8.12. Soft capacity calculations in the uplink

Bit rate (kbps)	Channels per cell	Hard blocked capacity	Trunking efficiency	Soft blocked capacity	Soft capacity
12.2	60.5	50.1 Erl	83%	52.8 Erl	5%
16	39.0	30.1 Erl	77%	32.3 Erl	7%
32	19.7	12.9 Erl	65%	14.4 Erl	12%
64	12.5	7.0 Erl	56%	8.2 Erl	17%
144	6.4	2.5 Erl	39%	3.2 Erl	28%

management algorithms are based on the wideband interference, not on the throughput or the number of connections.

Similar soft capacity is also available in the WCDMA downlink as well as in GSM if interference-based radio resource management algorithms are applied.

8.3 Capacity and Coverage Planning

8.3.1 Iterative Capacity and Coverage Prediction

In this section detailed capacity and coverage planning are presented. In the detailed planning phase real propagation data from the planned area is needed, together with the estimated user density and user traffic. Also, information about the existing base station sites is needed in order to utilise the existing site investments. The output of the detailed capacity and coverage planning are the base station locations, configurations and parameters.

Since in WCDMA all users are sharing the same interference resources in the air interface, they cannot be analysed independently. Each user is influencing the others and causing their transmission powers to change. These changes themselves again cause changes, and so on. Therefore, the whole prediction process has to be done iteratively until the transmission powers stabilise. Also, the mobile speeds, multipath channel profiles, and bit rates and type of services used play a more important role than in second generation TDMA/FDMA systems. Furthermore, in WCDMA fast power control in both uplink and downlink, soft/softer handover and orthogonal downlink channels are included, which also impact on system performance. The main difference between WCDMA and TDMA/FDMA coverage prediction is that the interference estimation is already crucial in the coverage prediction phase in WCDMA. In the current GSM coverage planning processes the base station sensitivity is typically assumed to be constant and the coverage threshold is the same for each base station. In the case of WCDMA the base station sensitivity depends on the number of users and used bit rates in all cells, thus it is cell and service specific. Note also that in third generation networks the downlink can be loaded higher than the uplink or vice versa.

8.3.2   Planning Tool

In second generation systems, detailed planning concentrated strongly on coverage planning. In third generation systems, a more detailed interference planning and capacity analysis than simple coverage optimisation is needed. The tool should aid the planner to optimise the base station configurations, the antenna selections and antenna directions and even the site locations, in order to meet the quality of service and the capacity and service requirements at minimum cost. To achieve the optimum result the tool must have knowledge of the radio resource algorithms in order to perform operations and make decisions, like the real network. Uplink and downlink coverage probability is determined for an specific service by testing the service availability in each location of the plan. A detailed description of the planning tool can be found in [8].

The actual detailed planning phase does not differ very much from second generation planning. The sites and sectors are placed in the tool. The main difference is the importance of the traffic layer. The proposed detailed analysis methods (see the following sections) use discrete mobile stations in the WCDMA analysis. The mobile station density in different cells should be based on actual traffic information. The hotspots should be identified as an input for accurate analysis. One source of information concerning user density would be the data from the operator's second generation network or later from the third generation.

The planning tool described here differs from the dynamic simulator introduced in Section 10.6.2. The planning tool is a static simulator that is based on average conditions, and snapshots of the network can be taken. The dynamic simulator includes the traffic and the mobility models which make it possible to develop and test the real-time radio resource management (RRM) algorithms. The dynamic simulations can be used to study the performance of the RRM algorithms in realistic environments, and the results of those simulations can be used as an input to this network planning tool. For example, the practical performance of handover algorithms with measurement errors and delays can be tested in the dynamic tool and the results fed into the network planning tool. Testing of RRM algorithms requires accurate modelling of WCDMA link performance, and therefore a time resolution corresponding to power control frequency of 1.5 kHz is used in the dynamic simulator. Such a high accuracy makes the dynamic simulation tool complex and the simulations still too slow—using current top-line high speed workstations—for practical network planning purposes. The accurate dynamic simulation tool can be used to verify and develop the more simple performance modelling in the network planning tool. When enough results from large-scale WCDMA networks are available, those results will be used in the calibration of the network planning tool.

8.3.2.1   Uplink and Downlink Iterations

The target in the uplink iteration is to allocate the simulated mobile stations' transmission powers so that the interference levels and the base station sensitivity values converge. The base station sensitivity level is corrected by the estimated uplink interference level (noise rise) and therefore is cell specific. The impact of the uplink loading on the sensitivity is taken into account with a term $-10 \cdot \log_{10}(1 - \eta_{UL})$, where η_{UL} is given by Equation (8.12). In the uplink iteration the transmission powers of the mobile stations are estimated based on the sensitivity level of the best server, the service, the speed and the link losses. Transmission powers are then compared to the maximum allowed transmission power of the mobile stations, and mobile stations exceeding this limit are put to outage. The interference can then be re-estimated and new loading values and sensitivities for each base station assigned. If the uplink load factor is higher than the set limit, the mobile stations are randomly moved from the highly loaded cell to another carrier (if the spectrum allows) or to outage.

The aim of the downlink iteration is to allocate correct base station transmission powers to each mobile station until the received signal at the mobile station meets the required E_b/N_0 target.

8.3.2.2   Modelling of Link Level Performance

In radio network dimensioning and planning it is necessary to make simplifying assumptions concerning the multipath propagation channel, transmitter and receiver. A traditional model is to use the average received E_b/N_0 ensuring the required quality of service as the basic number, which includes the effect of the power delay profile. In systems using fast power control the average received E_b/N_0 is not enough to characterise the influence of the radio channel on network performance. Also, the transmission power distribution must be taken into account when modelling link-level performance in network-level calculations. An appropriate approach is presented in [1] for the WCDMA uplink. It has been demonstrated that, due to the fast power control in the multipath fading environment, in addition to the average received E_b/N_0 requirement, an average transmission power rise is needed in interference calculations. The power rise is presented in detail in Section 9.2.1.2. Furthermore, a power control headroom must be included in the link budget estimation to allow power control to follow the fast fading at the cell edge.

Multiple links are taken into account in the simulator when estimating the soft handover gains in the average received and transmitted power and also in the required power control headroom. During the simulations the transmission powers are corrected by the voice activity factor, soft handover gain and average power rise for each mobile station.

8.3.3 Case Study

In this case study an area in Espoo, Finland, was planned, comprising roughly 12×12 km^2, shown in Figure 8.10. The network planning tool described in Section 8.3.2 was utilised in this case study.

The operator's coverage probability requirement for the 8 kbps, 64 kbps and 384 kbps services was set, respectively, to 95%, 80% and 50%, or better. The planning phase started with radio link budget estimation and site location selections. In the next planning step the dominance areas for each cell were optimised. In this context the dominance is related only to the propagation conditions. Antenna tilting, bearing and site locations can be tuned to achieve clear dominance areas for the cells. Dominance area optimisation is crucial for interference and soft handover area and soft handover probability control. The improved soft/softer handover and interference performance is automatically seen in the improved network capacity. The plan consists of 19 three-sectored macro sites, and the average site area is 7.6 km^2. In the city area the uplink loading limitation was set to 75%, corresponding to a 6 dB noise rise. In case the loading was exceeded, the necessary number of mobile stations were randomly set to outage (or moved to another carrier) from the highly loaded

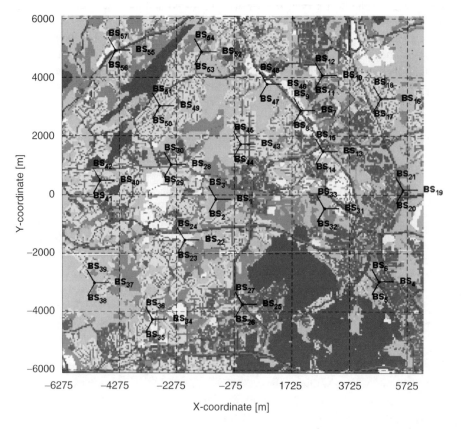

Figure 8.10. The network scenario. The area measures 12×12 km^2 and is covered with 19 sites, each with three sectors

cells. Table 8.13 shows the user distribution in the simulations and the other simulation parameters are listed in Table 8.14.

In all three simulation cases the cell throughput in kbps and the coverage probability for each service were of interest. Furthermore, the soft handover probability and loading results were collected. Tables 8.12 and 8.13 show the simulation results for cell throughput and coverage probabilities. The maximum uplink loading was set to 75% according to Table 8.14. Note that in Table 8.15 in some cells the loading is lower than 75%, and correspondingly the throughput is also lower than the achievable maximum value. The reason is that there was not enough offered traffic in the area to fully load the cells. The loading in cell 5 was 75%. Cell 5 is located in the lower right corner in Figure 8.10, and there is no other cells close to cell 5. Therefore, that cell can collect more traffic than the other cells. For example, the cells 2 and 3 are in the middle of the area and there is not enough traffic to fully load the cells.

Table 8.16 shows that mobile station speed has an impact on both throughput and coverage probability. When mobile stations are moving at 50 km/h, fewer can be served, the throughput is lower and the resulting loading is higher than when mobile stations are moving at 3 km/h. If the throughput values are normalised to correspond to the same loading value, the difference between the 3 km/h and 50 km/h cases is more than 20%. The better capacity with the slower-moving mobile stations can be explained by the better E_b/N_0 performance. The fast power control is able to follow the fading signal and the required E_b/N_0 target is reduced. The lower target value reduces the overall interference level and more users can be served in the network.

Comparing coverage probability, the faster-moving mobile stations experience better quality than the slow-moving ones, because for the latter a headroom is needed in the mobile

Table 8.13. The user distribution

Service in kbps	Users per service
8 kbps	1735
64 kbps	250
384 kbps	15

Table 8.14. Parameters used in the simulator

Uplink loading limit	75%
Base station maximum transmission power	20 W (43 dBm)
Mobile station maximum transmission power	300 mW (= 25 dBm)
Mobile station power control dynamic range	70 dB
Slow (log-normal) fading correlation between base stations	50%
Standard deviation for the slow fading	6 dB
Multipath channel profile	ITU Vehicular A
Mobile station speeds	3 km/h and 50 km/h
Mobile/base station noise figures	7 dB / 5 dB
Soft handover addition window	−6 dB
Pilot channel power	30 dBm
Combined power for other common channels	30 dBm
Downlink orthogonality	0.5
Activity factor speech/data	50% / 100%
Base station antennas	65° / 17 dBi
Mobile antennas speech/data	Omni / 1.5 dBi

Table 8.15. The cell throughput, loading and soft handover (SHO) overhead. UL = uplink, DL = downlink

Cell ID	Throughput UL (kbps)	Throughput DL (kbps)	UL loading	SHO overhead
Basic loading: mobile speed 3 km/h, served users: 1805				
cell 1	728.00	720.00	0.50	0.34
cell 2	208.70	216.00	0.26	0.50
cell 3	231.20	192.00	0.24	0.35
cell 4	721.60	760.00	0.43	0.17
cell 5	1508.80	1132.52	0.75	0.22
cell 6	762.67	800.00	0.53	0.30
MEAN (all cells)	519.20	508.85	0.37	0.39
Basic loading: mobile speed 50 km/h, served users: 1777				
cell 1	672.00	710.67	0.58	0.29
cell 2	208.70	216.00	0.33	0.50
cell 3	226.67	192.00	0.29	0.35
cell 4	721.60	760.00	0.50	0.12
cell 5	1101.60	629.14	0.74	0.29
cell 6	772.68	800.00	0.60	0.27
MEAN	531.04	506.62	0.45	0.39
Basic loading: mobile speed 50 km/h and 3 km/h, served users: 1802				
cell 1	728.00	720.00	0.51	0.34
cell 2	208.70	216.00	0.29	0.50
cell 3	240.00	200.00	0.25	0.33
cell 4	730.55	760.00	0.44	0.20
cell 5	1162.52	780.92	0.67	0.33
cell 6	772.68	800.00	0.55	0.32
MEAN	525.04	513.63	0.40	0.39

transmission power to be able to maintain the fast power control—see Section 8.2.1. The impact of the speed can be seen especially if the bit rates used are high, because for low bit rates the coverage is better due to a larger processing gain. The coverage is tested in this planning tool by using a test mobile after the uplink iterations have converged. It is assumed that this test mobile does not affect the loading in the network.

The downlink coverage probability analysis is different from the uplink one. In the uplink direction the limiting factor is the mobile station maximum transmission power. In the downlink direction the limitations are dependent on the used radio resource management algorithms. One limitation in the downlink direction is the total base station transmission power. In addition to that, another limit can be taken into use: the power limit per radio link. Figure 8.11 shows an example downlink coverage analysis for the speech service. It can be seen that if the power per link limitation is selected correctly, the downlink coverage probability can be set to the same value as the uplink coverage probability. Thus, the uplink and the downlink service areas can be balanced. The required

Table 8.16. The coverage probability results

Basic loading: mobile speed 3 km/h	Test mobile speed:	
	3 km/h	50 km/h
8 kbps	96.6%	97.7%
64 kbps	84.6%	88.9%
384 kbps	66.9%	71.4%
Basic loading: mobile speed 50 km/h	Test mobile speed:	
	3 km/h	50 km/h
8 kbps	95.5%	97.1%
64 kbps	82.4%	87.2%
384 kbps	63.0%	67.2%
Basic loading: mobile 3 and 50 km/h	Test mobile speed:	
	3 km/h	50 km/h
8 kbps	96.0%	97.5%
64 kbps	83.9%	88.3%
384 kbps	65.7%	70.2%

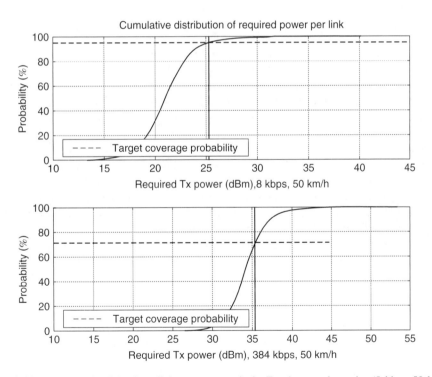

Figure 8.11. An example of the downlink coverage analysis. For the speech service (8 kbps, 50 km/h) the limit for the radio link was set to 25 dBm to achieve the 95% coverage probability. In the case of the 384 kbps and 71% coverage probability requirement the limit per radio link would be 35 dBm

powers per link in Figure 8.11 are the average powers and do not include the fast fading margin.

This example case demonstrates the impact of the user profile, i.e. the service used and the mobile station speed, on network performance. It is shown that the lower mobile station speed provides better capacity: the number of mobile stations served and the cell throughput are higher in the 3 km/h case than in the 50 km/h case. Comparing coverage probability, the impact of the mobile station speed is different. The higher speed reduces the required fast fading margin and thus the coverage probability is improved when the mobile station speed is increased.

The presented network planning tool was shown to be useful for taking into account the dependency between capacity and coverage in WCDMA. Also, the coverage areas of different bit rates can be evaluated for the chosen base station sites, and parameters for RRM algorithms can be set.

8.3.4 Network Optimisation

Network optimisation is a process to improve overall network quality as experienced by the mobile subscribers and to ensure that network resources are used efficiently. Optimisation includes the analysis of the network and improvements in the network configuration and performance. The transition in Figure 8.1 from detailed capacity and coverage planning to network operation and optimisation is smooth. Statistics of the key performance indicators for the operational network are fed to the network status analysis tool, and the radio resource management parameters can be tuned for better performance. The radio resource management algorithms are described in Chapter 9. An example optimisation parameter is soft handover area optimisation. The network status analysis tool could be an integrated part of the radio network planning tool presented in Section 8.3.2. The traffic growth in the network requires continuous interaction of the planning tool and the operational network. The capability of the current network to support the forecast traffic growth is analysed, and the radio network plan can be further processed based on actual measured data.

The first phase of the optimisation process is to define the key performance indicators. These consist of measurements in the network management system and of field measurement data, or any other information which can be used to determine the quality of service of the network. With the help of the network management system it is possible to analyse the past, present and predicted future performance of the network.

The performance of the radio resource management algorithms and their parameters can be analysed using the key performance indicator results. The radio resource management algorithms include handovers, power control, packet scheduling, admission and load control.

The network quality analysis is designed to give an operator a view of network quality and performance. The quality analysis and reporting consists of planning the case, field measurements and network management system measurements. After the quality of service criteria have been specified and the data has been analysed, a survey report can be generated. For second generation systems, quality of service has consisted, for example, of dropped call statistics, dropped call cause analysis, handover statistics and measurement of successful call attempts. For third generation systems with a greater variety of services, new definitions of quality of service for quality analysis must be generated.

Automatic optimisation will be important in third generation networks, since there are more services and bit rates than in second generation networks and manual optimisation would be too time consuming. Automatic adjustment should provide a fast response to the changing traffic conditions in the network. It should be noted that at the start of third generation deployment, only some of the parameters can be automatically tuned, and therefore a second generation type optimisation process must still be maintained.

8.4   GSM Co-planning

Utilisation of existing base station sites is important in speeding up WCDMA deployment and in sharing sites and transmission costs with the existing second generation system. The feasibility of sharing sites depends on the relative coverage of the existing network compared to WCDMA. In this section we compare the relative uplink coverage of existing GSM900 and GSM1800 full-rate speech services and WCDMA speech and 64 kbps and 144 kbps data services. Table 8.17 shows the assumptions made and the results of the comparison of coverage. The maximum path loss of the WCDMA 144 kbps here is 3 dB greater than in Table 8.4. The difference comes because of a smaller interference margin, a lower base

Table 8.17. Typical maximum path losses with existing GSM and with WCDMA

	GSM900/ speech	GSM1800/ speech	WCDMA/ speech	WCDMA/ 64 kbps	WCDMA/ 144 kbps
Mobile transmission power	33 dBm	30 dBm	21 dBm	21 dBm	21 dBm
Receiver sensitivity[1]	−110 dBm	−110 dBm	−124 dBm	−120 dBm	−117 dBm
Interference margin[2]	1.0 dB	0.0 dB	2.0 dB	2.0 dB	2.0 dB
Fast fading margin[3]	2.0 dB	2.0 dB	2.0 dB	2.0 dB	2.0 dB
Base station antenna gain[4]	16.0 dBi	18.0 dBi	18.0 dBi	18.0 dBi	18.0 dBi
Body loss[5]	3.0 dB	3.0 dB	3.0 dB	—	—
Mobile antenna gain[6]	0.0 dBi	0.0 dBi	0.0 dBi	2.0 dBi	2.0 dBi
Relative gain from lower frequency compared to UMTS frequency[7]	7.0 dB	1.0 dB	—	—	—
Maximum path loss	**160.0 dB**	**154.0 dB**	**156.0 dB**	**157.0 dB**	**154.0 dB**

[1] WCDMA sensitivity assumes 4.0 dB base station noise figure and E_b/N_0 of 5.0 dB for 12.2 kbps speech, 2.0 dB for 64 kbps and 1.5 dB for 144 kbps data. These E_b/N_0 values are similar to the requirements in 3GPP TS 25.104. The required sensitivity for speech is −121 dBm in static conditions with one antenna which corresponds to about −124 dBm with antenna diversity. GSM sensitivity is assumed to be −110 dBm with receive antenna diversity.

[2] The WCDMA interference margin corresponds to 37% loading of the pole capacity: see Figure 8.2. An interference margin of 1.0 dB is reserved for GSM900 because the small amount of spectrum in 900 MHz does not allow large reuse factors.

[3] The fast fading margin for WCDMA includes the macro diversity gain against fast fading.

[4] The antenna gain assumes three-sector configuration in both GSM and WCDMA.

[5] The body loss accounts for the loss when the terminal is close to the user's head.

[6] A 2.0 dBi antenna gain is assumed for the data terminal.

[7] The attenuation in 900 MHz is assumed to be 7.0 dB lower than in UMTS band and in GSM1800 band 1.0 dB lower than in UMTS band.

station receiver noise figure, and no cable loss. Note also that the soft handover gain is included into the fast fading margin in Table 8.17 and the mobile station power class is here assumed to be 21 dBm.

Table 8.17 shows that the maximum path loss of the 144 kbps data service is the same as for speech service of GSM1800. Therefore, a 144 kbps WCDMA data service can be provided when using GSM1800 sites, with the same coverage probability as GSM1800 speech. If GSM900 sites are used for WCDMA and 64 kbps full coverage is needed, an 3 dB coverage improvement is needed in WCDMA. This comparison assumes that GSM900 sites are planned as coverage-limited. In densely populated areas, however, GSM900 cells are typically smaller to provide enough capacity. Section 11.2 analyses the uplink coverage of WCDMA and presents a number of solutions for improving WCDMA coverage to match GSM site density.

If high-power terminals with a transmission power of 24 dBm are used, the coverage will be 3 dB better than with 21 dBm output power.

The downlink coverage of WCDMA is discussed in Section 11.2.3 and is shown to be better than the uplink coverage. Therefore, it is possible to provide full downlink coverage even for bit rates higher than 144 or 384 kbps using GSM1800 sites.

Any comparison of the coverage of WCDMA and GSM depends on the exact receiver sensitivity values and on system parameters such as handover parameters and frequency hopping. That presented in Table 8.17 is an example calculation. Note also that the aim of this exercise is to compare the coverage of the GSM base station systems that have been deployed up to the present with WCDMA coverage in the initial deployment phase during 2001–2002. The sensitivity of the latest GSM base stations is better than the one assumed in Table 8.17.

8.5 Multi-operator Interference

8.5.1 Introduction

In this section, the effect of adjacent channel interference between two operators on adjacent frequencies is studied. Adjacent channel interference needs to be considered, because it will affect all wideband systems where large guard bands are not possible, and WCDMA is no exception. If the adjacent frequencies are isolated in the frequency domain by large guard bands, spectrum is wasted due to the large system bandwidth. Tight spectrum mask requirements for a transmitter and high selectivity requirements for a receiver, in the mobile station and in the base station, would guarantee low adjacent channel interference. However, these requirements have a large impact, especially on the implementation of a small WCDMA mobile station.

In this section, the Adjacent Channel Interference power Ratio (ACIR) is defined as the ratio of the transmission power to the power measured after a receiver filter in the adjacent channel(s). Both the transmitted and the received power are measured with a filter that has a Root-Raised Cosine filter response with roll-off of 0.22 and a bandwidth equal to the chip rate [9]. The adjacent channel interference is caused by transmitter non-idealities and imperfect receiver filtering. In both uplink and downlink, the adjacent channel performance is limited by the performance of the mobile. In the uplink the main source of adjacent channel interference is the non-linear power amplifier in the mobile station, which introduces adjacent channel leakage power. In the downlink the limiting factor for adjacent channel interference is the receiver selectivity of the WCDMA terminal. The requirements for adjacent channel performance are shown in Table 8.18 and apply to both uplink and downlink.

A difficult interference scenario where the adjacent channel interference could affect network performance is illustrated in Figure 8.12. This assumes that there are a large macro cell of operator 1 and small micro cells of operator 2. In Figure 8.12 operator 1's mobile is transmitting with high power to a distant macro cell, and at the same time is located very close to one of operator 2's micro cells. It needs to be further assumed that these macro and micro cells are on adjacent frequencies. Part of the transmission of the mobile is leaking to the adjacent carrier and possibly causing interference to the micro cell reception.

In the following sections the effect of the adjacent channel interference in this interference scenario is analysed by worst-case calculations and system simulations. It will be shown that the worst-case calculations give very bad results but also that the worst-case scenario is extremely unlikely to happen in real networks. Therefore, simulations are also used to study this interference scenario. Finally, conclusions are drawn regarding adjacent channel interference and implications for network planning are discussed.

8.5.2 Worst-Case Uplink Calculations

In this section the worst-case adjacent channel interference scenario is presented. The worst-case adjacent channel interference occurs when a mobile is transmitting on full power very close to a base station that is receiving on the adjacent carrier. A minimum coupling loss

Table 8.18. Requirements for adjacent channel performance [9]

Frequency separation	Required attenuation
Adjacent carrier (5 MHz separation)	33 dB
Second adjacent carrier (10 MHz separation)	43 dB

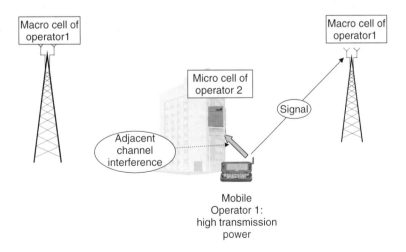

Figure 8.12. Adjacent channel interference in uplink from macro mobile to micro base station

of 50 dB is assumed here. The minimum coupling loss is defined as the minimum path loss between mobile and base station antenna connectors. The other assumptions in these worst-case calculations are shown in Table 8.19 and the results in Table 8.20.

The adjacent channel interference in the micro cell base station receiver is −62 dBm, which is 41 dB above the thermal noise level in the receiver. Such a high increase in the interference level would clearly affect the uplink coverage area of the micro cell. It is, however, extremely unlikely that this worst-case scenario would happen. It requires that the following conditions are fulfilled at the same time:

— Operator 1's mobile is transmitting with its full power of 21 dBm.

— Operator 1's mobile is located very close to operator 2's micro base station antenna. The minimum coupling loss of 50 dB occurs only if the base station antenna is very low and the mobile can get close to the antenna.

— Operator 1's mobile is transmitting on the adjacent carrier. If there is one carrier between the transmission and the reception, the adjacent channel interference power is attenuated by 10 dB more than on the adjacent carrier according to Table 8.19.

— The noise figure of the micro cell base station, including cable losses, is only 5 dB. If the sensitivity is not so good, the base station is less sensitive also to the adjacent channel interference.

Even if there is increased interference in the base station receiver, this does not necessarily lead to any dropped calls in the micro cell. It does lead, however, to a reduced coverage area, though this may not be a problem if the micro cell is not coverage limited.

Since the worst-case scenario discussed above is very unlikely, a simulation approach is used in Section 8.5.4 to study the practical effect of the adjacent channel interference on the WCDMA radio network performance.

8.5.3 Downlink Blocking

We need to note that operator 1's mobile is receiving adjacent channel interference in the downlink from operator 2's micro base station, and its connection will probably be

Table 8.19. Assumptions in uplink worst-case calculations of adjacent channel interference

Minimum coupling loss between mobile and micro cell in Figure 8.12	50 dB
Mobile transmission power	Maximum power 21 dBm
Micro base station noise figure	5 dB

Table 8.20. Results of uplink worst-case calculations

Thermal noise level with 3.84 Mcps	−108.2 dBm
Thermal noise level in the micro cell base station receiver	−108.2 dBm + 5 dB = −103.2 dBm
Interference from the adjacent channel	21 dBm − 50 dB − 33 dB (ACIR) = −62 dBm
Noise rise due to adjacent channel interference	−62 dBm − (−103.2 dBm) = 41.2 dB

dropped before that user is able to get extremely close to operator 2's base station. This happens because the selectivity of the mobile receiver is not perfect and the mobile is receiving interference from the micro cell base station on the adjacent carrier. The dropping of the downlink connection can even be considered desirable. It is preferable to drop one connection in downlink than to allow that mobile to interfere with all uplink connections of one cell. The dropping of the downlink connection depends on the maximum power that can be allocated to one connection from the macro cell base station to compensate the increased interference in the mobile reception.

If a mobile experiences adjacent channel interference in the downlink from a nearby base station, it can escape the interference by making an inter-frequency handover to the operator's other WCDMA carrier. Inter-frequency handovers can be used to avoid adjacent channel interference problems in the downlink.

This downlink blocking does not happen where there is interference between UTRA FDD and TDD systems, and therefore more difficult uplink interference situations can occur between FDD and TDD than within the FDD band. Interference between FDD and TDD systems is analysed in Section 12.3.

8.5.4   Uplink Simulations

The effect of adjacent channel interference from a macro cell mobile to the micro cell base station is studied in this section with a system simulator, which was presented in [10]. The worst-case calculation for that interference scenario was presented in Section 8.5.2, though that scenario is very unlikely to happen. In the simulation approach the effect of the adjacent channel interference can be analysed with a more realistic approach. The system simulator included macro cells and micro cells and all the mobiles connected to the cells. The minimum coupling loss in these simulations was 53 dB, which is a very low value and occurs only if the mobile can get very close to the base station antenna. The other simulation parameters are shown in Table 8.21.

The WCDMA nominal channel spacing is 5 MHz with a 3.84 Mcps chip rate. In these simulations the old chip rate of 4.096 Mcps has been used. The chip rate of 3.84 Mcps improves the adjacent channel performance by 0.5–0.7 dB compared to the results shown in this section [12]. In the uplink, all channels are assumed to be traffic channels of 8 kbps speech and no load due to control channels is calculated. In the downlink, the common control channels are modelled and contribute to the base station transmission power. The orthogonality factor was 0.4.

In these simulations the maximum allowed noise rise in the uplink was 20 dB in micro cells and 6 dB in macro cells. The noise rise is the increase in the wideband interference level

Table 8.21. Important simulation parameters for adjacent channel interference simulations. For other parameters see [11]

Parameter	Value
Minimum coupling loss	53 dB
Macro-to-macro base station distance	1000 m
Micro-to-micro base station distance	180 m
Mobile maximum power	21 dBm
Maximum allowed noise rise in micro cell	20 dB
Maximum allowed noise rise in macro cell	6 dB

over the thermal noise in the base station reception. For large macro cells a smaller noise rise is allowed because a higher noise rise decreases the coverage area. In the simulations this maximum loading was iteratively selected so that the predefined maximum noise rise values were not exceeded even if there was interference from the adjacent channel. This approach corresponds to the case where the real-time radio resource management algorithms, such as admission control, load control and packet scheduler, keep the uplink loading within the planned limits, and the coverage area is not affected by the adjacent channel interference. The effect of the adjacent channel interference can be seen as a reduced uplink capacity.

8.5.5   Simulation Results

In these simulations the coverage area of the micro and macro cells was kept within the planned limits even if there was adjacent channel interference, and the reduction in the maximum capacity was observed. The capacity loss is shown in Figure 8.13. The capacity is not sensitive to the adjacent channel interference if the attenuation between adjacent carriers (the ACIR) is higher than 20 dB. The minimum requirement for the ACIR is 33 dB: see Table 8.18. With a value of 33 dB the capacity loss in the micro cells is clearly below 1% and in the macro cell below 2%. Macro cell capacity is more sensitive to adjacent channel interference than micro cell capacity. The simulations show that micro cell capacity remains good while macro cell performance suffers more from adjacent channel interference. This is due to the fact that the number of interfering macro cell users is so small that the adjacent channel interference generated by them is negligible. On the other hand, the number of micro cell users is very high, thus they generate high adjacent channel interference to the macro cell base stations.

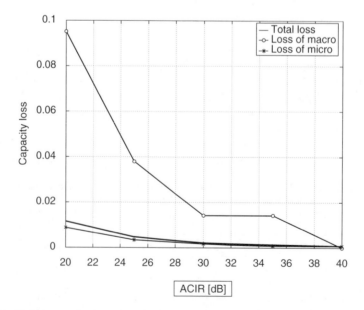

Figure 8.13. Uplink capacity loss due to adjacent channel interference for micro–macro scenario

8.5.6 Network Planning with Adjacent Channel Interference

This section presents a few network planning solutions that make sure that adjacent channel interference does not affect WCDMA network performance.

The selection of the base station antenna location and the antenna patterns affects the minimum coupling loss from the mobile to the base station. If that coupling loss is large, the adjacent channel interference can be avoided. In practice, the antenna should not be located so low that the mobile can easily get very close to the antenna.

It is possible also to reduce the sensitivity of the base station receiver, i.e., increase the noise figure of the base station RF parts. This approach is called desensitisation and can be used to make the base station receiver less sensitive to adjacent channel interference. At the same time the base station receiver also becomes less sensitive to the desired signal and the cell range is reduced. Therefore, this approach is suitable for small cells where uplink coverage is not the problem.

If the operators using adjacent frequency bands co-locate their base stations, either in the same sites or using the same masts, adjacent channel interference problems can be avoided, since the received power levels from the mobiles at both operators' base stations are then very similar. Since there are no large power differences, the adjacent channel attenuation of 33 dB is enough to prevent any adjacent channel interference problems. Also in the downlink the power levels received from both base stations are equal in both operators' mobiles. Note that co-location solves adjacent channel interference problems in UTRA FDD mode, but with UTRA TDD co-location can cause difficult interference situations. TDD interference scenarios are analysed in Section 12.3.

The nominal WCDMA carrier spacing is 5.0 MHz but can be adjusted with a 200 kHz raster according to the requirements of the adjacent channel interference. By using a larger carrier spacing, the adjacent channel interference can be reduced. If the operator has two carriers in the same base station, the carrier spacing between them could be as small as 4.0 MHz, because the adjacent channel interference problems are completely avoided if the two carriers use the same base station antennas. In that case a larger carrier spacing can be reserved between operators, as shown in Figure 8.14.

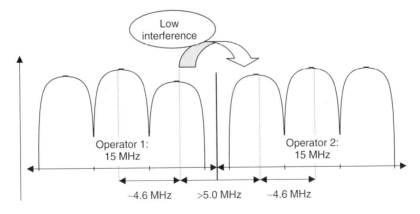

Figure 8.14. Selection of carrier spacings within operator's band and between operators

The network planning solutions to avoid adjacent channel interference are summarised below:

— selection of base station antenna locations
— desensitisation of the base station receiver
— co-location of the base stations with other operators
— adjustment of the carrier spacings
— inter-frequency handovers.

References

[1] Sipilä, K., Laiho-Steffens, J., Jäsberg, M. and Wacker, A., 'Modelling the Impact of the Fast Power Control on the WCDMA Uplink', *Proceedings of VTC'99*, Houston, Texas, May 1999, pp. 1266–1270.

[2] Ojanperä, T. and Prasad, R., *Wideband CDMA for Third Generation Mobile Communications*, Artech House, 1998.

[3] Saunders, S., *Antennas and Propagation for Wireless Communication Systems*, John Wiley & Sons, 1999.

[4] Wacker, A., Laiho-Steffens, J., Sipilä, K., and Heiska, K., 'The Impact of the Base Station Sectorisation on WCDMA Radio Network Performance', *Proceedings of VTC'99*, Amsterdam, The Netherlands, September 1999, pp. 2611–2615.

[5] Sipilä, K., Honkasalo, Z., Laiho-Steffens, J. and Wacker, A., 'Estimation of Capacity and Required Transmission Power of WCDMA Downlink Based on a Downlink Pole Equation', to appear in *Proceedings of VTC2000*, Spring 2000.

[6] Wang, Y.-P. and Ottosson, T., 'Cell Search in W-CDMA', *IEEE J. Select. Areas Commun.*, Vol. 18, No. 8, 2000, pp. 1470–1482.

[7] Lee, J. and Miller, L., *CDMA Systems Engineering Handbook*, Artech House, 1998.

[8] Wacker, A., Laiho-Steffens, J., Sipilä, K. and Jäsberg, M., 'Static Simulator for Studying WCDMA Radio Network Planning Issues', *Proceedings of VTC'99*, Houston, Texas, May 1999, pp. 2436–2440.

[9] 3GPP Technical Specification 25.101, UE Radio Transmission and Reception (FDD).

[10] Hämäläinen, S., Holma, H., and Toskala, A., 'Capacity Evaluation of a Cellular CDMA Uplink with Multiuser Detection', *Proceedings of ISSSTA'96*, Mainz, Germany, September 1996, pp. 339–343.

[11] 3GPP Technical Report 25.942, RF System Scenarios.

[12] 3GPP TSG RAN WG4 Tdoc 99/329, Impact of OHG Harmonisation Recommendation on UTRA/FDD, June 1999.

9

Radio Resource Management

Janne Laakso, Harri Holma and Oscar Salonaho

9.1   Interference-Based Radio Resource Management

Radio Resource Management (RRM) is responsible for utilisation of the air interface resources. RRM is needed to guarantee Quality of Service (QoS), to maintain the planned coverage area and to offer high capacity. RRM can be divided into handover, power control, admission control, load control and packet scheduling functionalities. Power control is needed to keep the interference levels at minimum in the air interface and to provide the required quality of service. WCDMA power control is described in Section 9.2. Handovers are needed in cellular systems to handle the mobility of the user when he or she is moving from the coverage area of one cell to another. Handovers are presented in Section 9.3. In third generation networks other radio resource management algorithms—admission control, load control and packet scheduling—are required to guarantee the quality of service and to maximise the system throughput with a mix of different bit rates, services and quality requirements. Admission control is presented in Section 9.5 and load control in Section 9.6. WCDMA packet scheduling is described in Chapter 10. The radio resource management algorithms can be based on the amount of hardware in the network or on the interference levels in the air interface. Hard blocking is defined as the case where the hardware limits the capacity before the air interface gets overloaded. Soft blocking is defined as the case where the air interface load is estimated to be above the planned limit. The difference between hard blocking and soft blocking is analysed in Section 8.2.3. It is shown that soft blocking based RRM gives higher capacity than hard blocking based RRM. If soft blocking based RRM is applied, the air interface load needs to be measured. The measurement of the air interface load is presented in Section 9.4. In IS-95 networks RRM is typically based on the available channel elements (hard blocking), but that approach is not applicable in the third generation WCDMA air interface, where various bit rates have to be supported simultaneously.

Typical locations of the RRM algorithms in a WCDMA network are shown in Figure 9.1.

WCDMA for UMTS, edited by Harri Holma and Antti Toskala
© 2001 John Wiley & Sons, Ltd

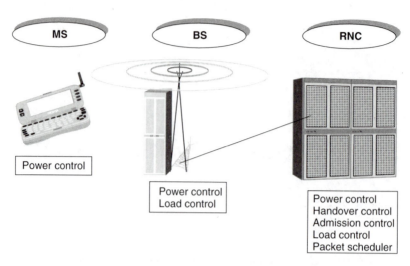

Figure 9.1. Typical locations of RRM algorithms in a WCDMA network

9.2 Power Control

Power control was introduced briefly in Section 3.5. In this chapter a few important aspects of WCDMA power control are covered. Some of these issues are not present in existing second generation systems, such as GSM and IS-95, but are new in third generation systems and therefore require special attention. In Section 9.2.1 fast power control is presented and in Section 9.2.2 outer loop power control is analysed. Outer loop power control sets the target for fast power control so that the required quality is provided.

In the following sections the need for fast power control and outer loop power control is shown using simulation results. Two special aspects of fast power control are presented in detail in Section 9.2.1: the relationship between fast power control and diversity, and fast power control in soft handover. Outer loop power control is presented in Section 9.2.2. It needs to estimate the received quality in order to adjust the target for fast power control. Estimation of quality is presented and outer loop power control algorithms are introduced. Third generation networks need to support high quality services and to multiplex several services on one connection. Those requirements affect also the outer loop. Finally, the differences between uplink and downlink outer loop algorithms are discussed.

9.2.1 Fast Power Control

In WCDMA fast power control with 1.5 kHz frequency is supported in both uplink and downlink. In GSM only slow (frequency approximately 2 Hz) power control is employed. In IS-95 fast power control with 800 Hz frequency is supported only in the uplink.

9.2.1.1 Gain of Fast Power Control

In this section, examples of the benefits of fast power control are presented. The simulated service is 8 kbps speech with BLER = 1% and 10 ms interleaving. Simulations are made with and without fast power control with a step size of 1 dB. Slow power control assumes that the average power is kept at the desired level and that the slow power control would be able to ideally compensate for the effect of path loss and shadowing, whereas fast power control can compensate also for fast fading. Two-branch receive diversity is assumed in the base station. ITU Vehicular A is a five-tap channel with WCDMA resolution, and ITU Pedestrian A is a two-path channel where the second tap is very weak. The required E_b/N_0 with and without fast power control are shown in Table 9.1 and the required average transmission powers in Table 9.2.

Fast power control gives clear gain, which can be seen from Tables 9.1 and 9.2. The gain from the fast power control is larger:

— for low mobile speeds than for high mobile speeds
— in required E_b/N_0 than in transmission powers
— for those cases where only a little multipath diversity is available, as in the ITU Pedestrian A channel. The relationship between fast power control and diversity is discussed in Section 9.2.1.2.

In Tables 9.1 and 9.2 the negative gains at 50 km/h indicate that an ideal slow power control would give better performance than the realistic fast power control. The negative gains are due to inaccuracies in the SIR estimation, power control signalling errors, and the delay in the power control loop.

Note that the gain from fast power control in Table 9.1 can be used to estimate the required fast fading margin in the link budget in Section 8.2.1. The fast fading margin is needed in the mobile station transmission power for maintaining adequate closed loop fast power control. The maximum cell range is obtained when the mobile is transmitting with full constant power, i.e. without the gain of fast power control.

9.2.1.2 Power Control and Diversity

In this section the importance of diversity is analysed together with fast power control. At low mobile speed the fast power control can compensate for the fading of the channel and keep the received power level fairly constant. The main sources of errors in the received powers arise from inaccurate SIR estimation, signalling errors and delays in the power control loop. The compensation of the fading causes peaks in the transmission power. The

Table 9.1. Required E_b/N_0 values with and without fast power control

	Slow power control	Fast 1.5 kHz power control	Gain from fast power control
ITU Pedestrian A 3 km/h	11.3 dB	5.5 dB	5.8 dB
ITU Vehicular A 3 km/h	8.5 dB	6.7 dB	1.8 dB
ITU Vehicular A 50 km/h	6.8 dB	7.3 dB	−0.5 dB

Table 9.2. Required relative transmission powers with and without fast power control

	Slow power control	Fast 1.5 kHz power control	Gain from fast power control
ITU Pedestrian A 3 km/h	11.3 dB	7.7 dB	3.6 dB
ITU Vehicular A 3 km/h	8.5 dB	7.5 dB	1.0 dB
ITU Vehicular A 50 km/h	6.8 dB	7.6 dB	−0.8 dB

received power and the transmitted power are shown as a function of time in Figures 9.2 and 9.3 with a mobile speed of 3 km/h. These simulation results include realistic SIR estimation and power control signalling. A power control step size of 1.0 dB is used. In Figure 9.2 very little diversity is assumed, while in Figure 9.3 more diversity is assumed in the simulation. Variations in the transmitted power are higher in Figure 9.2 than in Figure 9.3. This is due to the difference in the amount of diversity. The diversity can be obtained with, for example, multipath diversity, receive antenna diversity, transmit antenna diversity or macro diversity.

With less diversity there are more variations in the transmitted power, but also the average transmitted power is higher. Here we define power rise to be the ratio of the average transmission power in a fading channel to that in a non-fading channel when the received power level is the same in both fading and non-fading channels with fast power control. The power rise is depicted in Figure 9.4.

In Figure 9.5 theoretical power rises are calculated for two-path channels both with and without receive diversity as a function of the average power differences between the two multipath components. The stronger the second multipath component, i.e., the more multipath diversity is available, the smaller is the power rise. Antenna diversity reduces the power rise as well. If the two multipath components have the same average power, the power rise is 3 dB without antenna diversity. We obtain the same power rise also in a one-path channel with antenna diversity.

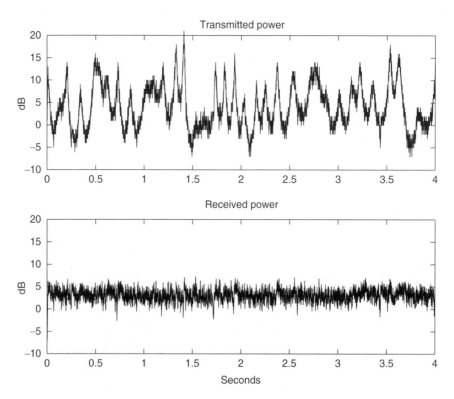

Figure 9.2. Transmitted and received powers in two-path (average tap powers 0 dB, −10 dB) Rayleigh fading channel at 3 km/h

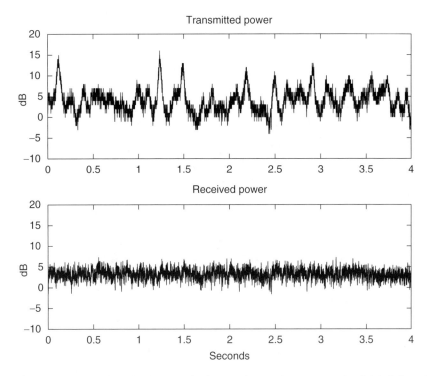

Figure 9.3. Transmitted and received powers in three-path (equal tap powers) Rayleigh fading channel at 3 km/h

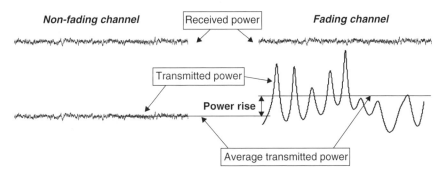

Figure 9.4. Power rise in fading channel with fast power control

In practice the power control is not ideal, so we resort to link-level simulations to find the real power rise. The link-level results for uplink power rise are presented in Table 9.3. The simulations are performed at different mobile speeds in a two-path ITU Pedestrian A channel with average multipath component powers of 0.0 dB and −12.5 dB. In the simulations the received and transmitted powers are collected slot by slot. The theoretical value for power rise in this ITU Pedestrian A multipath profile, according to Figure 9.5, is 2.3 dB with antenna diversity. In Table 9.3 we note that the simulated power rises at 3 km/h and 10 km/h are 2.1 dB and 2.0 dB, very close to the theoretical value of 2.3 dB.

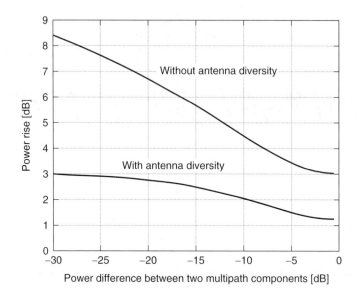

Figure 9.5. Theoretical power rise as a function of the power difference between the multipath components in a two-path Rayleigh fading channel

Table 9.3. Simulated power rises. Multipath channel ITU Pedestrian A, antenna diversity assumed

Mobile speed	Average power rise
3 km/h	2.1 dB
10 km/h	2.0 dB
20 km/h	1.6 dB
50 km/h	0.8 dB
140 km/h	0.2 dB

At high mobile speeds (>100 km/h) there is only very little power rise since the fast power control cannot compensate for the fading. At high mobile speeds the fast power control cannot follow the fast fading and a higher received power level is needed to obtain the required quality, as will be shown in Table 9.4. At high mobile speeds the diversity helps to keep the received power level constant; therefore a lower average received power level is adequate to provide the same quality of service.

More information about the uplink power control modelling can be found in [1].

Why is the power rise important for WCDMA system performance? In the downlink, the air interface capacity is directly determined by the required transmission power, since that determines the transmitted interference. Thus, to maximise the downlink capacity the transmission power needed by one link should be minimised. In the downlink, the received power level in the mobile does not affect the capacity.

In the uplink, the transmission powers determine the amount of interference to the adjacent cells, and the received powers determine the amount of interference to other users in the same cell. If, for example, there were only one WCDMA cell in one area, the uplink capacity of this cell would be maximised by minimising the required received powers, and the power rise would not affect the uplink capacity. We are, however, interested in cellular networks where the design of the uplink diversity schemes has to take into account both the transmitted and received powers. The lower the isolation between the adjacent cell in the network, the more emphasis must be put on the transmission power. The effect of received and transmission powers on network interference levels is illustrated in Figure 9.6.

9.2.1.3 Power Control in Soft Handover

Fast power control in soft handover has two major issues that are different from the single-link case: power drifting in the base station powers in the downlink, and reliable detection of the uplink power control commands in the mobile. These aspects are illustrated in Figure 9.7 and described in more detail in this section. A solution for improving the power control signalling quality is also presented in this section.

Downlink Power Drifting
The mobile sends a single command to control the downlink transmission powers; this is received by all base stations in the active set. The base stations detect the command independently, since the power control commands cannot be combined in RNC because it would cause too much delay and signalling in the network. Due to signalling errors in the air interface, the base stations may detect this power control command in a different way. It is possible that one of the base stations lowers its transmission power to that mobile while the other base station increases its transmission power. This behaviour leads to a situation where the downlink powers start drifting apart; this is referred to here as power drifting.

Power drifting is not desirable, since it mostly degrades the downlink soft handover performance. It can be controlled via RNC. The simplest method is to set relatively strict limits for the downlink power control dynamics. These power limits apply to the mobile's specific transmission powers. Naturally, the smaller the allowed power control dynamics,

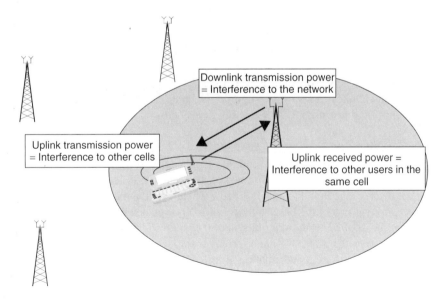

Figure 9.6. Effect of received and transmission powers on interference levels

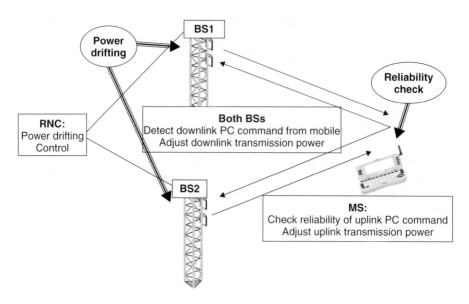

Figure 9.7. Fast power control in soft handover

the smaller the maximum power drifting. On the other hand, large power control dynamics typically improve power control performance, as shown in Table 9.2.

Another way to reduce power drifting is as follows. RNC can receive information from the base stations concerning the transmission power levels of the soft handover connections. These levels are averaged over a number of power control commands, e.g. over 500 ms or equivalently over 750 power control commands. Based on those measurements, RNC can send a reference value for the downlink transmission powers to the base stations. The soft handover base stations use that reference value in their downlink power control for that connection to reduce the power drifting. The idea is that a small correction is periodically performed towards the reference power. The size of this correction is proportional to the difference between the actual transmitted power and the reference power. This method will reduce the amount of power drifting. Power drifting can happen only if there is fast power control in the downlink. In IS-95 only slow power control is used in the downlink, and no method of controlling downlink power drifting is needed.

Reliability of Uplink Power Control Commands

All the base stations in the active set send an independent power control command to the mobile to control the uplink transmission power. It is enough if one of the base stations in the active set receives the uplink signal correctly. Therefore, the mobile station can lower its transmission power if one of the base stations sends a power-down command. Maximal ratio combining can be applied to the data bits in soft handover in the mobile station, because the same data is sent from all soft handover base stations, but not the power control bits because they contain different information from each of the base stations. Therefore, the reliability of the power control bits is not as good as for the data bits, and a threshold in the mobile is used to check the reliability of the power control commands. Very unreliable power control commands should be discarded because they are corrupted by interference.

Improved Power Control Signalling Quality

The power control signalling quality can be improved by setting a higher power for the dedicated physical control channel (DPCCH) than for the dedicated physical data channel (DPDCH) in the downlink if the mobile is in soft handover. This power offset between DPCCH and DPDCH can be different for different DPCCH fields: power control bits, pilot bits and TFCI. The power offset is illustrated in Figure 9.8.

The reduction of the mobile transmission power of a speech connection with downlink power offset is shown in Figure 9.9. The horizontal axis shows the difference in the attenuation from the mobile to the two soft handover base stations, 0 dB indicating that the attenuation is the same for both soft handover base stations. In this example 3 dB higher power was used for pilot and power control signalling. The reduction of the mobile transmission power is 0.4–0.6 dB with power offsets. This reduction is obtained because of the improved quality of the power control signalling.

9.2.2 *Outer Loop Power Control*

The outer loop power control is needed to keep the quality of communication at the required level by setting the target for the fast power control. The outer loop aims at providing the required quality: no worse, no better. Too high quality would waste capacity. The outer loop is needed in both uplink and downlink because there is fast power control in both uplink and downlink. In the following sections a few aspects of this control loop are described;

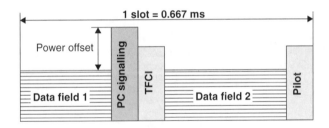

Figure 9.8. Power offset for improving downlink signalling quality

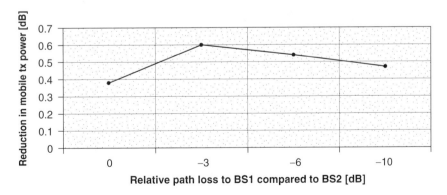

Figure 9.9. Gain in uplink transmission powers by using power offsets

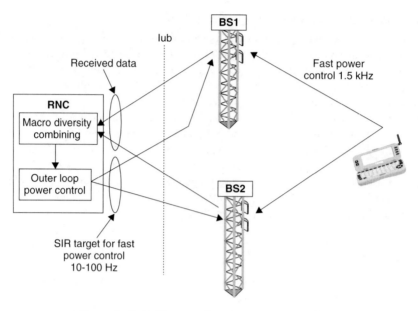

Figure 9.10. Uplink outer loop power control in RNC

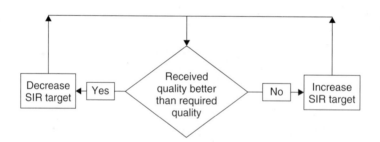

Figure 9.11. General outer loop power control algorithm

these apply to both uplink and downlink. In IS-95, outer loop power control is used only in the uplink because there is no fast power control in the downlink.

An overview of uplink outer loop power control is shown in Figure 9.10. The uplink quality is observed after macro diversity combining in RNC and the SIR target is sent to the base stations. The frequency of the fast power control is 1.5 kHz and frequency of the outer loop power control typically 10–100 Hz. A general outer loop power control algorithm is presented in Figure 9.11.

9.2.2.1 Gain of Outer Loop Power Control

In this section we analyse how much the SIR target needs to be adjusted when the mobile speed or the multipath propagation environment changes. The terms SIR target and E_b/N_0 target are used interchangeably in this chapter. Simulation results with AMR speech service and BLER $= 1\%$ are shown in Table 9.4 with outer loop power control. Three different multipath profiles are used: static channel corresponding to strong line-of-sight component,

Table 9.4. Average E_b/N_0 targets in different environments

Multipath	Mobile speed	Average E_b/N_0 target
Non-fading	—	**5.3 dB**
ITU Pedestrian A	3 km/h	5.9 dB
ITU Pedestrian A	20 km/h	6.8 dB
ITU Pedestrian A	50 km/h	6.8 dB
ITU Pedestrian A	120 km/h	**7.1 dB**
3-path equal powers	3 km/h	6.0 dB
3-path equal powers	20 km/h	6.4 dB
3-path equal powers	50 km/h	6.4 dB
3-path equal powers	120 km/h	6.9 dB

fading ITU Pedestrian A channel, and fading three-path channel with equal average powers of the multipath components. No antenna diversity is assumed here.

The lowest average E_b/N_0 target is needed in the static channel and the highest target in the ITU Pedestrian A channel with high mobile speed. This result indicates that the higher the variation in the received power, the higher the E_b/N_0 target needs to be to provide the same quality. If we were to select a fixed E_b/N_0 target of 5.3 dB according to the static channel, the frame error rate of the connection would be too high in fading channels and speech quality would be degraded. If we were to select a fixed E_b/N_0 target of 7.1 dB, the quality would be good enough but unnecessary high powers would be used in most situations. We can conclude that there is clearly a need to adjust the target of the fast closed loop power control by outer loop power control.

How fast should the outer loop power control adjust the target value? One example case could be a micro cellular environment where the mobile is first in line-of-sight to the base station and the average E_b/N_0 target of 5.3 dB provides the required quality. If the mobile turns around the corner, the line-of-sight component disappears and the multipath profile can change to ITU Pedestrian A. If the mobile is moving at 20 km/h, the E_b/N_0 target needs to be rapidly increased from 5.3 dB to 6.8 dB.

9.2.2.2 Estimation of Received Quality

A few different approaches to measuring the received quality are introduced in this section. A simple and reliable approach is to use the result of the error detection—cyclic redundancy check (CRC)—to detect whether there is an error or not. The advantages of using the CRC check are that it is a very reliable detector of frame errors, and it is simple. The CRC-based approach is well suited for those services where errors are allowed to occur fairly frequently, at least once every few seconds, such as the non-real-time packet data service where the block error rate (BLER) can be up to 10–20% before retransmissions, and the speech service where typically BLER = 1% provides the required quality. With Adaptive Multirate (AMR) speech codec the interleaving depth is 20 ms and BLER = 1% corresponds to one error on average every 2 seconds.

The received quality can also be estimated based on soft frame reliability information. Such information could be, for example:

— Estimated bit error rate (BER) before channel decoder, called raw BER or physical channel BER

— Soft information from Viterbi decoder with convolutional codes
— Soft information from Turbo decoder, for example BER or BLER after an intermediate
  decoding iteration
— Received E_b/N_0.

The problem with these quantities is that they may give an erroneous estimate of the received quality. Consider the use of the raw BER. The required raw BER to obtain a required final BLER after decoder is not constant but depends on the multipath profile, the mobile speed and the receiver algorithms. Soft information is needed for high quality services, see Section 9.2.2.4. Raw BER is used as soft information over Iub interface. The estimation of the quality is presented in Figure 9.12.

9.2.2.3   Outer Loop Power Control Algorithm

One possible outer loop power control algorithm is presented in [2]. The algorithm is based on the result of a CRC check of the data and can be characterised in pseudocode as shown in Figure 9.13.

If the BLER of the connection is a monotonically decreasing function of the E_b/N_0 target, this algorithm will result in a BLER equal to the FER target if the call is long enough. The step size parameter determines the convergence speed of the algorithm to the desired target and also defines the overhead caused by the algorithm. The principle is that the higher the step size the faster the convergence and the higher the overhead. Figure 9.14 gives an example of the behaviour of the algorithm with FER target of 1% and step size of 0.5 dB.

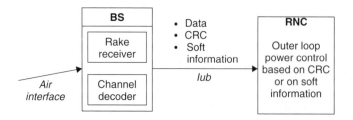

Figure 9.12. Estimation of quality in outer loop in RNC

```
IF CRC check OK
    Step_down = BLER_target * Step_size ;
    Eb/N0_target (n+1) = Eb/N0_target (n) - Step_down ;

ELSE
    Step_up = Step_size - BLER_target * Step_size ;
    Eb/N0_target (n+1) = Eb/N0_target (n) + Step_up ;
END

where

    Eb/N0_target (n) is the Eb/N0 target in frame n,
    BLER_target is the BLER target for the call and
    Step_size is a parameter, typically 0.3–0.5 dB
```

Figure 9.13. Pseudocode of one outer loop power control algorithm

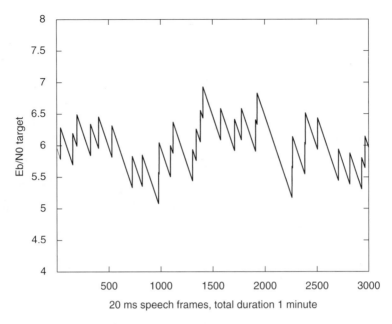

Figure 9.14. E_b/N_0 target in ITU Pedestrian A channel, AMR speech codec, BLER target 1%, step size 0.5 dB, speed 3 km/h

9.2.2.4 High Quality Services

High quality services with very low BLER ($<10^{-3}$) are required to be supported by third generation networks. In such services errors are very rare events. If the required BLER = 10^{-3} and the interleaving depth is 40 ms, an error occurs on average every $40/10^{-3}$ ms = 40 seconds. If the received quality is estimated based on the errors detected by CRC bits, the adjustments of the E_b/N_0 target is very slow and the convergence of the E_b/N_0 target to the optimal value takes a long time. Therefore, for high quality services the soft frame reliability information has advantages. Soft information can be obtained from every frame even if there are no errors.

9.2.2.5 Limited Power Control Dynamics

At the edge of the coverage area the mobile may hit its maximum transmission power. In that case the received BLER can be higher than desired. If we apply directly the outer loop algorithm of Figure 9.11, the uplink SIR target would be increased. The increase of the SIR target does not improve the uplink quality if the base station is already sending only power-up commands to the mobile. In that case the E_b/N_0 target might become unnecessarily high. When the mobile returns closer to the base station, the quality of the uplink connection is unnecessarily high before the outer loop lowers the E_b/N_0 target back to the optimal value. The situation in which the mobile hits its maximum transmission power is shown in Figure 9.15. In that example the AMR speech service with 20 ms interleaving is simulated with the outer loop power control algorithm from Figure 9.13. A BLER target of 1% and an outer loop step size of 0.5 dB are used here. With full power control dynamics an error should occur once every 2 seconds to provide an BLER of 1% with 20 ms interleaving. The maximum transmission power of the mobile station is 125 mW, i.e. 21 dBm.

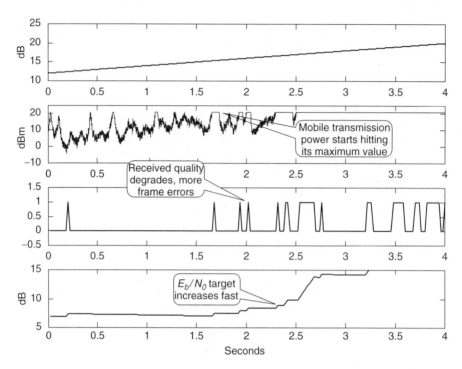

Figure 9.15. Increase of E_b/N_0 target when the mobile hits its maximum transmission power. Top: attenuation between mobile and base station; second figure: mobile transmission power (dBm); third figure: block errors (1 = error, 0 = no errors); bottom: uplink E_b/N_0 target

The same problem could also occur if the mobile hits its minimum transmission power. In that case the E_b/N_0 target would become unnecessarily low. The same problems can be observed also in the downlink if the power of the downlink connection is using its maximum or minimum value.

The outer loop problems from limited power control dynamics can be avoided by setting tight limits for the E_b/N_0 target or by an intelligent outer loop power control algorithm. Such an algorithm would not increase the E_b/N_0 target if the increase did not improve the quality.

9.2.2.6 Multiservice

One of the basic requirements of UMTS is to be able to multiplex several services on a single physical connection. Since all the services have a common fast power control, there can be only one common target for fast power control. This must be selected according to the service requiring the highest target as shown in Figure 9.16. There should be no large differences between the required targets if unequal rate matching has been applied on layer 1 to provide the different qualities.

9.2.2.7 Downlink Outer Loop Power Control

According to Section 9.2.2.1 there is clearly a need to adjust the target for fast closed loop power control. In the downlink the fastest adjustment of the downlink target is obtained

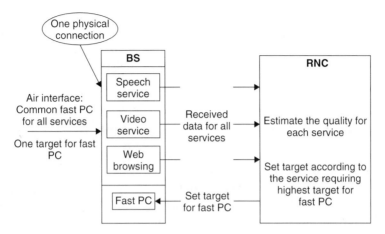

Figure 9.16. Uplink outer loop power control for multiple services on one physical connection

by having the outer loop power control within the mobile. Another approach would be a network-based downlink outer loop where the mobile reports the quality measurements to the network and the network then commands the mobile in adjusting the target value of the downlink fast power control. This network-based downlink outer loop power control would lead to increased signalling between the mobile station and RNC, and would also cause delays to the downlink outer loop power control. Therefore, the mobile-based outer loop power control is used in WCDMA.

The network can effectively control the downlink connections even if the downlink outer loop power control is running within the mobile. First, the network sets the quality target for each downlink connection; that target can be modified during the connection. Second, the base station does not need to increase the downlink power of that connection even if the mobile sends a power-up command. The network can control the quality of the different downlink connections very quickly by not obeying the power control commands from the mobile. This approach could be used, for example, during downlink overload to reduce the downlink power of those connections that have a lower priority, such as background-type services (see load control in Section 9.6). This reduction of downlink powers can take place at the frequency of fast power control, i.e. 1.5 kHz.

9.3   Handovers

9.3.1   Intra-frequency Handovers

9.3.1.1   Handover Algorithms

The algorithms to be presented shortly are the basic cdmaOne algorithm (IS-95A) [3,4,5] and the WCDMA soft handover algorithm [4,6]. Both algorithms use pilot channel CPICH E_c/I_0 as the handover measurement quantity, which is signalled to RNC by using layer 3 signalling (see Section 7.7).

The following terminology is used in the handover description:

Active set:

The cells in the active set form a soft handover connection to the mobile station.

Candidate set:

The candidate set is the list of cells that are not presently used in the soft handover connection, but whose pilot E_c/I_0 are strong enough to be added to the active set. The candidate set is not used in the WCDMA handover algorithm.

Neighbour set/monitored set:

The neighbour set or monitored set is the list of cells that the mobile station continuously measures, but whose pilot E_c/I_0 are not strong enough to be added to the active set.

Basic cdmaOne algorithm (IS-95A algorithm)

The functionality of the basic cdmaOne algorithm is depicted in Figure 9.17. In the IS-95A algorithm the handover threshold is a fixed value of received pilot E_c/I_0. The problem with the IS-95A algorithm is that some locations in the cell receive only weak pilots (requiring a lower threshold) and other locations receive a few strong and dominant pilots (requiring higher handover thresholds). Therefore an algorithm with dynamic thresholds is proposed [5]. In the modified cdmaOne algorithm (IS-95B) the only difference compared to the basic algorithm is the active set maintenance procedure. Candidate set maintenance is performed in a similar manner as in the basic algorithm.

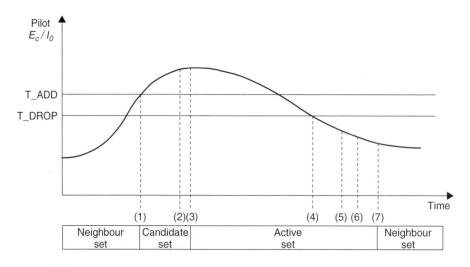

(1) Pilot strength exceeds *T_ADD*. Mobile station sends a Pilot Strength Measurement Message and transfers pilot to the candidate set.

(2) Base station sends a Handover Direction Message.

(3) Mobile station transfer pilot to the active set and sets a Handover Completion Message.

(4) Pilot strength drops below *T_DROP*. Mobile station starts the handover drop timer.

(5) Handover drop timer expires. Mobile station sends a Pilot Strength Measurement Message.

(6) Base station sends a Handover Direction Message.
   Mobile station moves pilot from the active set to the neighbour set and sends a Handover Completion Message.

Figure 9.17. Basic cdmaOne algorithm (IS-95A)

WCDMA handover algorithm

The WCDMA handover algorithm is described in Figure 9.18. For more detaile
see the 3GPP specification [6].

The soft handover algorithm as described in Figure 9.18 is as follows:

— If $Pilot\_E_c/I_0 > Best\_Pilot\_E_c/I_0 - Reporting\_range + Hysteresis\_event1A$ for a
   period of ΔT and the active set is not full, the cell is added to the active set. This
   event is called Event 1A or Radio Link Addition

— If $Pilot\_E_c/I_0 < Best\_Pilot\_E_c/I_0 - Reporting\_range - Hysteresis\_event1B$ for a
   period of ΔT, then the cell is removed from the active set. This event is called Event
   1B or Radio Link Removal

— If the active set is full and $Best\_candidate\_Pilot\_E_c/I_0 > Worst\_Old\_Pilot\_E_c/I_0 + Hysteresis\_event1C$ for a period of ΔT, then the weakest cell in the active set is
   replaced by the strongest candidate cell (i.e. strongest cell in the monitored set).
   This event is called Event 1C or Combined Radio Link Addition and Removal. The
   maximum size of the active set in Figure 9.18 is assumed to be two.

where

— *Reporting_range* is the threshold for soft handover
— *Hysteresis_event1A* is the addition hysteresis
— *Hysteresis_event1B* is the removal hysteresis
— *Hysteresis_event1C* is the replacement hysteresis

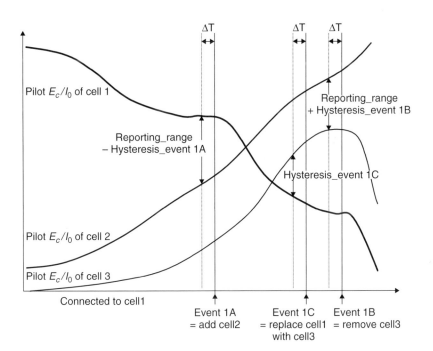

Figure 9.18. General scheme of the WCDMA soft handover algorithm

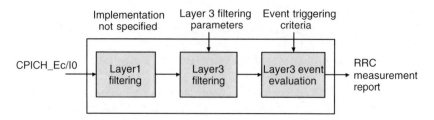

Figure 9.19. Handover measurement filtering and reporting

— ΔT is the time to trigger
— $Best\_Pilot\_E_c/I_0$ is the strongest measured cell in the active set
— $Worst\_Old\_Pilot\_E_c/I_0$ is the weakest measured cell in the active set
— $Best\_candidate\_Pilot\_E_c/I_0$ is the strongest measured cell in the monitored set
— $Pilot\_E_c/I_0$ is the measured and filtered quantity.

The WCDMA handover measurement reporting is described in Figure 9.19. Before the pilot E_c/I_0 is used by the active set update algorithm in the mobile, some filtering is applied. The measurement is filtered both on layer 1 and on layer 3. The layer 3 filtering can be controlled by the network. The measurements of the pilot E_c/I_0 are discussed in Section 9.3.1.2.

The handover measurement reporting from the mobile to RNC can be configured to be periodic, like in GSM, or event-triggered. According to [6] the event-triggered reporting provides the same performance as periodic reporting but with less signalling load.

9.3.1.2 Handover Measurements

The accuracy of the handover measurements, i.e. pilot E_c/I_0 measurements, is important for handover performance. The effect of the filtering on measurement accuracy is shown in Figure 9.20 with simulation results at 3 km/h and in Figure 9.21 at 50 km/h. The mobile travels the same distance in both figures. Path loss, shadowing and interference are not considered in these examples; only the effect of fast fading is shown. The target of the handover measurement is to obtain a measurement result where the effect of fast fading is averaged out. The example measurement here is done by taking one sample per 10 ms frame. The correct measurement value is 0 dB and the difference from that value is caused only by fast fading which is not completely averaged out. The assumed multipath profile here is one-path Rayleigh fading channel, which is the worst-case assumption. If multipath diversity is available and the mobile can measure with multiple fingers, fast fading causes fewer inaccuracies than in the one-path channel.

The filtering length of 100 ms causes very large measurement errors at 3 km/h, since fast fading cannot be filtered within such a short period, as shown in the uppermost part of Figure 9.20. Due to the measurement errors, unnecessary handovers take place, leading to increased handover signalling and short active set update periods. By increasing the filtering length to 1 s the measurement accuracy can be clearly improved. At low mobile speeds long filtering periods are advantageous.

At 50 km/h the filtering period of 100 ms gives reasonably good performance and only relatively small improvements can be achieved by increasing the filtering period. The drawback of long filtering periods is the delay caused to the handovers. At high mobile speeds

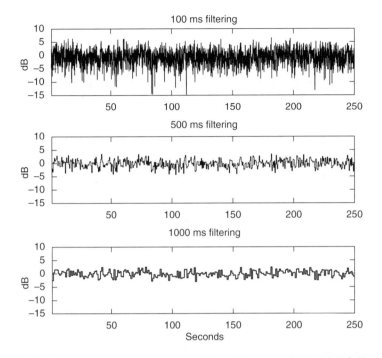

Figure 9.20. Handover measurement accuracy at 3 km/h in one-path Rayleigh fading channel

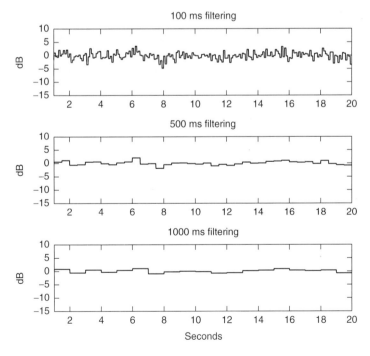

Figure 9.21. Handover measurement accuracy at 50 km/h in one-path Rayleigh fading channel

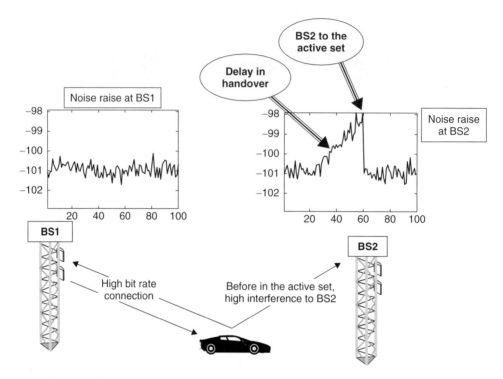

Figure 9.22. Noise rise peaks in the base station reception due to a delayed handover

fast handovers are important, especially in micro cellular networks where the path loss to the base stations can change quickly when a mobile is driven round a corner.

The effect of a very delayed handover with a fast mobile and a high bit rate connection is shown in Figure 9.22. As long as base station BS2 is not in the active set of the mobile, it cannot control the uplink transmission power, and noise rise peaks can be caused to the base station BS2. This problem can occur only if

— there are long delays in the handovers due to long averaging in the measurements or due to delays in the handover signalling, and
— the mobile is moving very fast, and
— the connection is using high bit rates.

Therefore, very long filtering periods cannot be used in the handover measurements. The optimal filtering period is a trade-off between measurement accuracy and handover delay.

The measurement period for handover measurements is 200 ms in 3GPP and additionally layer 3 filtering can be applied to top of the 200 ms period [8].

9.3.1.3 Soft Handover Gains

This section presents some examples of soft handover gains that have been obtained in simulations. These results are the gains of macro diversity combining against fast fading compared to ideal hard handover where the mobile would be connected to the base station

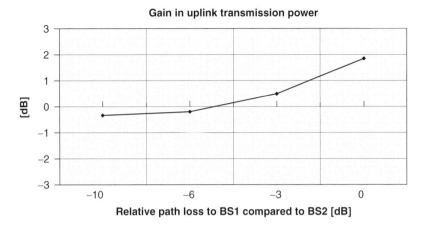

Figure 9.23. Soft handover gain in uplink transmission power (positive value = gain, negative value = loss)

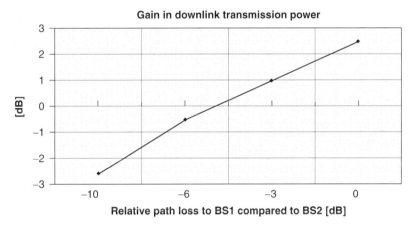

Figure 9.24. Soft handover gain in downlink transmission power (positive = gain, negative = loss)

where the path loss is smallest. The results were presented and discussed in more detail in [9]. The soft handover gains improve the coverage and capacity of the WCDMA network.

Figure 9.23 and Figure 9.24 show the simulation results of 8 kbps speech in ITU Pedestrian A channel, at 3 km/h, with soft handover containing two base stations in the active set. The relative path loss from the mobile to BS1 compared to BS2 was 0, −3, −6 or −10 dB. The highest gains are obtained when the path loss is the same to both base stations, i.e. the relative path loss difference is 0 dB. Figure 9.23 shows the soft handover gain in uplink transmission power with base station receive antenna diversity. Figure 9.24 shows the corresponding gains in downlink transmission power without transmit or receive antenna diversity. The gains are relative to the single-link case where the mobile would be connected only to the best base station. It should be noted that ITU Pedestrian A channel has only little multipath diversity, and thus the soft handover gains are relatively high. With more multipath diversity, the handover gains are lower.

In Figure 9.23 the maximum reduction of the mobile transmission power due to soft handover is 1.8 dB if the path loss is the same to both soft handover base stations. If the path loss difference is very large, the soft handover can cause an increase in the mobile transmission power. This increase is caused by the signalling errors of the uplink power control commands which are transmitted in the downlink. But, typically, the base station would not be in the active set of the mobile station if the path loss were 3–6 dB larger than the path loss to the nearest base station.

In the downlink the maximum soft handover gain is 2.3 dB (Figure 9.24), which is more than in the uplink (Figure 9.23). The reason is that no antenna diversity is assumed in the downlink, and thus in the downlink there is more need for macro diversity in soft handover.

In the downlink, soft handover causes an increase in the required downlink transmission power if the path loss difference is more than 4–5 dB in this example. In that case the mobile cannot receive effectively the signal from the more distant base station, and no additional diversity gain is provided.

These soft handover gains are example values only. The gains depend on the multipath profile, mobile speed, receiver algorithms and base station antenna configurations. The gains shown in this section are from the capacity point of view, while the soft handover gains for coverage are discussed in Section 11.2.1.4. The difference between these two aspects is that in the case of maximum coverage the mobile is transmitting on constant full power while in the present section fast power control is assumed.

9.3.1.4   Soft Handover Probabilities

Radio network planning is responsible for proper handover parameter setting and site planning so that the soft handover probability does not exceed the desired value. Typically, soft handover probability is required to keep below 30–40%, mainly because excessive soft handover probabilities could decreased the downlink capacity as shown in Figure 9.24. In the downlink each soft handover connection increases the transmitted interference to the network. When the increased interference exceeds the diversity gain, the soft handover does not provide any gain for system performance. Also, in the downlink the soft handover connections use more orthogonal codes than do only single-link connections. In both uplink and downlink the soft handovers require baseband resources in the base station, transmission capacity over the Iub interface, and RNC resources. It is the task of radio network planning and optimisation to keep the soft handover overhead below a desired threshold while still providing enough diversity in both uplink and downlink.

9.3.2   Inter-system Handovers Between WCDMA and GSM

WCDMA and GSM standards support handovers both ways between WCDMA and GSM. These handovers can be used for coverage or load balancing reasons. At the start of WCDMA deployment, handovers to GSM are needed to provide continuous coverage, and handovers from GSM to WCDMA can be used to lower the loading in GSM cells. This scenario is shown in Figure 9.25. When the traffic in WCDMA networks increases, it is important to have load reason handovers to both directions. The inter-system handovers are triggered in the source RNC/BSC, and from the receiving system point of view the inter-system handover is similar to inter-RNC or inter-BSC handover. The handover algorithms and triggers are not standardised. The handover procedure is standardised and presented in Section 7.7.3.

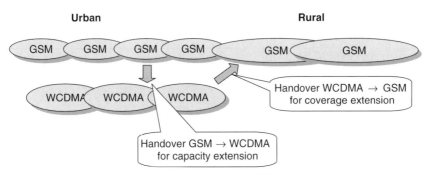

Figure 9.25. Inter-system handovers between GSM and WCDMA

Compressed Mode

WCDMA uses continuous transmission and reception and the mobile cannot make inter-system measurements with single receiver if there are no gaps generated to the WCDMA signals. Therefore, compressed mode is needed both for inter-frequency and for inter-system measurements. The compressed mode procedure is described in Chapter 6.

During the compressed mode gaps the fast power control cannot be applied and part of interleaving gain is lost. Therefore, a higher E_b/N_0 is needed during compressed frames leading to a capacity degradation. An example calculation for the capacity effect is shown in Table 9.5. In here it is assumed that the E_b/N_0 needs to be 2.0 dB higher during the compressed frames. It is further assumed that every third frame is compressed and 10% of the users are simultaneously using the compressed mode. In this case the interference level in the cell is increased 1.9%. In other words, the cell capacity is decreased by 1.9% if we want to keep the interference level constant.

Table 9.5 clearly shows that having all users in the compressed mode all the time would waste capacity, in this example 19% capacity degradation. Therefore, the compressed mode should be activated only when there is a need to execute an inter-system or inter-frequency handover. A typical inter-system handover procedure is as follows:

(1) Inter-system handover trigger is fulfilled in RNC, for example, mobile running out of WCDMA coverage area
(2) RNC commands the mobile to start inter-system measurements using compressed mode
(3) RNC selects the target GSM cell according to the mobile measurements
(4) RNC sends handover command to the mobile

Table 9.5. Effect of compressed mode to the capacity

Assumption	Average increase in the interference levels
Required E_b/N_0 is 2.0 dB higher during compressed frames	2.0 dB = 58% more interference during the compressed frame
Every third frame is compressed	58%/3 = 19% more interference from that connection
10% of the users in compressed mode at the same time	19%/10 = 1.9% more interference in the cell

Table 9.6. Effect of compressed mode to the coverage

Assumption	Coverage reduction
Required E_b/N_0 is 2.0 dB higher during compressed frames	2.0 dB
7-slot gap used in 15-slot frame	$10 * \log 10(15/(15 - 7)) = 2.7$ dB
Every second frame is compressed with 20 ms interleaving	$(2.0 \text{ dB} + 2.7 \text{ dB})/2 = 2.4$ dB reduced coverage

Compressed mode affects also the uplink coverage area of the real time services where the bit rate cannot be lowered during the compressed mode. An example effect to the coverage area is shown in Table 9.6 where the coverage is reduced by 2.4 dB. Therefore, the coverage reason inter-system handover procedure has to be initiated early enough at the cell edge to avoid any quality degradation during the compressed mode.

Inter-system handovers from GSM to WCDMA are initiated in GSM BSC. No compressed mode is needed for making WCDMA measurement from GSM because GSM uses discontinuous transmission and reception.

9.3.3 Inter-frequency Handovers within WCDMA

Most UMTS operators have two or three FDD carriers available. The operation can be started using one frequency and the second and the third frequency is needed later to enhance the capacity. Several frequencies can be used in two different ways as shown in Figure 9.26: several frequencies on the same sites for high capacity sites or macro and micro layers using different frequencies. Inter-frequency handovers between WCDMA carriers are needed to support those scenarios.

Compressed mode measurements are needed also in the inter-frequency handovers in the same way as in the inter-system handovers.

9.3.4 Summary of Handovers

The WCDMA handover types are summarised in Table 9.7. The most typical WCDMA handover is intra-frequency handover that is needed due to the mobility of the users.

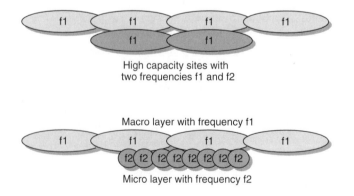

High capacity sites with two frequencies f1 and f2

Macro layer with frequency f1

Micro layer with frequency f2

Figure 9.26. Need for inter-frequency handovers between WCDMA carriers

Table 9.7. WCDMA handover types

Handover type	Handover measurements	Typical handover measurement reporting from mobile to RNC	Typical handover reason
WCDMA intra-frequency	Measurements all the time with matched filter	Event-triggered reporting	— Normal mobility
WCDMA → GSM inter-system	Measurements started only when needed, compressed mode used	Periodic during compressed mode	— Coverage — Load — Service
WCDMA inter-frequency	Measurements started only when needed, compressed mode used	Periodic during compressed mode	— Coverage — Load

The intra-frequency handover is controlled by those parameters shown in Figure 9.18. The intra-frequency handover reporting from the mobile to RNC typically is event-triggered, and RNC commands the handovers according to the measurement reports. In case of intra-frequency handover the mobile should be connected to the best base station(s) to avoid near–far problem, and RNC does not have any freedom in selecting the target cells.

Inter-frequency and inter-system measurements are typically initiated only when there is a need to make inter-system and inter-frequency handover to avoid unnecessary use of the compressed mode. Inter-frequency handovers are needed to balance loading between WCDMA carriers and cell layers, and to make a coverage reason handover from micro cell frequency to macro cells. Inter-system handovers to GSM are needed to extend the WCDMA coverage area, to balance load between systems and to direct services to the most suitable systems.

9.4   Measurement of Air Interface Load

If the radio resource management is based on the interference levels in the air interface, the air interface load need to be measured. The estimation of the uplink load is presented in Section 9.4.1 and the estimation of the downlink load is Section 9.4.2.

9.4.1   Uplink Load

In this section two uplink load measures are presented: load estimation based on wideband received power, and load estimation based on throughput. These are example approaches that could be used in WCDMA networks.

9.4.1.1   Load Estimation Based on Wideband Received Power

The wideband received power level can be used in estimating the uplink load. The received power levels can be measured in the base station. Based on those measurements, the uplink load factor can be obtained. The calculations are shown below.

The received wideband interference power, I_{total}, can be divided into the powers of own-cell (= intra-cell) users, I_{own}, other-cell (= inter-cell) users, I_{oth}, and background and receiver noise, P_N:

$$I_{total} = I_{own} + I_{oth} + P_N \tag{9.1}$$

The uplink noise rise is defined as the ratio of the total received power to the noise power:

$$\text{Noise rise} = \frac{I_{total}}{P_N} = \frac{1}{1 - \eta_{UL}} \tag{9.2}$$

This equation can be rearranged to give the uplink load factor η_{UL}:

$$\eta_{UL} = 1 - \frac{P_N}{I_{total}} = \frac{\text{Noise rise} - 1}{\text{Noise rise}} \tag{9.3}$$

where I_{total} can be measured by the base station and P_N is known beforehand.

The uplink load factor η_{UL} is normally used as the uplink load indicator. For example, if the uplink load is said to be 60% of the WCDMA pole capacity, this means that the load factor $\eta_{UL} = 0.60$.

Load estimation based on the received power level is also presented in [10] and [11].

9.4.1.2 Load Estimation Based on Throughput

The uplink load factor η_{UL} can be calculated as the sum of the load factors of the users that are connected to this base station:

$$\eta_{UL} = (1 + i) \cdot \sum_{j=1}^{N} L_j = (1 + i) \cdot \sum_{j=1}^{N} \frac{1}{1 + \dfrac{W}{(E_b/N_0)_j \cdot R_j \cdot \upsilon_j}} \tag{9.4}$$

where N is the number of users in the own cell, W is the chip rate, L_j is the load factor of the jth user, R_j is the bit rate of the jth user, $(E_b/N_0)_j$ is E_b/N_0 of the jth user, υ_j is the voice activity factor of the jth user, and i is the other-to-own cell interference ratio.

Note that Equation (9.4) is the same as the load factor calculation in radio network dimensioning in Section 8.2.2. In dimensioning, the average number of users N of a cell needs to be estimated, and average values for E_b/N_0, i and υ are used as input parameters. These values are typical for that environment and can be based on the measurements and simulations. In load estimation the instantaneous measured values for E_b/N_0, i, υ and the number of users N are used to estimate the instantaneous air interface load.

In throughput-based load estimation, interference from other cells is not directly included in the load but needs to be taken into account with the parameter i. Also, the part of own-cell interference that is not captured by the Rake receiver can be taken into account with the parameter i. If it is assumed that $i = 0$, then only own-cell interference is taken into account.

9.4.1.3 Comparison of Uplink Load Estimation Methods

Table 9.8 compares the above two load estimation methods. In the wideband power-based approach, interference from the adjacent cells is directly included in the load estimation because the measured wideband power includes all interference that is received in that carrier frequency by the base station. If the loading in the adjacent cells is low, this can be seen in the wideband power-based load measurement, and a higher load can be allowed in this cell, i.e. soft capacity can be obtained. The importance of soft capacity was explained in radio network dimensioning in Section 8.2.3.

The wideband power-based and throughput based load estimations are shown in Figure 9.27. The different curves represent a different loading in the adjacent cells. The

Table 9.8. Comparison of uplink load estimation methods

	Wideband received power	Throughput	Number of connections
What to measure	Wideband received power I_{total} per cell	Uplink E_b/N_0 and bit rates R for each connection	Number of connections
What needs to be assumed or measured separately	Thermal noise level (= unloaded interference power) P_N	Other-to-own cell interference ratio, i	Load caused by one connection
Other-cell interference	Included in measurement of wideband received power	Assumed explicitly in i	Assumed explicitly when choosing the maximum number of connections
Soft capacity	Yes, automatically	Not directly, possible via RNC	No
Other interference sources (= adjacent channel)	Reduced capacity	Reduced coverage	Reduced coverage

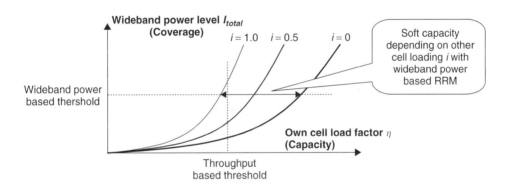

Figure 9.27. Wideband power based and throughput based load estimations

larger the value of i, the more interference from adjacent cells. The wideband power-based load estimation keeps the coverage within the planned limits and the delivered capacity depends on the loading in the adjacent cells (soft capacity). This approach effectively prevents cell breathing which would exceed the planned values.

The problem with wideband power-based load estimation is that the measured wideband power can include interference from adjacent frequencies. This could originate from another operator's mobile located very close to the base station antenna. Therefore, the interference-based method may overestimate the load of own carrier because of any external interference. The base station receiver cannot separate the interference from the own carrier and from other carriers by the wideband power measurements.

Throughput-based load estimation does not take interference from adjacent cells or adjacent carriers directly into account. If soft capacity is required, information about the adjacent cell loading can be obtained within RNC. The throughput-based RRM keeps the throughput of the cell at the planned level. If the loading in the adjacent cells is high, this affects the coverage area of the cell.

The third load estimation method in Table 9.8, in the right-hand column, is based simply on the number of connections in the base stations. This approach can be used in second generation networks where all connections use fairly similar low bit rates and no high bit rate connections are possible. In third generation networks the mix of different bit rates, services and quality requirements prevents the use of this approach. It is unreasonable to assume that the load caused by one 2-Mbps user is the same as that caused by one speech user.

9.4.1 Downlink Load

9.4.2.1 Power-Based Load Estimation

The downlink load of the cell can be determined by the total downlink transmission power, P_{total}. The downlink load factor, η_{DL}, can be defined to be the ratio of the current total transmission power divided by the maximum base station transmission power P_{max}:

$$\eta_{DL} = \frac{P_{total}}{P_{max}} \tag{9.5}$$

Note that in this load estimation approach the total base station transmission power P_{total} does not give accurate information concerning how close to the downlink air interface pole capacity the system is operating. In a small cell the same P_{total} corresponds to a higher air interface loading than in a large cell.

9.4.2.2 Throughput-Based Load Estimation

In the downlink, throughput-based load estimation can be effected by using the sum of the downlink allocated bit rates as the downlink load factor, η_{DL}, as follows:

$$\eta_{DL} = \frac{\displaystyle\sum_{i=1}^{N} R_i}{R_{max}} \tag{9.6}$$

where N is the number of downlink connections, including the common channels, R_j is the bit rate of the jth user, and R_{max} is the maximum allowed throughput of the cell.

It is also possible to weight the user bit rates with E_b/N_0 values as follows:

$$\eta_{DL} = \sum_{j=1}^{N} R_j \cdot \frac{\upsilon_j (E_b/N_0)_j}{W} \cdot [(1 - \overline{\alpha}) + \overline{i}] \tag{9.7}$$

where W is the chip rate, $(E_b/N_0)_j$ is the E_b/N_0 of the jth user, υ_j is the voice activity factor of the jth user, $\overline{\alpha}$ is the average orthogonality of the cell, and $\overline{i}$ is the average downlink other-to-own cell interference ratio of the cell. Note that Equation (9.7) is similar to the downlink radio network dimensioning (see Section 8.2.2).

The average downlink orthogonality can be estimated by the base station based on the multipath propagation in the uplink. The values of E_b/N_0 need to be assumed based on the typical values for that environment. The average interference from other cells can be obtained in RNC based on the adjacent cell loading.

9.5   Admission Control

9.5.1   Admission Control Principle

If the air interface loading is allowed to increase excessively, the coverage area of the cell is reduced below the planned values, and the quality of service of the existing connections cannot be guaranteed. Before admitting a new connection, admission control needs to check that the admittance will not sacrifice the planned coverage area or the quality of the existing connections. Admission control accepts or rejects a request to establish a radio access bearer in the radio access network. The admission control algorithm is executed when a bearer is set up or modified. The admission control functionality is located in RNC where the load information from several cells can be obtained. The admission control algorithm estimates the load increase that the establishment of the bearer would cause in the radio network. This has to be estimated separately for the uplink and downlink directions. The requesting bearer can be admitted only if both uplink and downlink admission control admit it, otherwise it is rejected because of the excessive interference that it would produce in the network. The limits for admission control are set by the radio network planning.

Several admission control schemes have been suggested in [12]–[16]. In [12]–[14] the use of the total power received by the base station is supported as the primary uplink admission control decision criterion. In [12] and [15] a downlink admission control algorithm based on the total downlink transmission power is presented.

9.5.2   Wideband Power-Based Admission Control Strategy

In the interference-based admission control strategy the new user is admitted by the uplink admission control algorithm if the new resulting total interference level is lower than the threshold value:

$$I_{total\_old} + \Delta I > I_{threshold} \qquad (9.8)$$

The threshold value $I_{threshold}$ is the same as the maximum uplink noise rise and can be set by radio network planning. This noise rise must be included in the link budgets as the interference margin: see Section 8.2.1. Wideband power-based admission control is shown in Figure 9.28. The uplink admission control algorithm estimates the load increase by using either of the two methods presented below. The uplink power increase estimation methods take into account the uplink load curve (see, e.g., [10], [11], [17] and [18]).

Two different uplink power increase estimation methods are shown below. They can be used in the interference-based admission control strategy. The idea is to estimate the increase ΔI of the uplink received wideband interference power I_{total} due to a new user. The admission of the new user and the power increase estimation are handled by the admission control functionality.

The first proposed method (the *derivative* method) is presented in Equation (9.11) and the second (the *integral* method) in Equation (9.12). Both take into account the load curve and are based on the derivative of uplink interference with respect to the uplink load factor

$$\frac{dI_{total}}{d\eta} \qquad (9.9)$$

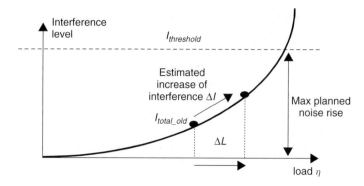

Figure 9.28. Uplink load curve and the estimation of the load increase due to a new user

which can be calculated as follows

$$\text{Noise rise} = \frac{I_{total}}{P_N} = \frac{1}{1-\eta} \Rightarrow$$

$$I_{total} = \frac{P_N}{1-\eta} \Rightarrow \qquad (9.10)$$

$$\frac{dI_{total}}{d\eta} = \frac{P_N}{(1-\eta)^2}$$

The change in uplink interference power can be obtained by Equation (9.11). This equation is based on the assumption that the power increase is the derivative of the old uplink interference power with respect to the uplink load factor, multiplied by the load factor of the new user ΔL:

$$\frac{\Delta I}{\Delta L} \approx \frac{dI_{total}}{d\eta} \Leftrightarrow$$

$$\Delta I \approx \frac{dI_{total}}{d\eta} \Delta L \Leftrightarrow \qquad (9.11)$$

$$\Delta I \approx \frac{P_N}{(1-\eta)^2} \Delta L \Leftrightarrow$$

$$\Delta I \approx \frac{I_{total}}{1-\eta} \Delta L \Leftrightarrow$$

The second uplink power increase estimation method is based on the integration method, in which the derivative of interference with respect to the load factor is integrated from the old value of the load factor ($\eta_{old} = \eta$) to the new value of the load factor ($\eta_{new} = \eta + \Delta L$) as follows:

$$\Delta I = \int_{\eta}^{\eta+\Delta L} dI_{total} \Leftrightarrow$$

$$\Delta I = \int_{\eta}^{\eta+\Delta L} \frac{P_N}{(1-\eta)^2} d\eta \Leftrightarrow$$

$$\Delta I = \frac{P_N}{1 - \eta - \Delta L} - \frac{P_N}{1 - \eta} \Leftrightarrow \tag{9.12}$$

$$\Delta I = \frac{\Delta L}{1 - \eta - \Delta L} \cdot \frac{P_N}{1 - \eta} \Leftrightarrow$$

$$\Delta I = \frac{I_{total}}{1 - \eta - \Delta L} \Delta L$$

In Equations (9.11) and (9.12) the load factor of the new user ΔL is the estimated load factor of the new connection and can be obtained as

$$\Delta L = \frac{1}{1 + \dfrac{W}{\upsilon \cdot E_b/N_0 \cdot R}} \tag{9.13}$$

where W is the chip rate, R is the bit rate of the new user, E_b/N_0 is the assumed E_b/N_0 of the new connection and υ is the assumed voice activity of the new connection.

The downlink admission control strategy is the same as in the uplink, i.e. the user is admitted if the new total downlink transmission power does not exceed the predefined target value:

$$P_{total\_old} + \Delta P_{total} > P_{threshold} \tag{9.14}$$

The threshold value $P_{threshold}$ is set by radio network planning. The load increase ΔP_{total} in the downlink can be estimated based on the initial power. The initial power depends on distance from the base station and is determined by the open loop power control algorithm.

9.5.3 Throughput-Based Admission Control Strategy

In throughput-based admission control strategy the new requesting user is admitted into the radio access network if

$$\eta_{UL} + \Delta L > \eta_{UL\_threshold} \tag{9.15}$$

and the same in downlink:

$$\eta_{DL} + \Delta L > \eta_{DL\_threshold} \tag{9.16}$$

where η_{UL} and η_{DL} are the uplink and downlink load factors before the admittance of the new connection and are estimated as shown in Section 9.4. The load factor of the new user ΔL is calculated as in Equation (9.13).

Finally, we need to note that different admission control strategies can be used in the uplink and in the downlink.

9.6 Load Control (Congestion Control)

One important task of the radio resource management functionality is to ensure that the system is not overloaded and remains stable. If the system is properly planned, and the admission control and packets scheduler work sufficiently well, overload situations should be exceptional. If overload is encountered, however, the load control functionality returns the system quickly and controllably back to the targeted load, which is defined by the radio network planning.

The possible load control actions in order to reduce load are listed below:

— Downlink fast load control : Deny downlink power-up commands received from the mobile
— Uplink fast load control : Reduce the uplink E_b/N_0 target used by the uplink fast power control
— Reduce the throughput of packet data traffic
— Handover to another WCDMA carrier
— Handover to GSM
— Decrease bit rates of real-time users, e.g. AMR speech codec
— Drop calls in a controlled fashion.

The first two in this list are fast actions that are carried out within a base station. These actions can take place within one timeslot, i.e. with 1.5 kHz frequency, and provide fast prioritisation of the different services. The instantaneous frame error rate of the non-delay-sensitive connections can be allowed to increase in order to maintain the quality of those services that cannot tolerate retransmission. These actions only cause increased delay of packet data services while the quality of the conversational services, such as speech and video telephony, is maintained.

The other load control actions are typically slower. Packet traffic is reduced by the packet scheduler: see Chapter 10.

One example of a real-time connection whose bit rate can be decreased by the radio access network is Adaptive Multirate (AMR) speech codec: for further information see Section 2.3. Inter-frequency and inter-system handovers can also be used as load balancing and load control algorithms. The final load control action is to drop real-time users (i.e. speech or circuit switched data users) in order to reduce the load on the system. This action is taken only if the load on the system remains very high even after other load control actions have been effected in order to reduce the overload. The third generation WCDMA air interface and the expected increase of non-real-time traffic in third generation networks give a large selection of possible actions to handle overload situations, and therefore the need to drop real-time users to reduce overload should be very rare.

References

[1] Sipilä, K., Laiho-Steffens, J., Wacker, A. and Jäsberg, M., 'Modelling the Impact of the Fast Power Control on the WCDMA Uplink', *Proceedings of VTC'99 Spring*, Houston, TX, 16–19 May 1999, pp. 1266–1270.
[2] Sampath, A., Kumar, P. and Holtzman, J., 'On Setting Reverse Link Target SIR in a CDMA System', *Proceedings of VTC'97*, Arizona, 4–7 May 1997, Vol. 2, pp. 929–933.
[3] TIA/EIA/IS-95-A, 'Mobile Station-Base Station Compatibility Standard for Dual-Mode Wideband Spread Spectrum Cellular System', Telecommunications Industry Association, Washington, DC, May 1995.
[4] Laiho-Steffens, J., Jäsberg, M., Sipilä, K., Wacker, A. and Kangas, A., 'Comparison of Three Diversity Handover Algorithms by Using Measured Propagation Data', *Proceedings of VTC'99 Spring*, Houston, TX, 16–19 May 1999, pp. 1370–1374.

[5]   Qualcomm Corporation, 'Diversity-Handover Method and Performance', *ETSI SMG2 Wideband CDMA Concept Group Alpha Meeting*, Stockholm, Sweden, September 1997.

[6]   3rd Generation Partnership Project, Technical Specification Group RAN, Working Group 2 (WG2), 'Radio Resource Management Strategies', 3G TR 25.922, Ver. 0.5.0, September 1999.

[7]   Hiltunen, K., Binucci, N. and Bergström, J., 'Comparison Between the Periodic and Event-Triggered Intra-Frequency Handover Measurement Reporting in WCDMA', *Proceedings of IEEE WCNC 2000*, Chicago, 23–28 September 2000, pp. xxx-xxx.

[8]   3G TS 25.133 Requirements for Support of Radio Resource Management (FDD).

[9]   Salonaho, O. and Laakso, J., 'Flexible Power Allocation for Physical Control Channel in Wideband CDMA', *Proceedings of VTC'99 Spring*, Houston, TX, 16–19 May 1999, pp. 1455–1458.

[10]   Shapira, J. and Padovani, R., 'Spatial Topology and Dynamics in CDMA Cellular Radio', *Proceedings of 42nd IEEE VTS Conference*, Denver, CO, May 1992, pp. 213–216.

[11]   Shapira, J., 'Microcell Engineering in CDMA Cellular Networks', *IEEE Transactions on Vehicular Technology*, Vol. 43, No. 4, November 1994, pp. 817–825.

[12]   Dahlman, E., Knutsson, J., Ovesjö, F., Persson, M. and Roobol, C., 'WCDMA—The Radio Interface for Future Mobile Multimedia Communications', *IEEE Transactions on Vehicular Technology*, Vol. 47, No. 4, November 1998, pp. 1105–1118.

[13]   Huang, C. and Yates, R., 'Call Admission in Power Controlled CDMA Systems', *Proceedings of VTC'96*, Atlanta, GA, May 1996, pp. 1665–1669.

[14]   Knutsson, J., Butovitsch, T., Persson, M. and Yates, R., Evaluation of Admission Control Algorithms for CDMA System in a Manhattan Environment', *Proceedings of 2nd CDMA International Conference*, CIC '97, Seoul, South Korea, October 1997, pp. 414–418.

[15]   Knutsson, J., Butovitsch, P., Persson, M. and Yates, R., 'Downlink Admission Control Strategies for CDMA Systems in a Manhattan Environment', *Proceedings of VTC'98*, Ottawa, Canada, May 1998, pp. 1453–1457.

[16]   Liu, Z. and Zarki, M. "SIR Based Call Admission Control for DS-CDMA Cellular System", *IEEE Journal on Selected Areas in Communications*, Vol. 12, 1994, pp. 638–644.

[17]   Holma, H. and Laakso, J., 'Uplink Admission Control and Soft Capacity with MUD in CDMA', *Proceedings of VTC'99 Fall*, Amsterdam, Netherlands, 19–22 September 1999, pp. 431–435.

[18]   Ojanperä, T. and Prasad, R., *Wideband CDMA for Third Generation Mobile Communications*, Artech House, 1998.

10

Packet Access

Mika Raitola and Harri Holma

This chapter presents packet access with WCDMA. It is organised as follows. Packet traffic characteristics are discussed in Section 10.1. An overview of WCDMA packet access is given in Section 10.2 and the transport channels for packet data are compared in Section 10.3. Example packet scheduling algorithms are presented in Section 10.4. The interaction between packet scheduling and other RRM algorithms is discussed in Section 10.5, and packet data performance simulation results are presented in Section 10.6.

10.1   Packet Data Traffic

The quality of service classes that are mainly considered in this chapter are interactive and background services: see Section 2.3. The conversational and streaming classes are assumed to be transmitted as real-time connections over the air interface. It is still possible to have all-IP for all service classes in the core network.

As an example of packet data traffic, an ETSI packet data model [1] is described below. A packet service session contains one or several packet calls depending on the application. During a packet call several packets may be generated, so that the packet call constitutes a bursty sequence of packets. The burstiness during the packet call is a characteristic feature of packet transmission. For example, in a web-browsing session a packet call corresponds to the downloading of a document. After the document is entirely received by the terminal, the user takes a certain amount of time to study the information. This time interval is called the reading time. It is also possible that the session contains only one packet call. The typical behaviour of the packet data traffic is illustrated Figure 10.1.

The following parameters define the characteristics of the packet data traffic [1]:

— session arrival process
— number of packet calls per session
— reading time between packet calls
— number of packets within a packet call

WCDMA for UMTS, edited by Harri Holma and Antti Toskala
© 2001 John Wiley & Sons, Ltd

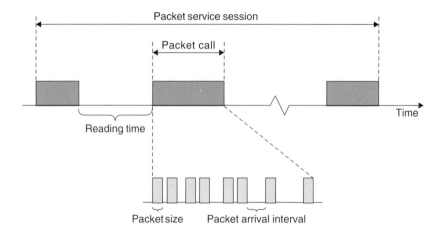

Figure 10.1. Characteristic of a packet service session

— time interval between two packets inside a packet call
— packet size.

The arrival of session setups to the network can be modelled as a Poisson process. Reading time starts when the last packet of the packet call is completely received by the user and ends when the user makes a request for the next packet call.

The model presented here is only one example made for web-browsing. The traffic model should be able to catch the various characteristic features that are possible in future traffic. For this reason different statistical distributions can be used.

The properties that are typical for non-real-time packet services from the air interface point of view are listed below:

— Packet data is bursty. The required bit rate can change rapidly from zero to hundreds of kilobits per second.
— Packet data tolerates longer delay than real-time services. Therefore, packet data is controllable traffic from the radio access network point of view. In interactive services the user must get a request inside a reasonable time, but in background-type services data can be transmitted when there is free radio interface capacity.
— Packets can be retransmitted by the radio link control (RLC) layer. This allows the use of worse radio link quality and much higher frame error ratio than in the case of real-time services.

It is also possible to transmit real-time services—conversational and streaming quality of service classes—over packet networks. An example is transmission of voice over IP. The transmission of real-time services is not considered in this chapter.

10.2 Overview of WCDMA Packet Access

In this section the functions of WCDMA radio packet access are briefly introduced. Packet allocations in WCDMA are controlled by the packet scheduler (PS). Its functions are to:

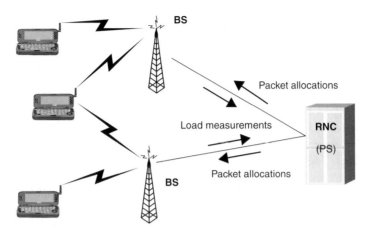

Figure 10.2. Packet access in WCDMA network

— divide the available air interface capacity between the packet data users
— decide the transport channel to be used for each user's packet data transmission
— monitor the packet allocations and the system load.

WCDMA packet access allows non-real-time bearers to use common, dedicated or shared channels dynamically. The usage of different channels is controlled by the packet scheduler (PS). The packet scheduler allocates a bit rate for a bearer and possibly changes this bit rate during an active connection.

The packet scheduler is typically located in RNC where the scheduling can be done efficiently for multiple cells, also taking into account the soft handover connections. The base station provides the measurements of the air interface load for the packet scheduler: see Figure 10.2. If the load exceeds the target, the packet scheduler can decrease the load by decreasing the bit rates of packet bearers; if the load is less than the target, it can increase the load by allocating more data. The packet scheduler is also a part of the network load control because it can increase or reduce the network load.

10.3   Transport Channels for Packet Data

In WCDMA there are three types of transport channels that can be used to transmit packet data: common, dedicated and shared transport channels. These channels are described in Chapter 6 and in this section their properties and feasibility for packet data are discussed. In downlink the transport channel for packet data is selected by the packet scheduling algorithm. In uplink the transport channel is selected by the mobile based on the parameters set by the packet scheduler.

10.3.1   Common Channels

Common channels are the random access channel (RACH) in the uplink, and the forward access channel (FACH) in the downlink. Both can carry signalling data but also user data in WCDMA. The advantage of common channels is their low setup time. Because they are also used for signalling before a connection is set up, they can be used to send packets immediately without any long setup time. There are typically only one or a few RACH or FACH channels per sector.

Common channels do not have a feedback channel, and therefore cannot use fast closed loop power control but only open loop power control or fixed power. Nor can these channels use soft handover. Therefore, the link-level performance of the common channels is worse than that of the dedicated channels, and more interference is generated than with dedicated channels. The gain of fast power control is analysed in Section 9.2 and the gain of soft handover in Section 9.3.

Now, it is easy to see that common channels are most suitable for transmitting small individual packets. Ideal applications to be used in common channels would be short message services and short text-only emails. Also, sending of a single request of a web page could fit well into common channels. In the case of larger data amounts, common channels suffer from poor radio performance.

The GPRS air interface does not support the use of common channels for transmission of data. The possibility of using common channels in WCDMA gives additional flexibility for packet data transmission compared to GPRS.

10.3.2 Dedicated Channels

The dedicated channels have the advantage that they can use fast power control and soft handover. These features improve their radio performance, and consequently less interference is generated than with common channels. On the other hand, setting up a dedicated channel takes more time than accessing common channels. Dedicated channels can have bit rates from a few kbps up to 2 Mbps. The bit rate can be changed during transmission. If the bit rate is changed within the transport format set, the downlink orthogonal code must be allocated according to the highest bit rate. Therefore, the variable bit rate dedicated channels consume downlink orthogonal codes.

10.3.3 Shared Channels

Shared channels are targeted to transfer bursty packet data. The idea is to share a single physical channel, i.e. orthogonal code, between many users in a time division manner. This saves the limited number of downlink orthogonal codes because several users share the code. If the dedicated channel were used instead, the orthogonal code would be reserved according to the maximum bit rate, and the efficiency of code usage would be low. Shared channels can be used in parallel with a lower bit rate dedicated channel. The dedicated channels carry the physical control channel, including the signalling for fast power control. It should be noted also that shared channels cannot use soft handover.

10.3.4 Common Packet Channel

The common packet channel (CPCH) is similar to the common and shared channels. It is used in the uplink and is accessed similarly to the RACH. Many users share this channel in a time division manner; in this sense it is similar to the uplink shared channel. The bit rate of the CPCH can be high or low and there can be many CPCHs per cell, each having a different bit rate. The CPCH can have fast power control after the access procedure, but it cannot use soft handover. The CPCH is ideal for small and medium-sized packets with bursty data.

Table 10.1. WCDMA channel types and their properties for packet data

	Dedicated channels	Common channels			Shared channels	
	DCH	FACH	RACH	CPCH	DSCH	USCH
Uplink/ Downlink	Both	Downlink	Uplink	Uplink	Downlink	Uplink, only in TDD
Code usage	According to maximum bit rate	Fixed codes per cell	Fixed codes per cell	Fixed codes per cell	Code shared between users	Code shared between users
Fast power control	Yes	No	No	Yes	Yes	No
Soft handover	Yes	No	No	No	No	No
Suited for	Medium or large data amounts	Small data amounts	Small data amounts	Small or medium data amounts	Medium or large data amounts	Medium or large data amounts
Suited for bursty data	No	Yes	Yes	Yes	Yes	Yes

10.3.5 Selection of Channel Type

The transport channels for packet data are summarised in Table 10.1. The packet scheduler in RNC selects the channels to be used for data transfer. This selection is based on:

— Service type or bearer parameters, for example delay requirements
— Data amount
— Load of the common channels and shared channels
— Interference levels in the air interface
— Radio performance of different transport channels.

10.4 Example Packet Scheduling Algorithms

10.4.1 Introduction

The packet scheduling function shares the available air interface capacity between packet users. The packet scheduler can decide the allocated bit rates and the length of the allocation. In WCDMA this can be done in two ways, in a code or time division manner (see Figure 10.3). In the code division approach a large number of users can have a low bit rate channel available simultaneously. When the number of users wanting capacity increases, the bit rate, which can be allocated for a single user, decreases. In time division scheduling the capacity is given to one user or only a few users at each moment of time. Thus, a user can have a very high bit rate but can use it only very briefly. In WCDMA the highest time resolution is a 10-ms frame. When the number of users increases in the time division approach, each user has to wait longer for transmission.

In Sections 10.4.2 and 10.4.3 the advantages and disadvantages of code division and time division scheduling are discussed. These two approaches are extreme examples. In practice,

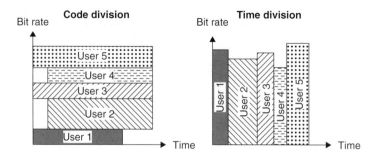

Figure 10.3. Code and time division scheduling principles

packet scheduling is a combination of time and code division approaches. Section 10.4.4 shows how to use the information about the required transmission power to improve the throughput of the packet scheduling.

10.4.2 Time Division Scheduling

When the packet scheduler allocates the packet data bit rates, radio performance should be taken into account. High bit rates typically require less energy per transmitted bit: see Section 11.2.1.1. Therefore, time division scheduling has the advantage of lower E_b/N_0 values compared to code division scheduling. Simulation results for packet data with various bit rates in the uplink are shown in Figure 10.4. The difference between 8 kbps and 256 kbps is about 2 dB, i.e. increasing the bit rate from 8 kbps to 256 kbps gives an increased throughput of $10^{(2\text{dB}/10)} = 58\%$ in the air interface capacity.

Since the bit rate is higher in the time division approach, the average delay is shorter than with the code division approach. The shorter delay is another reason to use time division scheduling, in addition to the lower E_b/N_0.

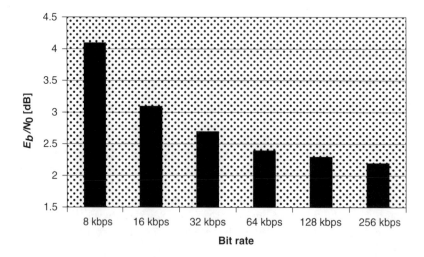

Figure 10.4. Packet data link-level simulation results with different bit rates in the uplink. Mobile speed 3 km/h in ITU Pedestrian A with FER target 10%

One disadvantage of the time division approach is the short transmission time. The establishment and release of a connection takes time, even as much as several frames, depending on the implementation. During this time the physical resources of the base station, such as channel units, are tied to a connection that is not in active use. Similarly, spreading codes are reserved at the same time. Also, the signalling links over radio and Iub interfaces must be set up. Thus, in time division scheduling the percentage of time when physical resources are not used is higher than in code division scheduling.

Use of time division allocation is constrained by the limited uplink range of high bit rates due to the mobile station's limited transmission power: see the discussion of uplink coverage in Section 11.2.1.1. In the downlink no similar range limitations exist: see the discussion of downlink coverage in Section 11.2.3. Also, the maximum supported bit rates of the mobile stations must be taken into account in the bit rate allocations.

Finally, we note that time division scheduling uses high bit rates and generates bursty traffic, which leads to higher variations in the interference levels than in code division scheduling.

Time division is used typically with shared channels but can also be used with dedicated channels. The bit rate allocations with shared channels are shown in Figure 10.5. A low bit rate dedicated channel is typically used together with the downlink shared channel.

10.4.3 Code Division Scheduling

In code division scheduling all users are allocated a channel when they need it. The bit rate is much lower than in time division scheduling if there are several packet users requesting capacity. Due to the lower bit rate, the transmission delay caused to each user is longer in code division scheduling than in time division scheduling.

With code division scheduling, establishment and release delays cause smaller losses in capacity due to the lower bit rate and longer time of transmission. Due to the lower bit rates, allocation of resources takes longer in code division scheduling than in time division scheduling. This makes the air interface interference levels more predictable and can be seen as an advantage for code division scheduling.

The code division scheduling can be static or dynamic. In static scheduling the allocated bit rate is kept fixed throughout the connection. This approach resembles circuit-switched

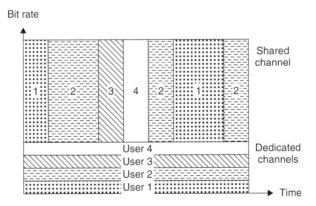

Figure 10.5. Principle of shared channel scheduling

connections and requires accurate estimation of the allocated bit rate. Unfortunately, packet traffic is often bursty and unpredictable. A connection may contain only one short document or many consecutive documents. To be able to estimate accurately the bit rate needed, large buffers are required, or the traffic has to be fairly constant. In many cases this is not practical or not true, and some way to modify the already allocated bit rate is needed, i.e. code division should be dynamic. It should be possible to increase or decrease the allocated bit rate. The change of bit rate can be based on the amount of data in the buffers.

Since the code division approach uses lower bit rates, it sets lower requirements for the mobile's capabilities. Also, the lower bit rates lead to a longer uplink range—see Section 11.2.1.1 for the uplink range of different bit rates.

Table 10.2 summarises the differences between time and code division scheduling.

10.4.4 Transmission Power-Based Scheduling

The allocated packet data bit rate could be based on the required transmission power of the connection: a higher bit rate for a user requiring less transmission power per transmitted bit. This approach would minimise the average required transmission power per bit, and the generated interference to the network, and increase the average cell throughput compared to equal bit rate scheduling. In practice, the users close to the base station would get a higher bit rate than those at the cell edge, and user throughput would depend on location if transmission power-based scheduling were used. In GPRS the link adaptation typically gives higher throughput close to the base station than at the cell edge, i.e. the GPRS bit rate depends on location. All the packet data simulation results in this section assume equal bit rate scheduling.

To enable transmission power-based scheduling, the packet scheduler must know or estimate the transmission power of each user when it is allocating the bit rates. In both uplink and downlink the received pilot E_c/I_0 measurements can be used to estimate the transmission power.

Transmission power-based scheduling gives more gain in the average throughput in downlink than in uplink compared to equal bit rate scheduling. In the uplink, typically at least 50% of the interference originates from the other users within the same cell, and that interference

Table 10.2. Comparison of time and code division scheduling strategies

	Time division	Code division
Number of simultaneous packet transmissions per cell in the air interface	Small (a few)	Large ($\sim$20–50)
Instantaneous bit rate per packet user	High ($>$100 kbps)	Low ($<$50 kbps)
Advantages	• Shorter total delay	• Longer range in uplink due to lower bit rate
	• Better E_b/N_0	• Fewer requirements for mobile capabilities
		• Less interference variation because more users

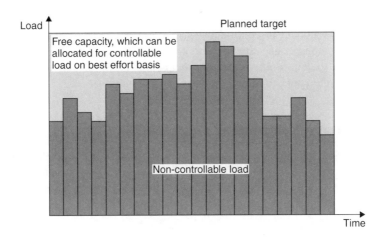

Figure 10.6. Capacity division between non-controllable and controllable traffic

does not depend on the transmission power but only on the received powers: see Figure 9.6. In the downlink the transmission power determines the air interface capacity directly, and transmission power-based scheduling can clearly increase the average downlink throughput.

10.5   Interaction between Packet Scheduler and Other RRM Algorithms

The operation of the packet scheduler is closely connected to other RRM functions. These interconnections are described below.

10.5.1   Packet Scheduler and Handover Control

If the mobile is in soft handover, the packet scheduler must take into account the air interface load and the physical resources in all base stations of the active set. The dedicated channels are the only transport channels that can use soft handover.

10.5.2   Packet Scheduler and Load Control (Congestion Control)

The packet scheduler and load control are closely connected, because the packet scheduler itself is an important part of load control. Since the packet scheduler does not guarantee the delays of the non-real-time connections, the load of non-real-time packet traffic can be controlled. If the load of real-time users gets too high, the packet scheduler can decrease the load of the controllable non-real-time users, as shown in Figure 10.6. Therefore, the non-real-time users and advanced packet scheduling algorithms together help to keep the system load at the desired level. Other load control actions besides the packet scheduler are introduced in Section 9.6.

10.5.3   Packet Scheduler and Admission Control

The admission control needs to estimate the load caused by the non-controllable real-time connections, because the controllable load from the non-real-time connections can be decreased if needed. If, for example, a video connection is requested, the admission control estimates the amount of controllable packet traffic that can be reduced, and determines whether the video connection can be admitted by reducing the packet data. The admission control also defines the connection setup parameters, including available bit rates that can be used in the connection.

10.6   Packet Data Performance

In this section packet scheduling performance is analysed. First, link-level performance of packet data is discussed in Section 10.6.1. In Section 10.6.2, the system-level simulator modelling principles are described, and the results of the dynamic system-level simulations with real packet scheduler and other RRM algorithms are presented.

10.6.1   Link-Level Performance

The effect of the frame error rate (FER) and retransmissions on throughput is studied at the link level and an optimal FER target level is proposed in Section 10.6.1.1. The effect of the interleaving length on packet data throughput is presented in Section 10.6.1.2.

10.6.1.1   Frame Error Rate Target

Packet data performance is studied in the ITU Pedestrian A multipath channel using mobile speeds of 3, 20 and 120 km/h in the uplink. The FER as a function of E_b/N_0 is shown in Figure 10.7. The low mobile speeds give the best performance because the fast power control is able to compensate for the fading channel. The required E_b/N_0 values are different in than in Table 11.12 because different bit rates and base station receiver algorithms have been assumed in these cases.

The higher the FER, the more retransmissions are needed to deliver error-free data. On the other hand, less power, or lower E_b/N_0, is needed for higher FER levels. What is the optimal FER operation point that requires the lowest energy per correctly received bit when the retransmissions are taken into account? To find out the optimal FER point, we use the

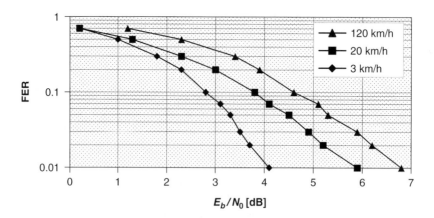

Figure 10.7. FER as a function of E_b/N_0 for 32 kbps packet service in the uplink

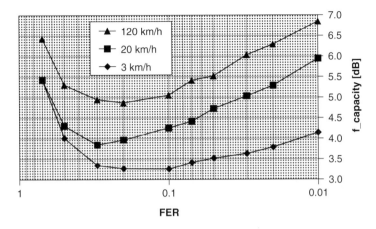

Figure 10.8. FER vs. capacity function for 32 kbps packet service in the uplink

definition of the cell capacity or throughput from [3]:

$$\text{Throughput}_{\text{cell}} = k \frac{1 - FER}{E_b/N_0} \tag{10.1}$$

where k is a constant that depends on the bandwidth and the propagation model but not on the link-level FER and E_b/N_0. To optimise the throughput we need only to optimise the latter part of Equation (10.1). The latter part is the capacity function. In this study the inverse of the capacity function is used as follows:

$$f_{\text{capacity}} = \frac{E_b/N_0}{1 - FER} \tag{10.2}$$

By minimising f_{capacity} the capacity of a cell is maximised. This capacity function can be seen as an effective E_b/N_0 when the retransmissions are taken into account. The FER vs. capacity function f_{capacity} is shown in Figure 10.8. The relationship between E_b/N_0 and FER is taken from Figure 10.7. The optimal FER operation point is between 10% and 30% depending on the mobile speed. If FER is lower, capacity is wasted because the retransmissions are not efficiently utilised to gain from the additional time diversity. If FER is higher, there are too many retransmissions, causing additional interference. With higher FER also the average delay will be longer due to retransmissions and the quality of the signalling will be reduced. A higher FER also consumes more downlink orthogonal codes because the code must be reserved for a longer time for the retransmissions.

10.6.1.2 Interleaving Length

A simulated example of E_b/N_0 performance for packet data as a function of interleaving length (transmission time interval) is shown in Figure 10.9. At 3 km/h the E_b/N_0 can be improved by 0.8 dB by increasing the interleaving from 10 ms to 80 ms. The difference of 0.8 dB corresponds to an increase in the interference-limited capacity of 20% ($10^{0.8 \text{ dB}/10} = 1.20$). At 20 km/h the difference between 10 ms and 80 ms interleaving is 1.35 dB, corresponding to a difference of 36% in the capacity.

The drawback of longer interleaving is a longer delay in packet transmission. For high quality of service a minimum delay is desired; this can be achieved with short interleaving.

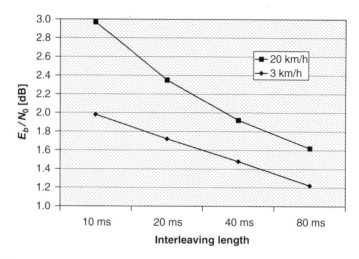

Figure 10.9. Required E_b/N_0 as a function of interleaving depths for FER = 10% packet data

The other drawback of longer interleaving is that the granularity is higher with shorter interleaving. For a 32 kbps packet connection, for example, the number of bits per 10 ms interleaving period is 320 (= 32 kbps × 10 ms) and the number per 80 ms interleaving period is 2560, and no smaller number of bits can be transmitted with 32 kbps without using padding. The selection of the packet data interleaving length is a trade-off between maximum throughput, delay and flexibility. Also, the mobile capabilities to support long interleaving must be taken into account.

10.6.2 System-Level Performance

This section introduces a dynamic system-level simulator, and the system-level performance of WCDMA packet scheduling is studied using this simulator. A more detailed description of the system simulator can be found in [4].

10.6.2.1 Difference between System-Level and Link-Level Simulators

Typically, radio network simulations can be classified as either link-level or system-level simulations. A single simulator approach would be preferable, but the complexity of such a simulator—including everything from transmitted waveforms to multi-cell network—is far too high for the required simulation resolutions and simulation times. For accurate receiver performance evaluation, a chip-level or symbol-level simulation model is needed, typically with 3.84 Mcps time resolution. On the other hand, at system level the traffic models and the mobility models require simulations of at least 10–20 minutes with a large number of mobiles and base stations. Therefore, separate link-level and system-level simulators are needed. The link-level simulators usually operate at symbol or chip frequency, while the system-level simulators typically operate with the fast power control frequency. In WCDMA the fast closed loop power control is operating at 1.5 kHz frequency and in this system simulator the frequency of 1.5 kHz is used.

The link-level simulator is needed for the system simulator to build a receiver model that can predict the receiver FER/BER performance, taking into account channel estimation, interleaving and decoding. The system-level simulator is needed to model a system with a large number of mobiles and base stations, and algorithms operating in such a system.

Because the simulation is divided into two parts, a method of interconnecting the two simulators has to be defined. Conventionally, the information obtained from the link-level tool is linked to the system simulation by using a so-called average value interface that describes the BER/FER performance by average E_b/N_0 requirements. The average value interface is not accurate if there are rapid changes in interference due to, for example, high bit rate packet users. This kind of approach is well suited to static snapshot simulations but cannot be used when simulating systems with fast power control and high bit rate packet data. With the simulator presented here, a so-called actual value interface (AVI) is used that provides accurate modelling of receiver performance with fast power control and high bit rate packet data [5]. The division of work between link-level and system-level simulators is summarised in Table 10.3.

10.6.2.2 Traffic Modelling

In the simulator the users are making calls and transmitting data according to the traffic models. Call generation for real-time services, such as speech and video, is implemented according to a Poisson process [6]. For speech, voice activity and discontinuous transmission have to be considered. For circuit-switched data services, the traffic model is a constant bit rate model, with 100% of activity. The traffic model of Figure 10.1 is used for packet data in this simulator.

10.6.2.3 Mobility Modelling

In this dynamic simulator the users are moving in the simulation area according to the mobility model. In [6] a separate mobility model is developed for micro cellular and macro cellular environments. When new users are generated in the macro cell simulation they are uniformly distributed over the simulation area. The direction in which a new user is moving is randomly selected when a new user is created. The direction of movement is updated for a user after every decorrelation length.

Table 10.3. Link-level and system-level simulators (AVI = Actual Value Interface)

	Link level	System level
Time resolution	1 sample/chip or 1 sample/symbol	1 sample/slot (1.5 kHz)
Number of mobiles	1	>100
Number of base stations	1–3	>10
Fast fading	Yes	Yes
Receiver channel estimation	Yes	Via AVI
Interleaving, channel coding	Yes	Via AVI
Fast power control	Yes	Yes
Packet retransmission	—	Yes
Path loss	—	Yes
Slow fading	—	Yes
Interference	Gaussian noise	Real transmitters
RRM algorithms	—	Yes
Mobility model	—	Yes
Traffic model	—	Yes
Simulated time span	1–5 min	20–60 min

10.6.2.4 System-Level Simulation Results

In this section the performance of the packet scheduler is studied using the system simulator described above. The target of these simulations is to find out how accurately the packet scheduler can utilise the air interface capacity. It is also important that the variations caused by the packet data are reasonably small so that the quality of the real-time services is not sacrificed. In these simulations a large number of users are placed into an 18-cell network with three-sector base stations. The possible bit rates are between 32 and 1024 kbps and the performance of the packet scheduling algorithms is observed. We want to observe how well the combination of RRM algorithms can keep the system load at the target level and what kind of bit rates users can have.

The environment in these simulations is macro cellular with 18 cells and a base station distance of 2 km. Mobile speed is 3 km/h and antenna diversity is used in uplink. The mobile maximum power is 125 mW (= 21 dBm). The noise floor, including the receiver noise figure, is -102.9 dBm in the uplink and -99.9 dBm in the downlink. The packet scheduler in these simulations is code division based. The FER target of the outer loop power control is 20%.

In the simulations the distribution of the total transmitted power and the total received interference of one base station in the middle of the base station grid are recorded. The average total transmitted power in the downlink (Figure 10.10) is 8.5 W, and the average noise rise over the thermal noise level in the uplink (Figure 10.11) is 5.7 dB. Only in a few cases is a transmission power of 12 W in the downlink or a noise rise of 7 dB in the uplink exceeded. Example time variations of the transmitted and received powers are

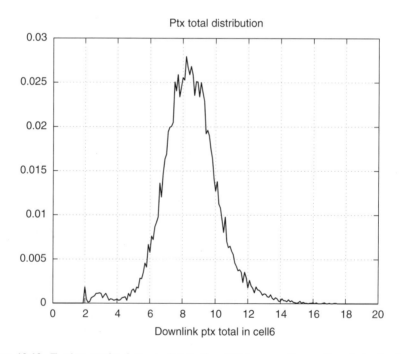

Figure 10.10. Total transmitted power distribution (W) in one cell over the whole simulation

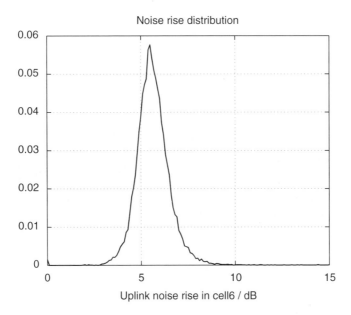

Figure 10.11. Distribution of the total received interference (dB) over thermal noise level (noise rise) over the whole simulation

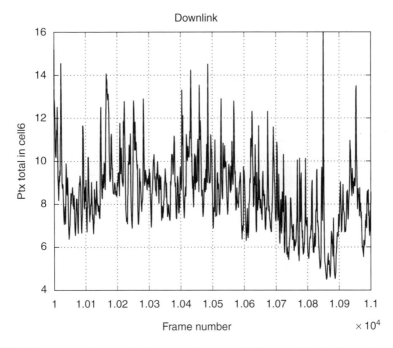

Figure 10.12. Example total transmitted powers in one cell. The *x*-axis is in 10-ms frames, i.e. the whole axis spans 10 seconds

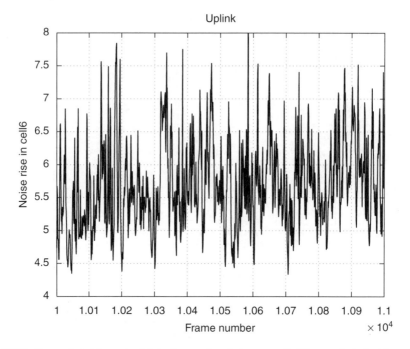

Figure 10.13. Example total received interference levels in one cell. The *x*-axis is in 10-ms frames, i.e. the whole axis spans 10 seconds

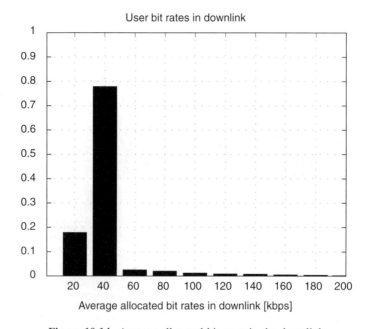

Figure 10.14. Average allocated bit rates in the downlink

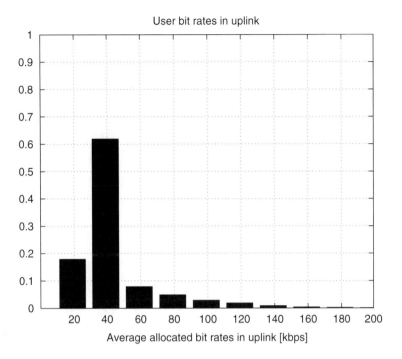

Figure 10.15. Average allocated bit rates in the uplink

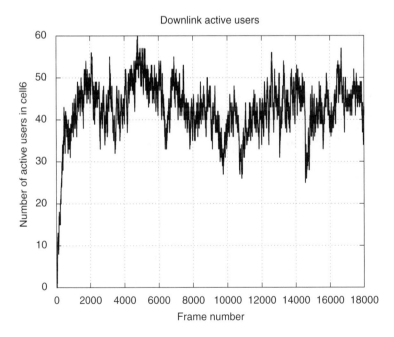

Figure 10.16. Average number of active connections (including soft handovers) in the downlink

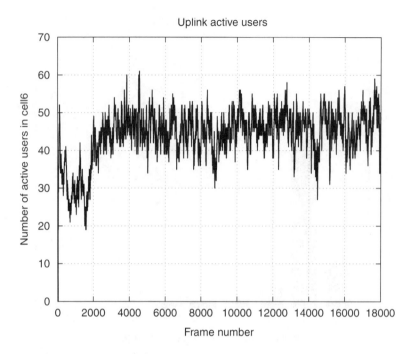

Figure 10.17. Average number of active connections (including soft handovers) in the uplink

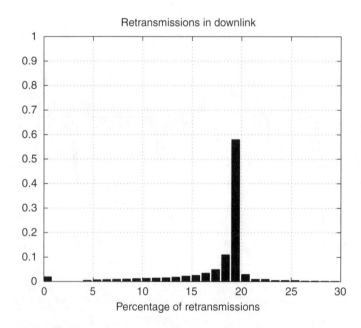

Figure 10.18. Average percentage of retransmissions in the downlink

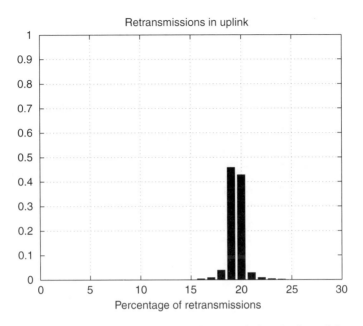

Figure 10.19. Average percentage of retransmissions in the uplink

shown in Figures 10.12 and 10.13. The variations in the power levels are reasonably low and the non-real-time packet data would not cause any degradation of quality of the real-time services. The air interface capacity would be fully utilised in these examples.

In these simulations the number of users was very high, almost 1400 per cell. A high number of users was used to ensure that there is always enough data in the buffers to test the performance of the packet scheduler under high load. Due to the high load, the resulting bit rates are low, as seen in Figures 10.14 and 10.15. However, the number of active connections per cell is fairly constant—about 50, including soft handover connections, except at the beginning of the simulation, where the offered load was lower. The soft handover overhead was 32% in these simulations. The number of active connections in the downlink is shown in Figure 10.16 and in the uplink in Figure 10.17.

The link quality can be studied by observing the number of retransmissions, which are shown in Figure 10.18 for the downlink and in Figure 10.19 for the uplink. The outer loop power control can keep the FER and the percentage of retransmissions close to the target value of 20%. The distribution of the downlink retransmissions shows a few cases where the proportion of retransmissions was less than 20%, even 0%. The reason is the limited downlink power control dynamics of 20 dB in these simulations. If the connection is using its minimum power and cannot power down any more, the quality will be better than the FER target value of 20%.

The dynamic simulations with code division packet allocation have shown that WCDMA is well suited for the transmission of packet data. The air interface capacity can be fully utilised, yet the variations in the interference levels remain low. The user bit rates here were low, but on the other hand the number of simultaneous users was very high.

References

[1] ETSI, Technical Report UMTS 30.06, UMTS Terrestrial Radio Access (UTRA); Concept Evaluation, Version 3.0.0, December 1997.

[2] Ghosh, A., Cudak, M. and Felix, K., 'Shared Channels for Packet Data Transmission in W-CDMA', *Proceedings of VTC'99 Fall*, Amsterdam, Netherlands, 19–22 September 1999, pp. 943–947.

[3] Christer, B. and Johansson, V., 'Packet Data Capacity in Wideband CDMA System', *Proceedings of VTC'98*, Ottawa, Canada, 18–21 May 1998, pp. 1878–1883.

[4] Hämäläinen, S., Holma, H. and Sipilä, K., 'Advanced WCDMA Radio Network Simulator', *Proceedings of PIMRC'99*, Osaka, Japan, September 1999, pp. 951–955.

[5] Hämäläinen, S., Slanina, P., Hartman, M., Lappeteläinen, A., Holma, H. and Salonaho, O., 'A Novel Interface between Link and System Level Simulations', *Proceedings of ACTS Summit 1997*, Aalborg, Denmark, October 1997, pp. 509–604.

[6] 'Universal Mobile Telecommunications System (UMTS); Selection procedures for the choice of radio transmission technologies of the UMTS', TR 101 112 V3.1.0 (1997–11), UMTS 30.03.

11

Physical Layer Performance

Harri Holma, Markku Juntti and Juha Ylitalo

11.1 Introduction

This chapter presents the effect of the propagation environment, base station solutions and WCDMA physical layer parameters on the coverage and capacity. The base station solutions include both baseband and antenna techniques. Coverage is important, especially in the initial stages of network deployment; WCDMA network coverage is analysed in Section 11.2. The importance of capacity will increase after the initial deployment when the amount of traffic increases; WCDMA capacity is presented in Section 11.3. In this chapter, we present the WCDMA air interface capacity which is limited by the interference. We assume that there are enough baseband hardware resources in the base station, transmission network and radio network controller to support the capacity. In Section 11.4 special attention is drawn to the performance of the high bit rate services up to 2 Mbps. Finally, Section 11.5 presents the possible performance enhancements that are supported by the 3GPP standard, including adaptive antenna structures and multi-user detection with advanced baseband processing.

The radio network planning and the optimisation of the radio resource management algorithms also affect coverage and capacity. Their effect is presented in Chapters 8 and 9.

11.2 Coverage

Coverage is important when the network is not limited by capacity, such as at the time of initial network deployment, and typically in rural areas. Macro cell coverage is determined by the uplink range, because the transmission power of the mobile is much lower than that of the macro base station. The output power of the mobile is typically 21 dBm (125 mW) and that of the macro cell base station 40–46 dBm (10–40 W) per sector. Therefore, in this section the uplink coverage is considered. Also, in Section 8.2.2 the macro cell coverage is shown to be uplink limited.

WCDMA for UMTS, edited by Harri Holma and Antti Toskala
© 2001 John Wiley & Sons, Ltd

The effect of the improvements in the link budget, ΔL, on the relative cell radius, ΔR, can be calculated assuming a propagation model, for example the Okumura–Hata model from Section 8.2. In this example the path loss exponent is 3.52 which results in

$$\Delta L = 35.2 \log_{10} \left(1 - \frac{\Delta R}{R} \right) \tag{11.1}$$

The ratio of the new cell area A_Δ to the original cell area A can be calculated as

$$\frac{A_\Delta}{A} = \left(1 - \frac{\Delta R}{R} \right)^2 = \left(10^{\frac{\Delta L}{35.2}} \right)^2 \tag{11.2}$$

The required relative base station site density with a given improvement in the link performance is calculated in Table 11.1. The number of base station sites is inversely proportional to the cell area. For example, with a link performance improvement of 5.3 dB, the base station density can be reduced by about 50%.

Typically, the radio access network represents 70% of the total UMTS network investment, and most part of the radio access network costs are base station site related costs. Therefore, a reduction of the number of sites is important in reducing the required investment to the UMTS network.

The factors affecting the maximum path loss can be seen from the link budget—see Section 8.2—and are shown in Figure 11.1. The effect of the base station solutions and the bit rate is described in this chapter. The relationship between uplink loading and coverage has been discussed in Section 8.2.2 and the power control headroom in Section 9.2.1.

11.2.1   Uplink Coverage

In this section we evaluate the effect of the physical layer parameters and the base station solutions on the WCDMA uplink coverage.

11.2.1.1   Bit Rate

The coverage of different bit rates is affected by the following two factors:

(1)   For higher bit rates the processing gain is lower, and the coverage is smaller: see the link budgets in the tables in Section 8.2, row l.

Table 11.1. Reduction in the base station site density with an improved link budget

Improvement in the link budget ΔL	Relative number of sites $1/(A_\Delta/A)$
0.0 dB = Reference case	100%
1.0 dB	88%
2.0 dB	77%
3.0 dB	68%
4.0 dB	59%
5.0 dB	52%
6.0 dB	46%
10.0 dB	27%

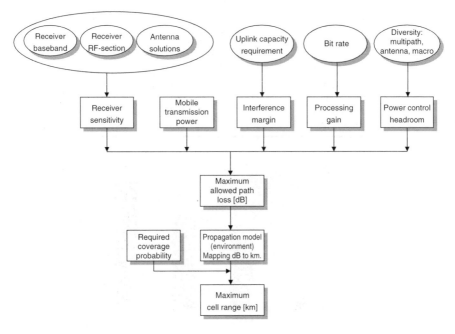

Figure 11.1. Factors affecting the uplink coverage

(2)   For higher bit rates the required E_b/N_0 tends to be lower, compensating for the reduced coverage of high bit rates. Typical E_b/N_0 values for different bit rates are shown in Figure 10.4. The lower the E_b/N_0 requirement, the less power is needed for the same performance and the larger the cell radius that can be obtained.

In this section we evaluate the coverage of different bit rates taking into account these two factors. The main reason why the required E_b/N_0 depends on the bit rate is that the DPCCH (dedicated physical control channel) is needed to keep the physical layer connection running and it contains reference symbols for channel estimation and power control signalling bits. The E_b/N_0 performance depends on the accuracy of the channel and SIR estimation algorithms. Those estimates are based on the reference symbols on DPCCH. The more power that can be allocated for DPCCH, the better the channel estimation. On the other hand, DPCCH is only overhead because it does not transmit any user data, and its power should therefore be minimised. The power difference between DPCCH and DPDCH (dedicated physical data channel) can be adjusted and is controlled by the network. The power of DPCCH is lower than the power of DPDCH when there is data in DPDCH. The value of the power difference is quantised into four bits, i.e. 15 values for power differences between −23.5 dB and 0.0 dB and one bit combination for no DPDCH when there is no data to be transmitted. Typical values for the power differences are shown in Table 11.2.

The relative received power levels of DPCCH with different bit rates are shown in Figure 11.2. The power differences between DPCCH and DPDCH are taken from Table 11.2. It is assumed that E_b/N_0 is the same for all bit rates. The received power of DPCCH is higher for higher bit rates. The more received power there is for DPCCH, the more accurate is the channel estimation and the better the E_b/N_0 performance.

Table 11.2. Typical power differences between DPCCH and DPDCH

Bit rate	Typical power difference between DPCCH and DPDCH
12.2 kbps speech	−3.0 dB
144 kbps data	−6.0 dB
384 kbps data	−9.0 dB
1024 kbps data	−12.0 dB

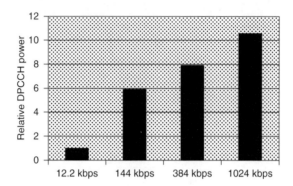

Figure 11.2. Relative received power of DPCCH with the same E_b/N_0

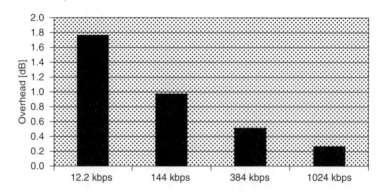

Figure 11.3. Overhead from DPCCH

The overhead from DPCCH for 144 kbps can be calculated as

$$DPCCH\_overhead = 10 \log_{10} \left(1 + 10^{\frac{-6\ \text{dB}}{10+}} \right) = 1.0 \text{ dB} \qquad (11.3)$$

The overhead for different bit rates is shown in Figure 11.3. For example, with the 144 kbps data service 20% of the transmission power is used to carry physical layer control information and 80% to carry data. The overhead from DPCCH is included in all E_b/N_0 values in this book.

We have seen now that for higher bit rates the power of DPCCH is higher, enabling more accurate channel estimation, and the overhead of DPCCH is still lower. Both these factors improve the E_b/N_0 performance.

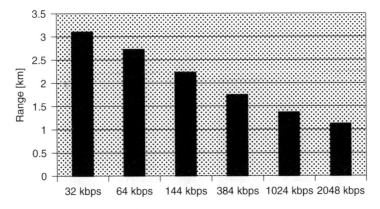

Figure 11.4. Uplink range of different data rates in suburban area

When we take into account the reduced processing gain and the improved E_b/N_0 performance for higher bit rates, we can calculate the coverage of the different bit rates in Figure 11.4. The same maximum power of the mobile station is assumed for all bit rates, a better E_b/N_0 performance is assumed for higher bit rates, and a suburban propagation model is assumed. In this example, the uplink range of 2 Mbps is 50% of the range of 144 kbps and 40% of the range of 64 kbps. If the cell is planned for 2 Mbps full uplink coverage instead of 144 kbps, the base station site density must be increased by a factor of $(1/0.5)^2 = 4.0$.

Wide area uplink coverage for high bit rate services will be challenging in UMTS, as shown in Figure 11.4, and providing full 2 Mbps uplink coverage requires high base station site density. These results also point out the importance of the solutions that improve uplink coverage in third generation systems. In second generation systems, coverage issues are less challenging since only low bit rate services are offered. The coverage of WCDMA data services is compared to GSM900 and GSM1800 speech coverage in Section 8.4.

Finally, we note that the bit rate of the uplink transmission can be decreased during the connection to improve the coverage when the mobile hits its maximum transmission power. The reduction of the bit rate is possible for non-real-time packet data services, which can tolerate delays, and for AMR speech service, which supports different bit rates from 4.75 kbps to 12.2 kbps. The coverage of AMR speech service is discussed in the next section.

11.2.1.2   Adaptive Multirate Speech Codec

With Adaptive Multirate (AMR) speech codec it is possible to switch to a lower bit rate if the mobile is moving out of the cell coverage area. The AMR speech codec is introduced in Section 2.3. The gain in the link budget by reducing the AMR bit rate can be calculated as follows:

$$
\text{Coverage\_gain} = 10 \cdot \log_{10} \left(\frac{\text{DPDCH}(12.2 \text{ kbps}) + \text{DPCCH}}{\text{DPDCH}(\text{AMR\_bit\_rate[kbps]}) + \text{DPCCH}} \right)
$$

$$
= 10 \cdot \log_{10} \left(\frac{12.2 + 12.2 \cdot 10^{\frac{-3 \text{ dB}}{10}}}{\text{AMR\_bit\_rate[kbps]} + 12.2 \cdot 10^{\frac{-3 \text{ dB}}{10}}} \right) \quad (11.4)
$$

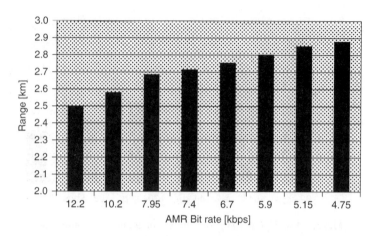

Figure 11.5. Relative uplink range of different AMR speech codec bit rates

where the power difference between DPCCH and DPDCH is assumed to be -3.0 dB for 12.2 kbps AMR speech. For lower AMR bit rates the DPCCH power is kept the same while the power of the DPDCH is changed according to the bit rate. The reduction of the total transmission power is calculated in Equation (11.4) and can be used to provide a larger uplink cell range. The coverage gain by reducing the bit rate from 12.2 kbps to 7.95 kbps is 1.1 dB, and the gain by reducing the bit rate from 12.2 kbps to 4.75 kbps is 2.3 dB. The relative cell ranges with different AMR bit rates are shown in Figure 11.5.

11.2.1.3 Multipath Diversity

We can study the effect of the multipath diversity on the uplink coverage by looking at example simulation results for the E_b/N_0 performance in two different multipath profiles: ITU Pedestrian A with little multipath diversity and ITU Vehicular A with more multipath diversity. ITU Vehicular A is a five-tap channel with WCDMA 3.84 Mcps resolution and ITU Pedestrian A is a two-path channel where the second tap is very weak [1]. The required E_b/N_0 for 8 kbps speech service with 10 ms interleaving, receive antenna diversity and full constant transmission power at 3 km/h is shown in Table 11.3. The 8 kbps speech simulation results in this section are obtained with 10 ms interleaving, but the AMR speech codec uses 20 ms interleaving, and therefore its performance is better than the simulation results herein.

In this example, the multipath diversity gain is 2.8 dB in uplink coverage. In general, the more multipath diversity that is available, the better the coverage. The degree of available multipath diversity depends on the environment but also on the transmission bandwidth. With wideband CDMA more multipath diversity can be received than with narrowband CDMA in the same environment: see Section 3.3.

Table 11.3. Required E_b/N_0 for FER $= 1\%$ for 8 kbps with full constant power

	E_b/N_0
ITU Pedestrian A (less multipath diversity)	11.3 dB
ITU Vehicular A (more multipath diversity)	8.5 dB
Multipath diversity gain	2.8 dB

Table 11.4. Required E_b/N_0 for FER = 1% for 8 kbps with full constant power

E_b/N_0	ITU Pedestrian A	ITU Vehicular A
Single link	11.3 dB	8.5 dB
Macro diversity result Equal powers / 3 dB difference	7.3 dB / 8.6 dB	6.3 dB / 7.7 dB
Macro diversity gain (Soft handover gain) Equal powers / 3 dB difference	4.0 dB / 2.7 dB	2.2 dB / 0.8 dB

11.2.1.4 Macro Diversity (Soft Handover)

During soft handover the uplink transmission from the mobile is received by two or more base stations. Since during soft handover there are at least two base stations trying to detect the mobile transmission, the probability of a correctly detected signal increases and macro diversity gain can be obtained. Example macro diversity gains for the uplink coverage are shown in Table 11.4 at 3 km/h with two base stations and 10 ms interleaving. Two cases are shown: when the path loss is the same to both base stations, and when there is a 3 dB difference in the path loss between the base stations. The first case gives the highest macro diversity gain. When the difference in the path loss increases, the macro diversity gain decreases and the mobile would not be in soft handover but connected to only one base station. The results show that the less multipath diversity that is available, the larger is the macro diversity gain. In this example, the best-case macro diversity gain in the ITU Pedestrian A channel is 4.0 dB and in ITU Vehicular A 2.2 dB.

The macro diversity gains for capacity are shown in Section 9.3.1.3. The capacity gains are lower than the coverage gains because diversity is more important for the uplink coverage. The reason is that at the edge of the coverage area the mobile is transmitting on full power and the diversity is important because the fast power control cannot compensate for the fast fading.

During softer handover the uplink transmission from the mobile is received by two sectors of one base station. The signals from two sectors are maximal ratio combined in the base station baseband Rake receiver, see Section 3.6. In soft handover selection combining is used in RNC. The maximal ratio combining of softer handover provides a better performance than the selection combining of soft handover. The soft and softer handover gains with equal power to both sectors and base stations are shown in Table 11.5. Softer handover provides 0.9–1.3 dB more gain than soft handover.

Table 11.5. Soft and softer handover gains for FER = 1% for 8 kbps with full constant power

E_b/N_0	ITU Pedestrian A	ITU Vehicular A
Soft handover gain (selection combining)	4.0 dB	2.2 dB
Softer handover gain (maximal ratio combining)	5.3 dB	3.1 dB
How much more gain from softer handover?	1.3 dB	0.9 dB

11.2.1.5 Receive Antenna Diversity

A 3 dB coverage gain can be obtained with receive antenna diversity even if antenna diversity branches have fully correlated fading. The reason is that the signals from two antennas can be combined coherently while the receiver thermal noise is combined non-coherently. The 3 dB gain assumes ideal channel estimation in the coherent combining. This 3 dB gain is achieved because there are more receiver branches collecting energy—but at the expense of increased hardware in the base station receiver. Additionally, antenna diversity provides gain against fast fading, since fast fading typically correlates poorly between the diversity antennas. Receive diversity antennas are shown in Figure 11.6. Antenna diversity can be obtained by space or polarisation diversity. The advantage of polarisation diversity is that the diversity branches do not need separation but can be located in one physical antenna housing. The performance of polarisation diversity in GSM is presented in [2], [3] and [4].

The uplink diversity reception can be extended beyond two-branch reception to four-branch reception. The four-branch antenna configuration can be obtained using two antennas with polarisation diversity with a separation of 2–3 meters. Those two antennas can also be placed very close to each other, even in a single radome, to make the visual impact lower. Those two four-branch antenna options are shown in Figure 11.7. The antenna gain of the single radome solution is typically 0.2–0.4 dB lower than the solution with separate antennas as shown in the measurement part.

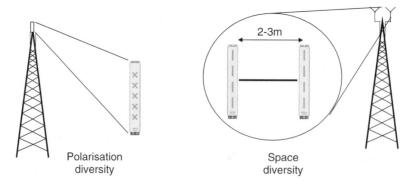

Figure 11.6. Polarisation and space diversity antennas

Figure 11.7. Four-branch receive antenna configurations

Table 11.6. Required E_b/N_0 for 8 kbps with full constant power

E_b/N_0	ITU Pedestrian A	ITU Vehicular A
With one receiver antenna	18.8 dB	12.8 dB
With two-branch receiver diversity	11.3 dB	8.5 dB
With four-branch receiver diversity	5.3 dB	5.4 dB

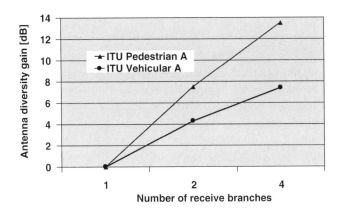

Figure 11.8. Antenna diversity gain with one, two and four-branch reception

Simulation results one branch receiver, with two-branch and with four-branch diversity with constant full power are shown in Table 11.6. The diversity gains are summarised in Figure 11.8. These results assume separate antennas in four-branch reception. The receive antenna diversity gain in the ITU Pedestrian A channel is 7.5 dB. In the ITU Vehicular A channel, the diversity gain is smaller, 4.3 dB, because there is more multipath diversity. The gain of four-branch diversity over two-branch diversity in ITU Vehicular A is 3.1 dB. The antenna diversity gain gets smaller when the number of antenna branches increases as shown in Figure 11.8.

The more diversity that is available, the smaller is the diversity gain from an additional diversity. This rule applies to antenna diversity and to all different sources of diversity. Therefore, there are no *a priori* values for any diversity gains, because the gains depend on the degree of other diversity sources.

Field Measurements of Four-branch Receive Antenna Diversity

The field performance of four-branch reception was tested in the WCDMA network in Espoo, Finland. The measurements area is in the middle of Figure 8.10. The measurement environment is urban and sub-urban type. The measurement routes are shown in Table 11.7.

In the field measurements the mobile transmission power was recorded slot-by-slot with three different base station antenna configurations:

Table 11.7. Measurement routes

Route A	up to 40 km/h in Leppävaara/Lintuvaara
Route B	up to 70 km/h on Ring road I
Route C	below 10 km/h in Mäkkylä

Table 11.8. Measured average mobile transmission powers

Route	Antenna separation	2-branch reception	4-branch reception	4-branch gain over 2-branch
Route A	1 m separation	6.95 dBm	4.44 dBm	2.5 dB
	no separation	6.95 dBm	4.83 dBm	2.1 dB
Route B	1 m separation	7.90 dBm	4.59 dBm	3.3 dB
	no separation	7.90 dBm	4.86 dBm	3.1 dB
Route C	1 m separation	5.63 dBm	2.54 dBm	3.0 dB

1. Two-branch reception with one polarisation diversity antenna
2. Four-branch reception with two polarisation diversity antennas separated by 1 m
3. Four-branch reception with two polarisation diversity antennas side-by-side (emulates single radome solution)

Several iterations were driven with each configuration. The different measurement drives are made comparable using differential Global Positioning System, GPS. The average transmission power over the measurement route is calculated from dBm values. These measured mobile transmission powers are shown in Table 11.8.

The multipath propagation in the measured environment is closer to ITU Vehicular A than to ITU Pedestrian A. Therefore, we compare the measurement results to the simulation results in ITU Vehicular A. The simulated gain of four-branch reception over two-branch reception in Table 11.6 is 3.1 dB with separate antennas. The average measured gain with 1 m separation is 3.0 dB in Table 11.8.

The measured difference of separate antennas and single radome solution is 0.2–0.4 dB. The effect of the antenna separation, i.e. space diversity, is small because the amount of other diversity sources is large: multipath and polarisation diversity.

Receiver antenna diversity is an effective approach to increase the uplink coverage area. A 3-dB improvement in the uplink performance reduces the required site density by about 30% according to Table 11.1.

11.2.1.6 Base Station Baseband Algorithms

The accuracy of the channel and SIR estimation is important for the receiver E_b/N_0 performance. Channel estimation in the receiver can be improved, for example, as follows.

— By averaging the estimation over several groups of pilot symbols on DPCCH. For low mobile speeds, in particular, it is possible to average the channel estimates over several timeslots. For higher mobile speeds, the averaging weights should adapt to the fading rate, i.e. adaptive channel estimation filters should be used.

— By using the modulated symbols on both DPCCH and DPDCH with decision feedback as additional pilot symbol. If the decisions of the channel bits are correct, those symbols can be regarded as additional pilot symbols. The typical error rate of the uncoded symbols in WCDMA is 5–20%. An error rate of roughly 10% or less can usually yield an observable channel estimation gain.

The largest improvement by the advanced receiver baseband algorithms can be obtained for the low bit rates, because there is only a little energy in DPCCH for channel estimation, as shown in Figure 11.2.

11.2.2 Random Access Channel Coverage

The uplink coverage discussion in Section 11.2.1 applies to the uplink dedicated channels (DCH) as well as to common channels such as the random access channel (RACH). If we want to improve the coverage of the WCDMA cells, we need to check whether the dedicated or the common channels are the limiting factor in the coverage. If, in any locations, the uplink dedicated channel can provide the required quality, it should also be possible to get the RACH message through to be able to start the connection. In this section the coverage area of the uplink dedicated and RACH channels are compared. The reasons for the differences in coverage performance between the dedicated channels and RACH are shown in Table 11.9. The random access procedure is described in Section 6.6.

The minimum number of bits that has to be transmitted in the initial RACH message is assumed to be 20 octets = 160 bits, which corresponds to the bit rate of 16 kbps with 10 ms interleaving and 8 kbps with 20 ms interleaving. Both 10 ms and 20 ms RACH are supported by the standard for the RACH message size of 20 octets. The interleaving period of the RACH message is indicated on BCCH to the mobile.

The high bit rate dedicated channels clearly have a smaller coverage than RACH, and therefore the RACH coverage needs to be checked against low bit rate dedicated channels, such as AMR speech codec with bit rates of 4.75–12.2 kbps.

Soft handover is not possible with common channels and no macro diversity gain can be obtained with RACH. Also, the reception of the short RACH burst is more difficult than that of the continuous dedicated channel. These factors make the RACH coverage smaller than the DCH coverage with the same bit rate. On the other hand, a higher FER can be allowed with RACH. The drawback of a high FER is that there is a longer delay in the call setup. The average FER of RACH can be controlled by the parameters given on BCCH.

Table 11.9. Reasons for different coverage of DCH and RACH

	DCH	RACH
Bit rate	Minimum AMR bit rate: 4.75 kbps Maximum packet data: 2.0 Mbps	Initial RACH message 20 octets which corresponds to 16 kbps (10 ms interleaving) 8 kbps (20 ms interleaving)
Soft handover (macro diversity gain)	0.8–4.0 dB: see Table 11.4	Soft handover not possible
E_b/N_0 performance	– Continuous transmission makes optimised reception easier than with RACH – Coding rate 1/3	– Short 10 or 20 ms burst makes optimised reception difficult – Coding rate 1/2
FER requirements	Speech 1%	Preferably 10% or below; higher FER causes longer delay in call setup

Table 11.10. RACH options to match with the coverage of dedicated channels

DCH bit rate	RACH option
AMR $\leq$7.95 kbps	20 ms / FER > 10%
AMR 12.2 kbps	20 ms / FER $\leq$ 10%
Bit rate > 20 kbps	10 ms / FER $\leq$ 10%

Suitable RACH options to match the coverage of the dedicated channels are presented in Table 11.10, taking into account all the differences listed above. Those options are based on the results from [5].

The 20 ms RACH option should be used only in large cells to improve the RACH coverage. The E_b/N_0 performance is worse for 20 ms RACH than for 10 ms RACH, and therefore the shorter 10 ms RACH is better for the uplink capacity. The reason for the degradation of the performance with longer RACH is that the power control cannot be done during RACH but the power is set only by preamble before the RACH message part. The channel can change during the 20 ms RACH more than during 10 ms RACH.

The coverage of RACH messages is important if the network is planned to provide continuous coverage only for low bit rate services. On the other hand, if the network is planned to provide continuous coverage for high bit rate services, the coverage of RACH is not a limiting factor, and 10 ms RACH can be used.

11.2.3 Downlink Coverage

In the downlink more power can be allowed for one connection than in the uplink because the base station output power can be higher than the mobile station output power. Therefore, better coverage can be given for high bit rate services in downlink than in uplink. See also the relation between the downlink capacity and coverage in Section 8.2.2. The uplink range of different bit rates is shown in Figure 11.4.

Here, a comparison of 12.2 kbps speech and 1 Mbps data coverage in downlink is shown. The assumptions in that comparison are:

— Cell size is determined by uplink coverage for speech
— Uplink reception in the base station has 6 dB better sensitivity than downlink reception, due to lower noise figure and antenna diversity
— 1 Mbps data has 3 dB lower E_b/N_0 than speech
— Speech terminal has 3 dB higher body loss than 1 Mbps data terminal
— Speech terminal transmission power is 21 dBm
— The amount of other cell interference is the same with both speech and 1 Mbps.

The required average downlink transmission power for speech is calculated to be 27 dBm, and for 1 Mbps connection about 40 dBm (= 10 W), to obtain full coverage in the downlink if the cell size is planned according to the uplink speech. The calculation is shown in Table 11.11. It is quite feasible to provide 10 W power in the downlink for one high bit rate user in a case when there are no other users requesting capacity in that cell, i.e., in the coverage-limited case.

Table 11.11. Required downlink transmission power for 1 Mbps full coverage

Mobile transmission power in uplink	21 dBm
Estimated transmission power for speech in downlink	21 dBm + 6 dB = 27 dBm
Difference in processing gain between 12.2 kbps and 1 Mbps	$10 * \log_{10}(1000/12.2) = 19.1$ dB
Lower E_b/N_0 for data terminal	3 dB
Body loss	Speech terminal: 3 dB
	Data terminal: 0 dB
Required transmission power for 1 Mbps full coverage	27 dBm + 19.1 dB − 3 dB − 3 dB = 40 dBm = 10 W

In practice the downlink coverage of high bit rates, 1–2 Mbps, depends on

— uplink dimensioning: the bit rate for which the uplink cell range is dimensioned
— downlink power amplifier rating
— adjacent cell loading.

If the cell is planned to provide high bit rates also in the uplink from the cell edge, the cell is smaller and then also the downlink coverage will be better. The loading of the adjacent cell affects the possibility of having a high bit rate connection at the cell edge. A large number of high bit rate users with continuous coverage requires a high capacity. The capacity aspects of a WCDMA network are considered in Section 11.3.

This discussion of downlink coverage has considered mainly macro cell base stations where the maximum output power is in the order of 43 dBm. With low-power micro or pico base stations the coverage can also be downlink-limited.

11.2.4 Coverage Improvements

Some ways of improving the uplink coverage of a WCDMA base station site are as follows:

— Reduce E_b/N_0 by improving the base station baseband algorithms or by increasing the number of receiver antennas
— Reduce the base station noise figure in the base station RF section
— Reduce the cable loss between the antenna and the base station low noise amplifier
— Reduce the interference margin, i.e. the maximum allowed capacity in the uplink
— Increase the antenna gain

The E_b/N_0 in the base station can be reduced with the techniques listed in Section 11.2.1. The most important methods are by increasing the number of receiver diversity branches and by optimising the baseband receiver algorithms in the base station. The base station noise figure can be lowered by improving the base station radio frequency (RF) parts. Cable loss can be reduced by using thicker cables or by using a mast head amplifier. A typical cable loss is 6 dB per 100 m.

In WCDMA the capacity and coverage are tied together as shown in Section 8.2. The uplink coverage can be improved by allowing a lower uplink capacity, which allows a lower interference margin. For example, lowering the maximum uplink loading from 50% to 30% reduces the required interference margin from 3.0 dB to 1.5 dB and thus gives a coverage gain of 1.5 dB in the link budget. The load is kept within the planned limits by the real-time radio resource management algorithms shown in Chapter 9. Another way to lower the interference margin is to use interference cancellation or multi-user detection in the base station, as shown in Section 11.5.2.

A higher antenna gain can be used to improve the coverage and can be obtained by increasing the number of sectors and narrowing the horizontal antenna pattern. Antenna gain can also be increased by narrowing the vertical antenna beam. Typically, the maximum antenna gain for a three-sector antenna is 18 dBi, assuming a vertical beam of 6°. The antenna gain can also be increased by adaptive antenna solutions, as shown in Section 11.5.1.

11.3   Capacity

The WCDMA downlink air interface capacity is shown to be less than the uplink capacity [6–8]. The main reason is that better receiver techniques can be used in the base station than in the mobile station. These techniques include receiver antenna diversity and multi-user detection. Additionally, in UMTS, the downlink capacity is expected to be more important than the uplink capacity because of asymmetric downloading type of traffic. Therefore, in this section the downlink capacity and its performance enhancements are considered.

In the following sections two aspects are presented that affect the downlink capacity and are different from the uplink: orthogonal codes in Section 11.3.1 and base station transmit diversity in Section 11.3.2.

11.3.1   Downlink Orthogonal Codes

11.3.1.1   Multipath Diversity Gain in Downlink

The effect of the downlink orthogonal codes on capacity performance is considered in this section. The short codes in the downlink within one scrambling code are orthogonal but only in a one-path channel. In case of multipath channel, the orthogonality is partly lost and the intra-cell users interfere with each other. In GSM, there is no intra-cell interference because the users are orthogonal in the time domain in different timeslots, regardless of multipath propagation. The downlink performance in the ITU Vehicular A and ITU Pedestrian A multipath profiles for 8 kbps speech with 10ms interleaving with 1% FER is presented below. The ITU Pedestrian A channel is close to a one-path channel and does not give much multipath diversity, while the ITU Vehicular A channel gives a significant degree of multipath diversity. The simulation scenario is shown in Figure 11.9. The required transmission power per speech connection ($= I_c$) as compared to the total base station power ($= I_{or}$) is shown on the vertical axis in Figure 11.10. For example, the value of -20 dB means that this connection takes $10^{(-20 \text{ dB}/10)} = 1\%$ of the total base station transmission power. The lower is the value on the vertical axis, the better is the performance. The horizontal axis shows the total transmitted power from this base station divided by the received interference from the other cells, including thermal noise ($= N_0$). This ratio is also known as the geometry factor, G . A high value of G is obtained when the mobile is close to the base station and a low value at the cell edge.

We can learn a few important lessons about downlink performance from Figure 11.10. At the cell edge, i.e. for low values of G , the multipath diversity in the ITU Vehicular A channel gives a better performance compared to less multipath diversity in the ITU Pedestrian A channel. At the cell edge the multipath diversity improves the downlink performance. Close

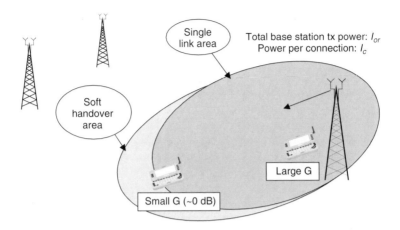

Figure 11.9. Simulation scenario for downlink performance evaluation

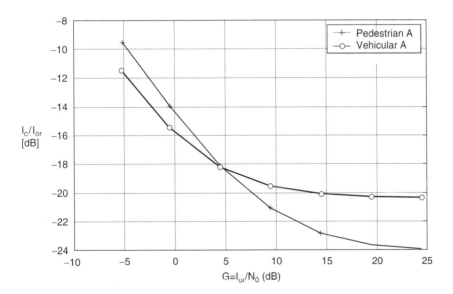

Figure 11.10. Effect of multipath propagation

to the base station the performance is better in the ITU Pedestrian A channel because the multipath propagation in the ITU Vehicular A channel reduces the orthogonality of the downlink codes. Furthermore, there is not much need for diversity close to the base station, since the intra-cell interference experiences the same fast fading as the desired user's signal. If signal and interference have the same fading, the signal to interference ratio remains fairly constant despite the fading.

The effect of soft handover is not shown in these simulations but it would improve the performance, especially in the ITU Pedestrian A channel at the cell edge by providing extra diversity—macro diversity. The macro diversity gain is presented in detail in Section 9.3.1.3.

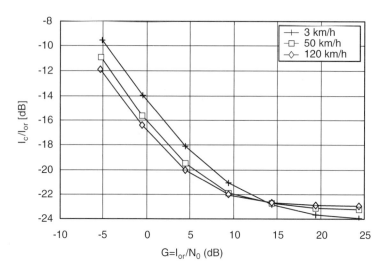

Figure 11.11. Effect of mobile speed in the ITU Pedestrian A channel

We note that in the downlink the multipath propagation is not clearly beneficial—it gives diversity gain but at the same time reduces orthogonality. It was shown in Section 11.2.1.3 that the multipath diversity improves uplink coverage but the multipath propagation does not necessarily improve downlink capacity because of the loss of orthogonality. But we need to note that the diversity reduces the variation of the transmission power, as shown in Section 9.2.1.3. The gain of the reduced transmission power variation is not shown in Figure 11.10, where only the average transmission powers are shown. For low bit rates the average transmission power is a good measure, since there are a large number of users and the variation of the total transmission power is small. The variation of the transmission power is important for high bit rate performance when there are only a few users. If the variation is large, a margin needs to be reserved in the radio resource management algorithms, such as in admission control, to guarantee quality of service of that connection. The effect of multipath diversity on high bit rate performance in the downlink is discussed further in Section 11.4.2.

The effect of the mobile speed on downlink performance in the Pedestrian A channel is shown in Figure 11.11. At the cell edge the best performance is obtained for high mobile speeds, while close to the base station low mobile speeds perform better. This behaviour can be explained by the fact that for high mobile speeds interleaving and channel coding, here convolutional code, provide time diversity and coding gain. In Figure 11.10 it was shown that diversity is important at the cell edge to improve the performance.

11.3.1.2 Downlink Capacity in Different Environments

In this section the WCDMA capacity formulas from Section 8.2.2 are used to evaluate the effect of orthogonal codes on the downlink capacity in macro and micro cellular environments. The downlink orthogonal codes make the WCDMA downlink more resistant to intra-cell interference than the uplink direction, and the effect of inter-cell interference from adjacent base stations has a large effect on the downlink capacity. The amount of interference from the adjacent cells depends on the propagation environment and the network planning. Here we assume that the amount of inter-cell interference is lower in micro cells where street corners isolate the cells more strictly than in macro cells. This cell isolation is represented in the formula by the other-to-own cell interference ratio i. We also assume that in micro cellular environments there is less multipath propagation, and thus a better orthogonality of the downlink codes. On the other hand, less multipath propagation gives less multipath diversity, and therefore we assume a higher E_b/N_0 requirement in the downlink in micro cells than in macro cells.

The assumed loading in uplink is allowed to be 60% and in downlink 80%. A lower loading is assumed in uplink than in downlink because the coverage is more challenging in uplink. A higher loading results into smaller coverage as shown in Section 8.2.2.

We assume that 20% of the downlink capacity is allocated for downlink common channels, for more information about those channels see Section 6.5.

The assumed downlink performance here does not include the effect of transmit antenna diversity while receive antenna diversity is assumed in uplink. We calculate the example data throughputs in macro and micro cellular environments in both uplink and downlink. The assumptions of the calculations are shown in Table 11.12 and the results in Table 11.13. Retransmissions are not taken into account in these throughputs. With FER = 10%, the user throughput will be 90% of the values shown Table 11.13.

In macro cells the uplink throughput is higher than the downlink throughput, while in micro cells the downlink and uplink capacities are quite balanced. The downlink capacity depends more on the propagation and multipath environment than does the uplink capacity. The reason is the application of the orthogonal codes.

The capacities in Table 11.13 assume that the users are equally distributed over the cell area. If the users are on average closer to the base station, the capacities will be higher. The capacities also show that supporting a 2 Mbps user in every cell is not possible if the 2 Mbps users can be anywhere in the cell area, including the cell edge.

11.3.1.3 Number of Orthogonal Codes

The number of downlink orthogonal codes is limited within one scrambling code. With a spreading factor of SF, the maximum number of orthogonal codes is SF. This code

Table 11.12. Assumptions in the throughput calculations

	Macro cell	Micro cell
Downlink orthogonality	0.6	0.95
Other-to-own cell interference ratio i	0.65	0.2
Uplink E_b/N_0	1.5 dB	1.5 dB
Uplink loading	60%	60%
Downlink E_b/N_0	5.5 dB	8.0 dB
Downlink loading	80%	80%

Table 11.13. Data throughput in macro and micro cell environments per sector per carrier

	Macro cell	Micro cell
Uplink	1040 kbps	1430 kbps
Downlink	660 kbps	1560 kbps

Table 11.14. Assumptions in the calculation of Table 11.15

Common channels	10 codes with SF = 128
Soft handover overhead	20%
Spreading factor (SF) for half-rate speech	256
Spreading factor (SF) for full-rate speech	128
Chip rate	3.84 Mcps
Modulation	QPSK (2 bits per symbol)
Average DPCCH overhead for data	10%
Channel coding rate for data	1/3 with 30% puncturing

Table 11.15. Maximum downlink capacity with one scrambling code per sector

Speech, full rate (AMR 12.2 kbps and 10.2 kbps)	128 channels	Number of codes with spreading factor of 128
	$*(128 - 10)/128$	Common channel overhead
	$/1.2$	Soft handover overhead
	= 98 channels	
Speech, half rate (AMR ≤ 7.95 kbps)	2*98 channels	Spreading factor of 256
	= 196 channels	
Packet data	3.84e6	Chip rate
	$*(128 - 10)/128$	Common channel overhead
	$/1.2$	Soft handover overhead
	$*2$	QPSK modulation
	$*0.9$	DPCCH overhead
	$/3$	1/3 rate channel coding
	$/(1 - 0.3)$	30% puncturing
	= 2.5 Mbps	

limitation can affect the downlink capacity if the propagation environment is favourable and the network planning and hardware support such a high capacity. In this section the achievable downlink capacity with one set of orthogonal codes is estimated. The assumptions in these calculations are shown in Table 11.14 and the results in Table 11.15. Part of the downlink orthogonal codes must be reserved for the common channel and for soft and softer handover overhead. These factors are taken into account in Tables 11.11 and 11.12. The maximum number of full-rate speech channels per sector is 98 with these assumptions, and the maximum data throughput is 2.5 Mbps per sector.

The number of orthogonal codes is not a hard-blocking limitation for the downlink capacity. If this number is not large enough, a second (or more) scrambling code can be taken into use in the downlink, which gives a second set of orthogonal short codes: see Section 6.3. These two sets of orthogonal codes are not orthogonal against each other. If the second scrambling code is used, those code channels under the second scrambling code cause more interference to those under the first scrambling code than do the other code channels under the first scrambling code.

The maximum capacity with one set of orthogonal codes can be lower than that shown in Table 11.15 if variable rate connections are used, because the orthogonal code must be reserved according to the maximum bit rate of the connection.

The second scrambling code will most likely be needed with downlink adaptive antennas, which improve the downlink performance above those values shown in Table 11.13.

11.3.2 Downlink Transmit Diversity

The downlink capacity could be improved by using receive antenna diversity in the mobile. For small and cheap mobiles it is not, however, feasible to use two antennas and receiver chains. Therefore, the WCDMA standard supports the use of base station transmit diversity. The downlink transmit diversity modes are described with physical layer procedures in Section 6.6. With transmit diversity the downlink signal is transmitted via two base station antenna branches. If we already have receive diversity in the base station and we duplex the downlink transmission to the receive antennas, there is no need for extra antennas for downlink diversity. In we could use both antennas for reception and for transmission. The downlink transmit diversity could utilise space or polarisation diversity antennas.

In this section we compare the downlink transmit diversity to the uplink receive diversity presented in Section 11.2.1.3 and analyse the performance gain from the downlink transmit diversity. The performance gain from either receive or transmit diversity can be divided into two parts:

(1)   Coherent combining gain
(2)   Diversity gain against fast fading.

The coherent combining gain can be obtained because the signal is combined coherently while interference is combined non-coherently. The gain from ideal coherent combining is 3 dB with two antennas. In the uplink it is the Rake receiver that makes the coherent combining of the signal from the two diversity antennas. Rake reception is introduced in Section 3.4. In the uplink the coherent combining gain is 2.5–3.0 dB, depending on the accuracy of the channel estimation which affects the accuracy of the coherent combining. Also, with downlink transmit diversity it is possible to obtain coherent combining in the mobile reception if the phases from the two transmission antennas are adjusted according to the feedback commands from the mobile. The coherent combining is not perfect, since the multipath components cannot be combined coherently; only the relative phasing of the signals from two antennas can be adjusted. The feedback loop has only discrete steps and has a delay that further degrades the coherent combining gain in the downlink. The delay effects are especially notable with high mobile speeds. Therefore, the coherent combining gives less gain with downlink transmit diversity than with uplink receive diversity. The downlink transmit diversity with feedback is depicted in Figure 11.12.

Both receive and transmit diversity provide gain against fast fading. This gain is larger when there is less multipath diversity. The relationship between power control and diversity is presented in detail in Section 9.2.1.2. With fast power control at low mobile speeds, the average transmitted power is lower, i.e. the power rise is smaller. At low mobile speeds this reduction of power rise with diversity, e.g. in ITU Pedestrian A, is 2.8 dB: see Figure 9.5. The differences between uplink receive diversity and downlink transmit diversity are summarised in Table 11.16.

It is important to note the difference between two sources of diversity in the downlink: multipath and transmit diversity. Multipath diversity reduces the orthogonality of the downlink codes, while transmit diversity keeps the downlink codes orthogonal in flat fading channels. In order to maximise the interference-limited downlink capacity, it would be beneficial to avoid multipath propagation to keep the codes orthogonal and to provide the diversity with transmit antenna diversity.

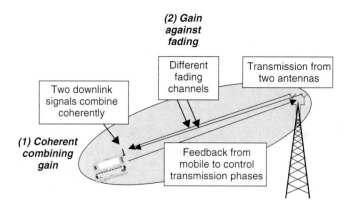

Figure 11.12. Downlink transmit diversity with feedback

Table 11.16. Comparison of uplink receive and downlink transmit diversity

	Uplink receive diversity	Downlink transmit diversity with feedback
(1) Coherent combining gain		
Gain from ideal coherent combining is 3.0 dB with two antennas		
How to obtain coherent combining	Rake receiver with channel estimation from pilot symbols	Feedback loop from mobile to base station to control the transmission phases to make received signals to combine coherently in mobile
Non-idealities in coherent combining	Inaccurate channel estimation in Rake receiver	Inaccurate channel estimation in mobile Discrete steps in feedback loop Delay in feedback loop Multipath propagation
Practical gain of coherent combining	2.5–3.0 dB gain	Gain is lower than with receive diversity
(2) Diversity gain against fading		
Diversity gain with fast power control	Diversity gain = reduction of power rise. Example values: ITU Pedestrian A: 2.8 dB ITU Vehicular A: 0.8 dB	
Total gain in reduction of transmission powers		
Total gain from antenna diversity	3.0–6.0 dB	0.0–5.0 dB

The uplink and downlink capacities are compared in Table 11.13. In that comparison receive diversity was assumed for uplink while no transmit diversity was assumed for downlink. If we add the transmit diversity gain to the downlink capacities in Table 11.13, the downlink capacity becomes approximately equal to the uplink capacity in a macro cell environment. In micro cells the downlink and uplink capacities are roughly equal without transmit diversity. With transmit diversity the interference-limited downlink capacity in a micro cell environment becomes clearly higher than the uplink capacity. This asymmetric

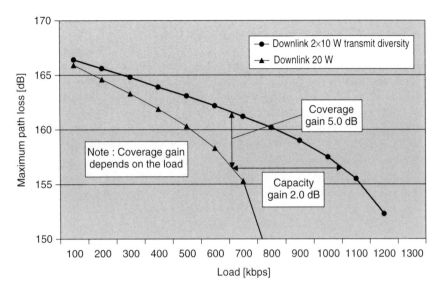

Figure 11.13. Downlink capacity and coverage gains with transmit diversity. A 2 dB link-level gain from transmit diversity is assumed

air interface capacity is beneficial, since the expected capacity requirement is higher in downlink than in uplink. The explanation for the higher downlink capacity is the orthogonal codes in the downlink.

The effect of the downlink transmit diversity gains on downlink capacity and coverage is illustrated in Figure 11.13. A 2 dB gain—including coherent combining gain and diversity gain against fading—is assumed. This improvement gives a capacity gain of 2 dB. If we allow, for example, a maximum path loss of 157 dB, the capacity can be increased from 650 kbps to 1030 kbps. The transmit diversity gain can be used alternatively to improve the downlink coverage while keeping the load unchanged. In the example in Figure 11.14 the maximum path loss could be increased by 5 dB, from 157 dB to 162 dB, if the load were kept at 650 kbps. The coverage gain is higher than the capacity gain because of the WCDMA load curve. It may not be possible to utilise the downlink coverage gains and extend the cell size with downlink transmit diversity if the uplink is the limiting direction in coverage. The coverage gain could be used alternatively to reduce the required base station transmission power. If we keep the load unchanged at 650 kbps and the maximum path loss unchanged at 157 dB, we could reduce the transmission power by 5 dB, from 20 W to 2×3.2 W.

11.3.3 Capacity Improvements

Some ways to improve the downlink capacity per site are as follows:

— More frequencies = carriers
— Transmit diversity
— Sectorisation
— Lower bit rate codec, for example with AMR speech codec.

If the operator's frequency allocation allows, the operator can take another carrier into use. WCDMA supports efficient inter-frequency handovers and several carriers can be utilised to balance the loading and to enhance the capacity per site. It is possible to share one power amplifier between several carriers. In Section 8.2.2 it was shown that sharing a power amplifier between two carriers provides the most efficient use of the power amplifier, since the loading can be divided between two carriers and, when we come down in the WCDMA load curve, the required transmission power per user is reduced. The WCDMA load curve also shows that increasing the downlink transmission power may give only a very marginal capacity gain and is not an effective solution for improving downlink capacity.

Downlink transmit diversity improves downlink capacity to an extent that depends on the degree of multipath diversity in that environment. The less multipath diversity is available, the larger is the downlink capacity gain by using transmit diversity. Therefore, the highest capacity gains can be expected in small micro and pico cells where there is only limited multipath diversity. If the uplink reception already uses diversity, the downlink transmit diversity could be included in the network without modifications to the antenna structures.

Sectorisation can be used to increase the capacity per site. In an ideal case N sectors give N times higher capacity, but in practice the sectorisation efficiency is typically about 90%. This means that upgrading the site from an omni site to a three-sector site gives a capacity increase of about 2.7, and to a six-sector site a capacity increase of about 5.4. The increased number of sectors also brings improved coverage through a higher antenna gain. The drawback of upgrading the capacity by increasing the number of sectors is that the antennas must be replaced and the radio network planning and optimisation must be redone. The sectorisation is analysed in more detail in [6].

With AMR speech codec it is possible to increase the speech capacity in WCDMA by using a lower bit rate AMR mode. The AMR speech codec is introduced in Section 2.3. The total number of transmitted user bits is not increased by using lower AMR bit rates—the number of connections is increased while the bit rate per user is decreased. The AMR codec allows a trade-off between speech capacity and quality according to the operator's needs.

An example capacity upgrade path is shown in Section 8.2.4.

11.4 High Bit Rates

In this section the link-level performance of 512 kbps to 2 Mbps in multipath channels is evaluated [10,11]. The evaluation is based on link-level simulations with a Rake receiver in different multipath profiles. In WCDMA the higher bit rates are obtained with a lower processing gain—either with a variable spreading factor or with multicodes. The interference properties and the performance of those two solutions have been shown to be equal [12, 13]. Since the autocorrelation properties of the spreading codes are not ideal, multipath components interfere with each other. If the processing gain is large, these inter-path interference (IPI) and inter-symbol interference (ISI) terms are negligible, but with a low processing gain the inter-path interference clearly affects the performance. The processing gain for 2 Mbps is only 2.8 dB $(= 10 \cdot \log_{10}(3.84 \text{ Mcps}/2.0 \text{ Mbps}))$ and clearly part of the spread-spectrum properties of WCDMA are lost. In TDMA systems, like in GSM, the inter-symbol interference is tackled by an equaliser.

In link-level simulations the E_b/N_0 performance in different multipath profiles depends on both inter-path interference and multipath diversity gain. If we wish to observe only the effect of the inter-path interference, we need to know the effect of the diversity gain. Therefore, simulation results are also obtained using a simulation model where the inter-path interference between the multipath components is not modelled. These results emulate an ideal receiver structure that could perfectly eliminate the inter-path interference.

Since the multipath propagation causes inter-path interference but also provides multipath diversity, the total effect of multipath diversity depends on the bit rate. For very high bit rates the inter-path interference may cause more degradation than that gained from the multipath diversity. The multipath diversity gain is analysed in Section 11.4.2.

11.4.1   Inter-path Interference

11.4.1.1   Uplink 512 kbps

The uplink performance of 512 kbps is obtained in these simulations with a single code transmission supporting an efficient power amplifier operation in the mobile station. Half-rate convolutional code is used. The frame error rate of packets is used as a performance measure with a packet size of 320 user bits. According to 3GPP Release-99 specifications 1/3 rate Turbo codes could be used. Turbo codes would provide slightly better E_b/N_0 performance than convolutional codes but the effect of inter-path interference would not be affected. Receive antenna diversity is assumed in the base station.

The uplink performance of 512 kbps has been simulated both with a chip-level simulator and with a symbol-level simulator. In the chip-level simulator spreading and despreading are modelled with a chip-level time resolution of 0.24 μs ($= 1/4.096$ Mcps), while in the symbol-level simulator only symbol-level time resolution is used. The former chip rate of 4.096 Mcps has been used in these simulations. In the symbol-level simulator multipath components as well as I- and Q-branches are kept separate, and orthogonal, thus inter-path, interference is not modelled. The symbol-level model corresponds to an ideal receiver structure which could perfectly eliminate the inter-path interference.

The received E_b/N_0 requirements with chip-level and symbol-level models are shown in Figure 11.14 and Tables 11.17 and 11.8. The difference in performance between one-, two- and three-path channels with symbol-level model is less than 0.2 dB. This indicates that the multipath diversity does not cause differences, because fast power control can compensate for fast fading at slow mobile speeds. Therefore, we can assume that the reason for the difference between chip-level and symbol-level results is only the inter-path interference. In these simulations all multipath components have equal average powers.

The simulation results show that the degradation caused by inter-path interference is about 0.6 dB at FER = 10%. At lower FER, it is larger—about 1.2 dB at FER = 1%.

11.4.1.2   Downlink 2.3 Mbps

In these simulations downlink 2 Mbps is achieved with four parallel code channels, each with a spreading factor of four. In practice three parallel spreading codes should be used in the downlink instead of four, because part of the orthogonal codes must be reserved for the downlink common channels: see Table 6.3. If we use only three codes instead of four, we need to apply more puncturing. Neither transmit nor receive antenna diversity is assumed in these downlink results.

The downlink simulation results with the chip-level model are shown in Figure 11.15 and for FER = 10% in Table 11.19. The chip-level model includes the effect of the inter-path

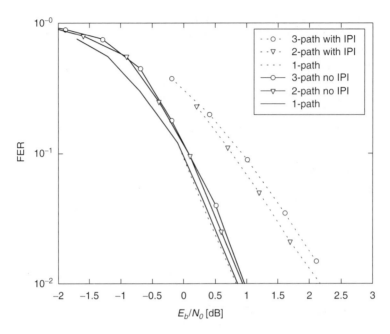

Figure 11.14. Uplink 512 kbps E_b/N_0 with chip-level and symbol-level simulation models at 3 km/h (IPI = Inter-path interference)

Table 11.17. Uplink 512 kbps performance at FER = 10% (IPI = inter-path interference)

	Required E_b/N_0 per antenna		
	Chip level with IPI	Symbol level, no IPI	Degradation due to IPI
1-path	0.0 dB	0.0 dB	No IPI
2-path	0.7 dB	0.1 dB	0.6 dB
3-path	0.8 dB	0.2 dB	0.6 dB

Table 11.18. Uplink 512 kbps performance at FER = 1% (IPI = inter-path interference)

	Required E_b/N_0 per antenna		
	Chip level with IPI	Symbol level, no IPI	Degradation due to IPI
1-path	0.9 dB	0.9 dB	No IPI
2-path	2.2 dB	0.9 dB	1.3 dB
3-path	2.2 dB	1.0 dB	1.2 dB

interference. The difference between the different multipath profiles is caused by inter-path interference.

With 2.3 Mbps in the downlink, the degradation due to inter-path interference can be up to 3.7 dB at FER = 10%. Inter-path interference causes an error floor, and very low FER cannot be achieved.

The degradation due to inter-path interference could be mitigated by using an advanced receiver structure where the impact of the channel would be equalised. The maximum gains

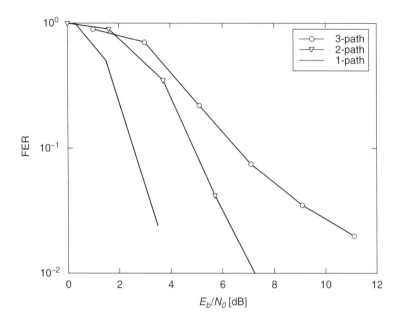

Figure 11.15. Downlink 2.3 Mbps E_b/N_0 with inter-path interference at 3 km/h

Table 11.19. Downlink 2.3 Mbps performance at FER = 10%

	Required E_b/N_0	
	Chip level with IPI	Degradation due to IPI
1-path	2.7 dB	No IPI
2-path	4.6 dB	1.9 dB
3-path	6.4 dB	3.7 dB

from such a receiver structure could be up 3.7 dB with 2 Mbps in a three-path channel. The implementation of such a receiver would not be excessively complex because it requires only a linear equaliser: see Section 11.5.2 for more discussion.

The downlink performance of high bit rates could be improved also by using the transmit diversity in the base station. The transmit diversity improves the downlink performance, as shown in Section 11.3.2, and makes the high bit rate transmission also more robust against inter-path interference.

11.4.2 Multipath Diversity Gain

The multipath propagation causes inter-path interference but also provides multipath diversity. At low mobile speeds this diversity gain can be seen in the transmission powers which are lower with diversity. In Figure 11.16 the downlink transmission powers are shown with 2 Mbps. At FER = 10%, two-path and three-path results are equal and the required transmission power is about 3 dB less than in the one-path channel. At FER = 1% the effect of inter-path interference is larger than at FER = 10%.

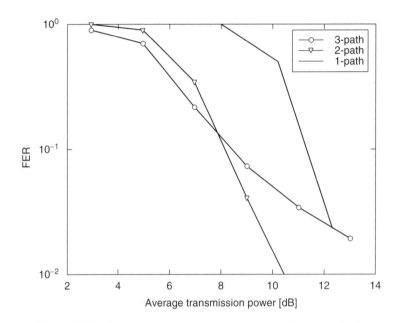

Figure 11.16. Downlink transmission power with 2 Mbps at 3 km/h

The difference between the required received power in Figure 11.15 and the required transmission power in Figure 11.16 is the power rise, which is explained in detail in Section 9.2.1.2. The power rise with ideal power control would be 3.0 dB in a two-path channel and 1.8 dB in a three-path channel. The simulated power rise in these 2 Mbps simulations with real power control seems to be very close to the theoretical values.

What is the difference between these 2 Mbps downlink results and the downlink results for speech service in Section 11.3.1? For speech service the multipath propagation causes interference between the users within one cell. For 2 Mbps there is no interference between own-cell users because the capacity does not allow more than one 2 Mbps user at the same time in one cell but the multipath propagation causes inter-path interference.

11.4.3 Feasibility of High Bit Rates

The feasibility of high bit rates in cellular environments is summarised in this section. The feasibility considers coverage, capacity and link-level performance of high bit rates up to 2 Mbps. Coverage is discussed in general in Section 11.2, and capacity in Section 11.3.

The downlink coverage is analysed in Section 11.2.3, and it is shown that providing full downlink coverage of 1 Mbps even in large macro cells is possible. Several assumptions affect the coverage. In practice the downlink coverage of 2 Mbps can be 50–100% of the cell area and therefore will not prevent the provision of high bit rate services with WCDMA. In general, the downlink coverage of high bit rates is better than the uplink coverage, since more power can be allocated per connection in downlink than in uplink. This is especially true for the high bit rates because there can be only a low number of high bit rate users per cell at the case time, and therefore a large proportion of the base station power can be allocated to a high bit rate user.

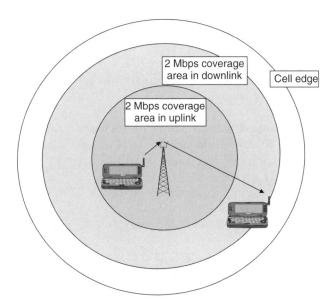

Figure 11.17. 2 Mbps coverage area in macro cells

The uplink coverage of high bit rates is smaller than the downlink coverage because the transmission power of the mobile is limited. The uplink coverage depends heavily on the cell size. Figure 11.4 showed the relative coverage of the different bit rates. In that example the cell was planned to provide full coverage for 144 kbps, and the range of the 2 Mbps service was 50% of the 144 kbps range. Example coverage areas of the 2 Mbps service are shown in Figure 11.17.

Typical capacities in different environments are shown in Table 11.13, from which it is seen that the average capacity per sector is smaller than 2 Mbps. In Table 11.13 it is assumed that the users are equally distributed over the cell area. If the loading in the adjacent cells is low or if the 2 Mbps user is closer to the base station, it is possible to support a 2 Mbps user in the downlink. If we required a 2 Mbps user in every cell, including at the cell edge, an improvement in capacity would be needed. Section 11.3.3 suggested a number of solutions for improving WCDMA capacity.

The results of simulation of link-level performance of high bit rates in multipath channels have been given in Section 11.4. It was shown that 2 Mbps is feasible with WCDMA in multipath channels.

To summarise, the WCDMA air interface can provide high bit rate services even in large cells. The air interface is ready for those services that can take advantage of these high bit rate capabilities.

11.5   Performance Enhancements

In this section WCDMA performance enhancements with advanced antenna structures and with baseband multi-user detection are described. Both solutions are considered here for the base station for improving uplink performance.

11.5.1 Antenna Solutions

WCDMA coverage is discussed in Section 11.2 and is shown to be typically uplink limited. One effective way of improving the uplink coverage is to increase the number of receiving antennas at the base station. The selection of the particular antenna configuration depends, for example, on the following issues:

— Radio channel characteristics (pico cell/micro cell/macro cell) which is affected by degree of multipath diversity in the channel
— Angular spread
— Typical mobile speeds
— Uplink and downlink capacity requirements
— Implementation complexity
— Environmental issues, visual impact of the antennas.

The basic choice is whether to maximise the number of diversity branches or to maximise the antenna gain. Figure 11.18 shows different options for base station antenna configurations. In a conventional arrangement each cell is divided into three sectors and the base station employs two antennas per sector. Each antenna pattern covers the entire sector. Two-branch diversity is usually applied in the uplink, while often only one of the two antennas is used for downlink transmission. Figure 11.18 also shows the higher-order diversity approach and the beam-forming approach (fixed beam) when four antennas per sector are utilised. In the case of higher-order diversity, multiple antennas, each covering the entire cell sector, are employed. Low correlation between the antennas is desired, which can be achieved by adequate separation between the antennas or by using antennas with orthogonal polarisation. The beam-forming approach can exploit a uniform linear array in which the inter-antenna spacing is in the order of one half of a carrier wavelength. The sector is covered with narrow beams which have an increased antenna gain compared to a conventional sector antenna.

If the radio channel has a high degree of multipath diversity, it is beneficial to employ relatively few antenna diversity branches. However, if the radio channel can be characterised as a flat Rayleigh fading (one-tap) channel, it is advantageous to increase the number of diversity branches. Macro cells typically comprise more multipath diversity but less angular

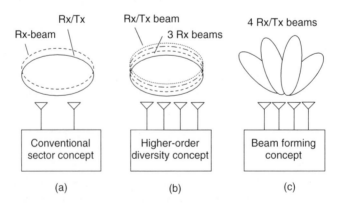

Figure 11.18. Different base station antenna concepts (a single sector is illustrated)

diversity than micro or pico cells. Angular diversity can be exploited by techniques. Obviously, there exists an optimum number of diversity branc coherent combining of the diversity signals requires accurate channel estin number of diversity branches increases the E_b/N_0 per diversity branch decreases. Therefore, the channel estimation for each branch becomes more critical.

In the following sections the coverage and capacity issues with adaptive antenna arrays are discussed separately. Here the focus is mainly on the uplink solutions. In the FDD mode of WCDMA it is rather difficult to achieve optimal downlink performance with adaptive antenna arrays, since the uplink and downlink use different frequencies, and fast fading is uncorrelated between the uplink and the downlink. Therefore, in the beam-forming approach the direction of arrival of the uplink signal paths has to be estimated. In addition, calibration of the antenna array is required continuously in order to optimise downlink performance. In the transmit diversity mode the direction of arrival does not need to be estimated because of the feedback from the terminal. The downlink transmit antenna diversity is presented in more detail in Section 11.3.2. A comprehensive review of the applications of antenna arrays to mobile communications can be found in [14] and [15].

11.5.1.1 Coverage Enhancement with Antenna Arrays

In the coverage-limited situation a typical assumption is that the interference can be characterised as temporally and spatially white noise and the loading can be presumed to be at low level, i.e. the ratio between multiple access interference and noise, I_0/N_0, is well below 0 dB. In these circumstances, the optimal receiver for a fading channel is a diversity receiver that employs maximal ratio combining, MRC [12]. If the amplitudes and phase angles of the diversity branches are perfectly estimated, the phase angle distortions due to the propagation channels can be compensated, and the antenna signals can be weighted proportionally to the signal-to-noise ratio of each antenna branch. This leads to coherent combining of the antenna signals. The Rake receiver in Figure 11.19 performs the coherent combining of the antenna signals by using the pilot symbols for channel estimation. Only a single Rake finger is depicted. In a case of frequency-selective fading, multiple signal paths have to be coherently combined and one Rake finger is required for each separable propagation path. The sum of the Rake fingers gives the composite signal, which is a coherent sum over the antennas and the multipaths. With maximal ratio combining the Rake receiver of Figure 11.19 multiplies a particular antenna signal with a weight w_m which is the complex conjugate of the channel impulse response estimated from that antenna signal. This calculation has to be performed in each Rake finger. In a line-of-sight situation without fading, the optimal receiver is a beam former that does the antenna phasing of the linear uniform antenna array in a way that directs the main beam to the direction of the desired signal. In the receiver of Figure 11.19 beam forming can be obtained by choosing the phase angles of antenna weights w_m so that the beam is directed in the direction that gives the largest desired signal power. If the multiple access interference can be modelled as spatially and temporally white noise, uniform antenna weighting can be applied. In the line-of-sight case, beam forming and maximal ratio combining both give the same optimal result if the direction of arrival of the desired signal is known in the first method and the channel coefficients are known in the latter method. For example, with an eight-element array and a flat fading channel, one parameter in the beam-forming case and eight parameters in the maximal ratio combining case have to be estimated. Accordingly, the relative performance between the

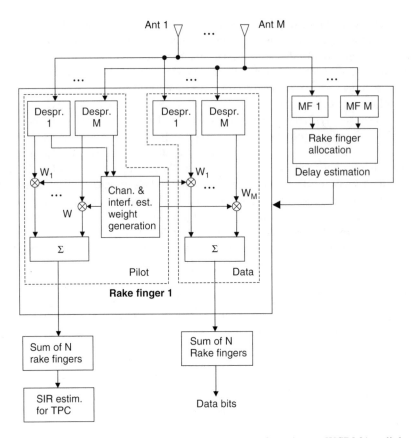

Figure 11.19. Principle of an antenna array receiver for coherent WCDMA uplink

beam forming and maximal ratio combining changes as the number of diversity branches increases since the maximal ratio combining suffers from higher estimation errors.

Figure 11.20 shows the simulation result with a WCDMA macro cell when the number of antennas per sector was increased from two to eight. Here the MRC and the fixed beam cases always assume a three-sector base station with a variable number of antennas per sector. However, sectorisation refers to a two-antenna diversity with a variable number of sectors. For example, a 12-antenna base station can be configured to a three-sector base station with four antennas or four beams per sector, or to a six-sector base station with two antennas per sector. The simulation parameters for the channel model are shown in Table 11.20. The performance of different antenna array approaches are depicted when the terminal speed was 50 km/h and the angular spread was 15°. The WCDMA speech service was assumed. In this case the multiple access interference was modelled as temporally and spatially white Gaussian noise. In the beam-forming and sectorisation approaches the desired user was located at the beam centre.

According to the simulation results of Figure 11.20 the coverage improves as the number of antennas is increased. The differences between the antenna solutions with the same number of receiver branches are small. With eight or more receiver branches, the diversity (MRC) option suffers from estimation errors.

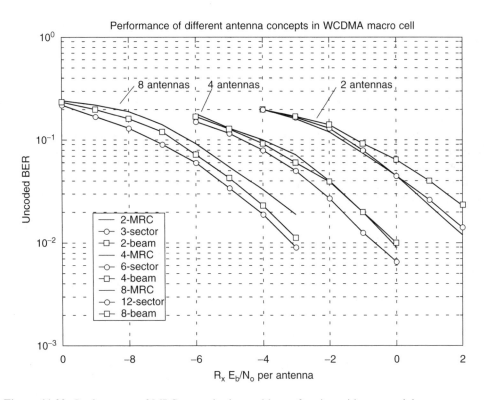

Figure 11.20. Performance of MRC, sectorisation and beam forming with two to eight antennas per 120° sector

Table 11.20. Simulation parameters describing the channel and interference models

Channel for desired signal	Three-path Rayleigh with tap powers of 0 dB, −2.75 dB and −7 dB
Interference modelling as random QPSK data	Spatially coloured interference
Interference channel	Two-path Rayleigh
Channel data rate	1024 kbps (∼300 kbps with channel coding)
Spreading factor	4
Load factor	0.5
I_0/N_0 (dB)	0
Number of interfering 1024 kbps users in Figure 11.22	3

11.5.1.2 Capacity Enhancement with Antenna Arrays

If increased capacity for low bit rate speech users is required, the antenna array solutions of Section 11.5.1.1 apply, since the interference with a high number of low bit rate users can be approximated as spatially and temporally white Gaussian noise. If there are only a very few high bit rate users at a time in a single cell or cell sector, the multiple access interference can often be characterised as spatially coloured noise and the interference scenario comes close to that of the TDMA systems. Accordingly, similar adaptive antenna algorithms as in TDMA can be applied for interference suppression. However, the power control of WCDMA sets the typical operation point to a level in which the I_0/N_0 ratio is much smaller than in the TDMA systems. For example, if the load factor is 50% the I_0/N_0 ratio is 0 dB [17]. Therefore, the achievable gains due to the optimum combining, or interference rejection combining (IRC), cannot reach such high figures as in TDMA systems [18]. The WCDMA receiver structure of Figure 11.19 allows an approach which adaptively maximises the signal-to-interference ratio. In addition to the channel estimation from each antenna and delay tap, also the interference pattern has to be estimated. One way to achieve this is to reconstruct the desired signal by using the pilot symbols and the estimated channel impulse response and to subtract the reconstructed signal from the received signal. This leads to optimum combining, which maximises the signal-to-interference ratio [19]. The optimum weight vector $\hat{\mathbf{w}}_n$ for each finger n can be expressed as

$$\hat{\mathbf{w}}_n = \hat{\mathbf{R}}_{uu,n}^{-1}\hat{\mathbf{h}}_n \qquad (11.5)$$

where $\hat{\mathbf{R}}_{uu,n}$ is the estimate of the spatially coloured interference and $\hat{\mathbf{h}}_n$ is the channel impulse response vector of the antenna signals.

Figure 11.21 illustrates the relative performance of IRC compared to MRC in a macro cellular environment. The simulation parameters of Table 11.20 describe the nature of the interference. A load factor of 50% and uncorrelated diversity antennas are assumed. The results indicate that the relative gain of the optimum combining compared to maximal ratio

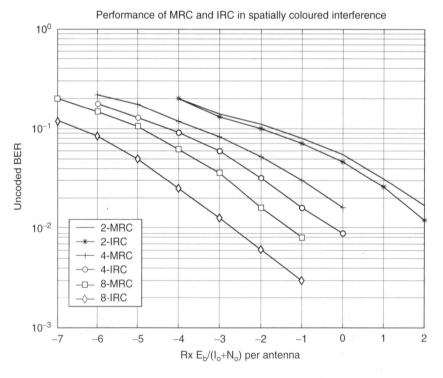

Figure 11.21. Performance of MRC and IRC in spatially coloured interference with two to eight antennas per 120° sector

combining (MRC) increases as the number of antennas increases. The gain of the IRC compared to the MRC is up to 1.6 dB with eight antennas at the uncoded BER of 10%. If the cell is more heavily loaded, the relative gains increase further. The results show that IRC can be used to improve the uplink capacity of high bit rate services if at least four receiver branches can be used. In theory, the maximum gain with the I_0/N_0 ratio of 3 dB (67% loading) is about 3–4 dB with 4–8 antennas, respectively [18].

11.5.2 Multi-user Detection

In this section the uplink performance improvements with base station multi-user detection are discussed. The target is to give an overview of different multi-user detection algorithms and references to more detailed information. The algorithms presented can be applied for both WCDMA FDD and TDD operation. Simulation results for WCDMA FDD with the most promising algorithms are shown. The application of the advanced receiver to the WCDMA TDD mode is discussed in Section 12.2.

CDMA systems are inherently interference-limited from both the receiver performance and system capacity points of view [20–22]. From the receiver perspective this means that, if the number of users is large enough, an increase in signal-to-noise ratio yields no improvement in bit or frame error rate. From the system capacity view it means that the larger the signal-to-interference-plus-noise ratio required for the desired quality of service, the fewer users can be accommodated in the communication channel.

The interference-limited nature of CDMA systems results from the receiver design. In CDMA systems, the core of the receiver is a spreading code matched filter (MF) or correlator [22]. Since the received spreading codes are usually not completely orthogonal, multiple access interference (MAI) is generated in the receiver. If the spreading factor is moderate, the received powers of users are equal (no near–far problem), and the number of interfering users is large (>10), by the central limit theorem the multiple access interference can be modelled as increased background noise with a Gaussian distribution. This approximation has led to the conclusion that the matched filter followed by decoding is the optimal receiver for CDMA systems in additive white Gaussian noise (AWGN) channels. In frequency-selective channels, the Rake receiver [23] can be considered optimal with corresponding reasoning.

Although multiple access interference can be approximated as AWGN, it inherently consists of received signals of CDMA users. Thus, multiple access interference is very structured, and can be taken into consideration in the receiver. This observation led Verdú [24] to analyse the optimal multi-user detectors (MUDs) for multiple access communications. Verdú was able to show that CDMA is not inherently interference-limited, but that is a limitation of the conventional matched filter receiver.

The optimal multi-user detectors [24] can use either maximum *a posteriori* (MAP) detection or maximum likelihood sequence detection (MLSD). In other words, techniques (including the Viterbi algorithm) similar to those applied in channels with inter-symbol interference [23] can be used to combat multiple access interference. The drawback of both the MLSD and MAP based multi-user detectors is that their implementation complexity is an exponential function of the number of users. Thus, they are not feasible for most practical CDMA receivers. This fact, together with Verdú's observation that CDMA with a MLSD receiver is not interference-limited, has triggered an avalanche of papers on suboptimal multi-user receivers. A brief summary of the suboptimal multi-user detection techniques is given below. For a more complete treatment, the reader is referred to the overview paper by Juntti and Glisic [25] and to the book by Verdú [26].

The existing suboptimal multi-user detection techniques can be categorised in several ways. One way is to classify the detection algorithms as centralised multi-user detection or decentralised single-user detection algorithms. The centralised algorithms perform real multi-user joint detection, i.e., they detect jointly each user's data symbols; they can be considered practical in base station receivers. The decentralised algorithms detect the data symbols of a single user based on the received signal observed in a multi-user environment containing multiple access interference; the single-user detection algorithms are applicable to both base station and terminal (mobile station) receivers.

In addition to one kind of implementation-based categorisation on multi-user and single-user detectors, the multi-user detectors can be classified based on the method applied. Two main classes in this category can be identified: linear equalisers and subtractive interference cancellation (IC) receivers. Linear equalisers are linear filters suppressing multiple access interference. The most widely studied equalisers include the zero-forcing (ZF) or decorrelating detector [27–29] and the minimum mean square error (MMSE) detector [30, 31]. The IC receivers attempt to explicitly estimate the multiple access interference component, after which it is subtracted from the received signal. Thus, the decisions become more reliable. Multiple access interference cancellation can be performed in parallel to all users, resulting in parallel interference cancellation (PIC) [32, 33]. Interference cancellation can also be performed in a serial fashion, resulting in serial interference cancellation (SIC) [34, 35].

Both linear equalisers and interference cancellation receivers can be applied in centralised receivers. The linear equalisers can also be implemented adaptively as single-user type decentralised detectors. This is possible if the spreading sequences of the users are periodic over a symbol interval so that multiple access interference becomes cyclostationary. Various adaptive implementations based on training sequences of the MMSE detectors have been studied [36–40]. So-called blind adaptive detectors not requiring training sequences have also been considered [39, 41, 42].

The choice of multi-user detection techniques for WCDMA base station receivers has been studied [43–46]. Both the receiver performance and implementation complexity have been considered. The conclusion of the studies is that a multi-user receiver based on multi-stage parallel interference cancellation (PIC) is currently the most suitable method to be applied in CDMA systems with a single spreading factor. The PIC receiver principle with one cancellation stage for a two-user CDMA system is illustrated in Figure 11.22. The parallel interference cancellation means that interference is cancelled from all users simultaneously, i.e., in parallel. The cancellation performance can be improved by reusing the decisions made after interference cancellation in a new IC stage. This results in a multi-stage interference cancellation receiver, which is illustrated in Figure 11.23.

The choice of multi-user detection for multi-service CDMA systems with a variable spreading factor access has been considered by Ojanperä [47]. For such a system, a group-wise serial interference cancellation (GSIC) [48–51] receiver seems to be the most appropriate choice with the current level of receiver technology. In the GSIC receiver, the users with a certain spreading factor are detected in parallel, after which multiple access interference caused by them is subtracted from the users with other spreading factors. A main reason for the GSIC being efficient is that the power of users depends on the spreading

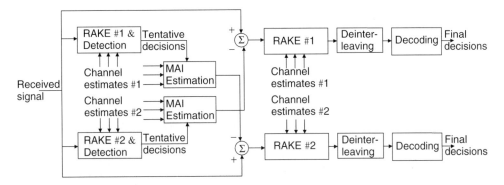

Figure 11.22. Parallel interference cancellation receiver for two users

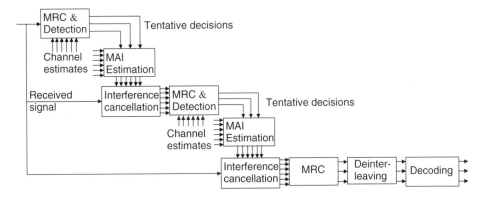

Figure 11.23. Multi-stage interference cancellation receiver

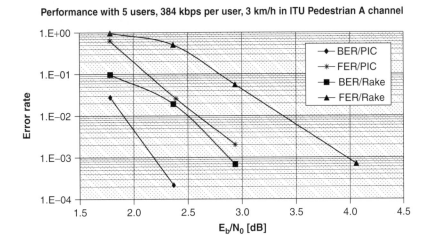

Figure 11.24. Performance of PIC receiver

factor. By starting the cancellation from the users with the lowest spreading factor, the highest power users (the most severe interferers) are cancelled first.

The performance of the interference cancellation receiver in a WCDMA system was studied using Monte Carlo computer simulations. The potential performance gains of the IC receivers are illustrated in Figure 11.24. Three users with data rate of 384 kbps in the ITU Pedestrian A channel are considered. The mobile speed is 3 km/h. Fast power control is applied for each user. The results show that the improvement in radio link performance is 0.7–1.0 dB. This performance gain could be used to improve the uplink capacity while keeping the uplink interference margin and the base station density unchanged. Another option would be to improve the coverage and reduce the base station density while keeping the uplink capacity unchanged. The coverage improvement can be seen in the link budget in Section 8.2 as a reduction of the interference margin.

The adaptive linear equalisers can be applied only if the spreading sequences of users are periodic over a relatively short time, such as over the symbol interval. Therefore, by using the short scrambling code option in a WCDMA uplink, the adaptive receivers could be utilised therein. In the WCDMA downlink, the spreading codes are periodic over one radio frame, whose duration is 10 ms. The period is so long that conventional adaptive receivers are practically useless. The problem can be partially overcome by introducing chip equalisers [52–57]. The idea here is to equalise the impact of the frequency-selective multipath channel on a chip-interval level. This suppresses inter-path interference (IPI) of the signals and also retains (at least partially) the orthogonality of spreading codes of users within one cell. The latter impact is possible, since synchronous transmission with orthogonal signature waveforms are applied in the downlink. In other words, multiple access interference in the downlink is caused by the multipath propagation, which can now be compensated for by the equaliser. The effect of multipath propagation on downlink performance without any interference suppression receivers is presented in Section 11.3.1.1.

Interference cancellation and suppression techniques are promising methods for improving receiver performance, as well as system capacity and coverage, in both uplink and downlink. In this section, uplink performance improvements are considered. In the uplink, explicit interference estimation-subtraction based receivers appear to be the most promising ones for practical implementation. In the downlink, on the other hand, linear equalisers can be applied.

References

[1]   UMTS, Selection Procedures for the Choice of Radio Transmission Technologies of the UMTS, ETSI, v.3.1.0, 1997.
[2]   Laiho-Steffens, J. and Lempiäinen, J., 'Impact of the Mobile Antenna Inclinations on the Polarisation Diversity Gain in DCS1800 Network', *Proceedings of PIMRC'97*, Helsinki, Finland, September 1997, pp. 580–583.
[3]   Lempiäinen, J. and Laiho-Steffens, J., 'The Performance of Polarisation Diversity Schemes at a Base-Station in Small/Micro Cells at 1800MHz', *IEEE Trans. Vehic. Tech.*, Vol. 47, No. 3, August 1998, pp. 1087–1092.
[4]   Sorensen, T.B., Nielsen, A.O., Mogensen, P.E., Tolstrup, M. and Steffensen, K., 'Performance of Two-Branch Polarisation Antenna Diversity in an Operational GSM Network', *Proceedings of VTC'98*, Ottawa, Canada, 18–21 May 1998, pp. 741–746.
[5]   TSGR1 WG1 #8(99)f58, 3GPP/WG1 contribution "Proposal to Have Optional 20 ms RACH Message Length', Source: Nokia.

[6]   Westman, T. and Holma, H., 'CDMA System for UMTS High Bit Rate Services', *Proceedings of VTC'97*, Phoenix, AZ, May 1997, pp. 825–829.

[7]   Pehkonen, K., Holma, H., Keskitalo, I., Nikula, E., and Westman, T., 'A Performance Analysis of TDMA and CDMA Based Air Interface Solutions for UMTS High Bit Rate Services', *Proceedings of PIMRC'97*, Helsinki, Finland, September 1997, pp. 22–26.

[8]   Ojanperä, T. and Prasad, R., *Wideband CDMA for Third Generation Mobile Communications*, Artech House, 1998, 439 pp.

[9]   Wacker, A., Laiho-Steffens, J., Sipilä, K. and Heiska, K., 'The Impact of the Base Station Sectorisation on WCDMA Radio Network Performance', *Proceedings of VTC'99*, Houston, TX, May 1999, pp. 2611–2615.

[10] Heiska, K. and Holma, H., 'Performance of 2 Mbps Packet Data with WCDMA in Small Microcellular Environment', *Proceedings of WPMC'98*, Yokosuka, Japan, November 1998, pp. 64–69.

[11] Holma, H. and Heiska, K., 'Performance of High Bit Rates with WCDMA over Multipath Channels', *Proceedings of VTC'99*, Houston, TX, May 1999, pp. 25–29.

[12] Dahlman, E. and Jamal, K., 'Wideband Services in a DS-CDMA Based FPLMTS System', *Proceedings of VTC'96*, Atlanta, GA, 28 April-1 May 1996, pp. 1656–1660.

[13] Ramakrishna, S. and Holtzman, J., 'A Comparison between Single Code and Multiple Code Transmission Schemes in a CDMA System', *Proceedings of VTC'98*, Ottawa, Canada, 18–21 May 1998, pp. 791–795.

[14] Godara, L.C., 'Application of Antenna Arrays to Mobile Communications, Part I: Performance Improvement, Feasibility, and System Considerations', *Proc. IEEE*, Vol. 85, No. 7, 1997, pp. 1031–1060.

[15] Godara, L.C., 'Application of Antenna Arrays to Mobile Communications, Part II: Beam-Forming and Direction-of-Arrival Considerations', *Proc. IEEE*, Vol. 85, No. 8, 1997, pp. 1195–1245.

[16] Jakes, W.J. (ed.), *Microwave Mobile Communications*, IEEE Press, New Jersey, 1974.

[17] Muszynski, P., 'Interference Rejection Rake-Combining for WCDMA', *Proceedings of WPMC'98*, Yokosuka, Japan, November 1998, pp. 93–97.

[18] Winters, J.H., 'Optimum Combining in Digital Mobile Radio with Co-channel Interference', *IEEE Trans. Vehic. Tech.*, Vol. 33, No. 3, 1984, pp. 144–155.

[19] Monzingo, R.A. and Miller, T.W., *Introduction to Adaptive Arrays*, John Wiley& Sons, New York, 1980.

[20] Pursley, M.B., 'Performance Evaluation for Phase-Coded Spread-Spectrum Multiple-Access Communication—Part I: System Analysis', *IEEE Trans. Commun.*, Vol. 25, No. 8, 1977, pp. 795–799.

[21] Gilhousen, K.S., Jacobs, I.M., Padovani, R., Viterbi, A.J., Weaver, L.A. and Wheatley III, C. E., 'On the Capacity of a Cellular CDMA System', *IEEE Trans. Vehic. Tech.*, Vol. 40, No. 2, 1991, pp. 303–312.

[22] Viterbi, A.J., *CDMA: Principles of Spread Spectrum Communication, Addison-Wesley Wireless Communications Series*, Addison-Wesley, Reading, MA, 1995.

[23] Proakis, J.G., *Digital Communications*, 3rd edn, McGraw-Hill, New York, 1995.

[24] Verdú, S., 'Minimum Probability of Error for Asynchronous Gaussian Multiple-Access Channels', *IEEE Trans. Inform. Th.*, Vol. 32, No. 1, 1986, pp. 85–96.

[25] Juntti, M. and Glisic, S., 'Advanced CDMA for Wireless Communications', in *Wireless Communications: TDMA Versus CDMA*, ed. S. Glisic and P. Leppänen, Chapter 4, pp. 447–490, Kluwer, 1997.

[26] Verdú, S., *Multiuser Detection*, Cambridge University Press, Cambridge, UK, 1998.

[27] Lupas, R. and Verdú, S., 'Near-Far Resistance of Multiuser Detectors in Asynchronous Channels', *IEEE Trans. Commun.*, Vol. 38, No. 4, 1990, pp. 496–508.

[28] Klein, A. and Baier, P.W., 'Linear Unbiased Data Estimation in Mobile Radio Systems Applying CDMA', *IEEE J. Select. Areas Commun.*, Vol. 11, No. 7, 1999, pp. 1058–1066.

[29] Zvonar, Z., 'Multiuser Detection in Asynchronous CDMA Frequency-Selective Fading Channels', *Wireless Personal Communications*, Kluwer, Vol. 3, No. 3–4, 1996, pp. 373–392.

[30] Xie, Z., Short, R.T. and Rushforth, C.K., 'A Family of Suboptimum Detectors for Coherent Multiuser Communications', *IEEE J. Select. Areas Commun.*, Vol. 8, No. 4, 1990, pp. 683–690.

[31] Klein, A., Kaleh, G.K. and Baier, P.W., 'Zero Forcing and Minimum Mean-Square-Error Equalization for Multiuser Detection in Code-Division Multiple Access Channels', *IEEE Trans. Vehic. Tech.*, Vol. 45, No. 2, 1996, pp. 276–287.

[32] Varanasi, M.K. and Aazhang, B., 'Multistage Detection in Asynchronous Code-Division Multiple-Access Communications', *IEEE Trans. Commun.*, Vol. 38, No. 4, 1990, pp. 509–519.

[33] Kohno, R., Imai, H., Hatori, M. and Pasupathy, S., 'Combination of an Adaptive Array Antenna and a Canceller of Interference for Direct-Sequence Spread-Spectrum Multiple-Access System', *IEEE J. Select. Areas Commun.*, Vol. 8, No. 4, 1990, pp. 675–682.

[34] Viterbi, A.J., 'Very Low Rate Convolutional Codes for Maximum Theoretical Performance of Spread-Spectrum Multiple-Access Channels', *IEEE J. Select. Areas Commun.*, Vol. 8, No. 4, 1990, pp. 641–649.

[35] Patel, P. and Holtzman, J., 'Analysis of a Simple Successive Interference Cancellation Scheme in a DS/CDMA System', *IEEE J. Select. Areas Commun.*, Vol. 12, No. 10, 1994, pp. 796–807.

[36] Madhow, U. and Honig, M.L., 'MMSE Interference Suppression for Direct-Sequence Spread-Spectrum CDMA', *IEEE Trans. Commun.*, Vol. 42, No. 12, 1994, pp. 3178–3188.

[37] Rapajic, P.B. and Vucetic, B.S., 'Linear Adaptive Transmitter-Receiver Structures for Asynchronous CDMA Systems', *European Trans. Telecommun.*, Vol. 6, No. 1, 1995, pp. 21–27.

[38] Miller, S.L., 'An Adaptive Direct-Sequence Code-Division Multiple-Access Receiver for Multiuser Interference Rejection', *IEEE Trans. Commun.*, Vol. 43, No. 2/3/4, 1995, pp. 1746–1755.

[39] Latva-aho, M., 'Advanced Receivers for Wideband CDMA Systems', Vol. C125 of *Acta Universitatis Ouluensis, Doctoral thesis*, University of Oulu Press, Oulu, Finland, 1998

[40] Latva-aho, M. and Juntti, M., 'Modified LMMSE Receiver for DS-CDMA—Part I: Performance Analysis and Adaptive Implementations', *Proceedings of ISSSTA'98*, Sun City, South Africa, September 1998, pp. 652–657.

[41] Honig, M., Madhow, U. and Verdú, S., 'Blind Adaptive Multiuser Detection', *IEEE Trans. Inform. Th.*, Vol. 41, No. 3, 1995, pp. 944–960.

[42] Latva-aho, M., 'LMMSE Receivers for DS-CDMA Systems in Frequency-Selective Fading Channels', in *CDMA Techniques for 3rd Generation Mobile Systems*, ed. F. Swarts, P. van Rooyen, I. Oppermann and M. Lötter, Chapter 13, Kluwer, 1998.

[43] Juntti, M. and Latva-aho, M., 'Multiuser Receivers for CDMA Systems in Rayleigh Fading Channels', *IEEE Trans. Vehic. Tech.*, to appear, 2000.

[44] Correal, N.S., Swanchara, S.F. and Woerner, B.D., 'Implementation Issues for Multiuser DS-CDMA Receivers, *Int. J. Wireless Inform. Networks*, Vol. 5, No. 3, 1998, pp. 257–279.

[45] Juntti, M., 'Multiuser Demodulation for DS-CDMA Systems in Fading Channels', Vol. C106 of *Acta Universitatis Ouluensis, Doctoral thesis*, University of Oulu Press, Oulu, Finland, 1997.

[46] Ojanperä, T., Prasad, R. and Harada, H., 'Qualitative Comparison of Some Multiuser Detector Algorithms for Wideband CDMA', *Proceedings of VTC'98*, Ottawa, Canada, May 1998, pp. 46–50.

[47] Ojanperä, T., 'Multirate Multiuser Detectors for Wideband CDMA', *Ph.D. thesis*, Technical University of Delft, Delft, The Netherlands, 1999.

[48] Johansson, A.-L., 'Successive Interference Cancellation in DS-CDMA Systems', *Doctoral thesis*, Chalmers University of Technology, Göteborg, Sweden, 1998.

[49] Juntti, M., 'Multiuser Detector Performance Comparisons in Multirate CDMA Systems', *Proceedings of VTC'98*, Ottawa, Canada, May 1998, pp. 36–40.

[50] Wijting, C.S., Ojanperä, T., Juntti, M.J., Kansanen, K. and Prasad, R., 'Groupwise Serial Multiuser Detectors for Multirate DS-CDMA', *Proceedings of VTC'99*, Houston, TX, May 1999, pp. 836–840.

[51] Juntti, M., 'Performance of Multiuser Detection in Multirate CDMA Systems', *Wireless Pers. Commun.*, Kluwer, Vol. 11, No. 3, 1999, pp. 293–311.

[52] Werner, S. and Lillberg, J., 'Downlink Channel Decorrelation in CDMA Systems with Long Codes', *Proceedings of VTC'99*, Houston, TX, May 1999, pp. 836–840.

[53] Hooli, K., Latva-aho, M. and Juntti, M., 'Multiple Access Interference Suppression with Linear Chip Equalizers in WCDMA Downlink Receivers', *Proceedings of Globecom'99*, Rio de Janeiro, Brazil, December 1999, pp. 467–471.

[54] Hooli, K., Juntti, M. and Latva-aho, M., 'Inter-Path Interference Suppression in WCDMA System with Low Spreading Factors', *Proceedings of VTC'99*, Amsterdam, The Netherlands, September 1999, pp. 421–425.

[55] Komulainen, P. and Heikkilä, M., 'Adaptive Channel Equalization Based on Chip Separation for CDMA Downlink', *Proceedings of PIMRC'99*, Osaka, Japan, September 1999, pp. 1114–1118.

[56] Heikkilä, M., Komulainen, P. and Lilleberg, J., 'Interference Suppression in CDMA Downlink through Adaptive Channel Equalization', *Proceedings of VTC'99*, Amsterdam, The Netherlands, September 1999, pp. 978–982.

[57] Grant, P.M., Spangenberg, S.M., Cruickshank, G.M., McLaughlin, S. and Mulgrew, B., 'New Adaptive Multiuser Detection Technique for CDMA Mobile Receivers', *Proceedings of PIMRC'99*, Osaka, Japan, September 1999, pp. 52–54.

12

UTRA TDD Mode

Otto Lehtinen, Antti Toskala, Harri Holma and Heli Väätäjä

12.1 Introduction

The UTRA TDD is intended to operate in the unpaired spectrum, as shown in Figure 1.2 in Chapter 1, illustrating the spectrum allocations in various regions. As can be seen from Figure 1.2, there is no TDD spectrum available in all regions. The background of UTRA TDD was described in Chapter 4. Since the technology selection in ETSI in January 1998, the major parameters have been harmonised between UTRA FDD and TDD, including chip rate and modulation. The presentation in this chapter follows the technical specifications for the TDD mode in the 3rd Generation Partnership Project (3GPP), specifically documents TS 25.221–TS 25.224 and TS 25.102 [1–5].

This chapter first introduces TDD as a duplex method on a general level. The physical layer and related procedures of the UTRA TDD mode are introduced in Section 12.2. UTRA TDD interference issues are evaluated in Section 12.3.

12.1.1 Time Division Duplex (TDD)

Three different duplex transmission methods are used in telecommunications: frequency division duplex (FDD), time division duplex (TDD) and space division duplex (SDD). The FDD method is the most common duplex method in the cellular systems. It is used, for example, in GSM. The FDD method requires separate frequency bands for both uplink and downlink. The TDD method uses the same frequency band but alternates the transmission direction in time. TDD is used, for example, for the digital enhanced cordless telephone (DECT). The SDD method is used in fixed-point transmission where directive antennas can be used. It is not used in cellular terminals.

Figure 12.1 illustrates the operating principles of the FDD and TDD methods. The term downlink or forward link refers to transmission from the base station (fixed network side) to the mobile terminal (user equipment), and the term uplink or reverse link refers to transmission from the mobile terminal to the base station.

WCDMA for UMTS, edited by Harri Holma and Antti Toskala
© 2001 John Wiley & Sons, Ltd

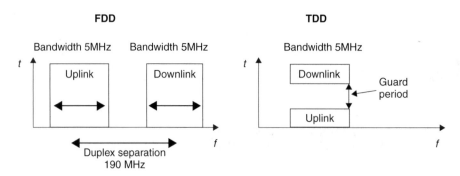

Figure 12.1. Principles of FDD and TDD operation

There are some characteristics peculiar to the TDD system listed below.

- Utilisation of unpaired band
  The TDD system can be implemented on an unpaired band while the FDD system always requires a pair of bands. In the future it is more likely that unpaired spectrum resources are cleared for UMTS as no pair is required for TDD operation.

- Discontinuous transmission
  Switching between transmission directions requires time, and the switching transients must be controlled. To avoid corrupted transmission, the uplink and downlink transmissions require a common means of agreeing on transmission direction and allowed time to transmit. Corruption of transmission is avoided by allocating a guard period which allows uncorrupted propagation to counter the propagation delay. Discontinuous transmission may also cause audible interference to audio equipment that does not comply with electromagnetic susceptibility requirements.

- Interference between uplink and downlink
  Since uplink and downlink share the same frequency band, the signals in those two transmission directions can interfere with each other. In FDD this interference is completely avoided by the duplex separation of 190 MHz. In UTRA TDD individual base stations are synchronised to each other at frame level to avoid this interference. This interference is further analysed in Section 12.3.

- Asymmetric uplink/downlink capacity allocation
  In TDD operation, uplink and downlink are divided in the time domain. It is possible to change the duplex switching point and move capacity from uplink to downlink, or vice versa, depending on the capacity requirement between uplink and downlink.

- Reciprocal channel
  The fast fading depends on the frequency, and therefore, in FDD systems, the fast fading is uncorrelated between uplink and downlink. As the same frequency is used for both uplink and downlink in TDD, the fast fading is the same in uplink and in downlink. Based on the received signal, the TDD transceiver can estimate the fast fading, which will affect its transmission. Knowledge of the fast fading can be utilised in power control and in adaptive antenna techniques in TDD.

12.2   UTRA TDD Physical Layer

The UTRA TDD mode uses a combined time division and code division multiple access (TD/CDMA) scheme that adds a CDMA component to a TDMA system. The different user signals are separated in both time and code domain. Table 12.1 presents a summary of the UTRA physical layer parameters. All the major RF parameters are harmonised within UTRA for FDD and TDD modes.

12.2.1   Transport and Physical Channels

UTRA TDD mode transport channels can be divided into dedicated and common channels. Dedicated channels (DCH) are characterised in basically the same way as in the FDD mode. Common channels can be further divided into common control channels (CCCH), the random access channel (RACH), the downlink shared channel (DSCH) in the downlink,

Table 12.1. Comparison of UTRA FDD and TDD physical layer key parameters

	UTRA TDD	**UTRA FDD**
Multiple access method	TDMA, CDMA (inherent FDMA)	CDMA (inherent FDMA)
Duplex method	TDD	FDD
Channel spacing	5 MHz (nominal)	
Carrier chip rate	3.84 Mcps	
Timeslot structure	15 slots/frame	
Frame length	10 ms	
Multirate concept	Multicode, multislot and orthogonal variable spreading factor (OVSF)	Multicode and OVSF
Forward error correction (FEC) codes	Convolutional coding $R = \frac{1}{2}$ or 1/3, constraint length $K = 9$, turbo coding (8-state PCCC $R = 1/3$) or service-specific coding	
Interleaving	Inter-frame interleaving (10, 20, 40 and 80 ms)	
Modulation	QPSK	
Detection	Coherent, based on midamble	Coherent, based on pilot symbols
Dedicated channel power control	Uplink: open loop; 100 Hz or 200 Hz Downlink: closed loop; rate $\leq$ 800 Hz	Fast closed loop; rate = 1500 Hz
Intra-frequency handover	Hard handover	Soft handover
Inter-frequency handover	Hard handover	
Channel allocation	Slow and fast DCA supported	No DCA required
Intra-cell interference cancellation	Support for joint detection	Support for advanced receivers at base station
Spreading factors	1 ... 16	4 ... 512

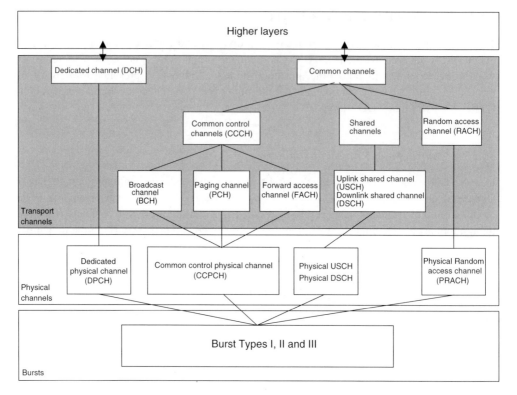

Figure 12.2. Mapping of the UTRA TDD transport channels to physical channels

and the uplink shared channel (USCH) in the uplink. Each of these transport channels is then mapped to the corresponding physical channel.

The physical channels of UTRA TDD are the dedicated physical channel (DPCH), common control physical channel (CCPCH), physical random access channel (PRACH), paging indicator channel (PICH) and synchronisation channel (SCH). For the SCH and PICH there do not exist corresponding transport channels. The mapping of the different transport channels to the physical channels and all the way to the bursts is shown in Figure 12.2. The physical channel structure is discussed in the following section in more detail.

12.2.2 Modulation and Spreading

The data modulation scheme in UTRA TDD is QPSK. The modulated data symbols are spread with a specific channelisation code of length 1–16. The modulated and spread data is finally scrambled by a pseudorandom sequence of length 16. The same type of orthogonal channelisation codes is used in the UTRA FDD system (see Section 6.3). Data spreading is followed by scrambling with a cell- or source-specific scrambling sequence; the scrambling process is chip-by-chip multiplication. The combination of multiplying with channelisation code and the cell-specific scrambling code is a user- or cell-specific spreading procedure. Finally, pulse shape filtering is applied to each chip at the transmitter: each chip is filtered with a root raised cosine filter with roll-off factor $\alpha = 0.22$, identical to UTRA FDD.

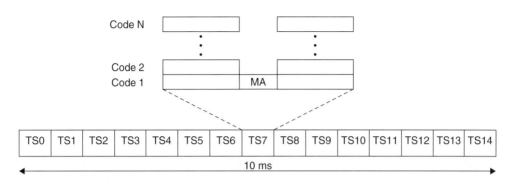

Figure 12.3. Frame structure of UTRA TDD. The number of code channels that may be used within a single timeslot varies depending on the propagation conditions (MA = midamble)

12.2.3 Physical Channel Structures, Slot and Frame Format

The physical frame structure is similar to that of the UTRA FDD mode. The frame length is 10 ms and it is divided into 15 timeslots each of 2560 chips, i.e., the timeslot duration is 666 μs. Figure 12.3 shows the frame structure.

Each of the 15 timeslots within a 10 ms frame is allocated to either uplink or downlink. Multiple switching points for different transmission directions per frame allow closed loop power control and a physical synchronisation channel (PSCH) in dedicated downlink slots to speed cell search. On the other hand, to be able to cover dynamic asymmetric services, the flexibility in slot allocation in the downlink/uplink direction guarantees efficient use of the spectrum. To maintain maximum flexibility while allowing closed loop power control whenever useful, the SCH has two timeslots per frame for downlink transmission in cellular usage. Figure 12.4 (a) shows such a maximum uplink asymmetry slot allocation (2:13). The PSCH is mapped to two downlink slots. For public systems a single-slot-per- frame SCH scheme could be applied. On the other hand, at least one timeslot has to be allocated to uplink transmission for the random access channel. Figure 12.4 (b) illustrates a maximum downlink asymmetry of 14:1.

Since the TDMA transmission in UTRA TDD is discontinuous the average transmission power is reduced by a factor of $10 \times \log_{10}(n/15)$, where n is the number of active timeslots per frame. For example, to provide the same coverage with UTRA TDD using a single timeslot for 144 kbps requires at least four times more base station sites than with UTRA FDD. This 12 dB reduction in the average power would result in typical macro cell environment to reduce the cell range more than into half, and thus, the cell area to a quarter. When utilising the same hardware in the UE the TDMA discontinuous transmission with

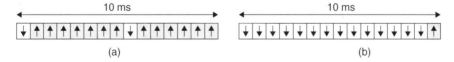

Figure 12.4. (a) Maximum uplink asymmetry of 2:13. Two slots per frame are allocated to downlink transmission; the synchronisation channel is assigned to these downlink slots. (b) Maximum downlink asymmetry (14:1); at least one uplink slot per frame is assigned for random access purposes

low duty cycle leads to reduced uplink range. With higher data rates the coverage difference to FDD reduces. Due to these properties, TDD should be used in small cell environment where power is not limiting factor and data rates used for the coverage planning are higher.

12.2.3.1 Burst Types

There are three bursts defined for UTRA TDD. All of them are used for dedicated channels while common channels typically use only a subset of them.

Burst types I and II are usable for both uplink and downlink direction with the difference between the Type I and II being the midamble length. Figure 12.5 illustrates the general downlink and Figure 12.6 the uplink burst structure with transmission power control (TPC) and transmission format combination indicator (TFCI). The burst types with two variants of midamble length can be used for all services up to 2 Mbps. The logical traffic channel (TCH), which contains user data, is mapped to a burst.

The burst contains two data fields separated by a midamble and followed by a guard period. The duration of a burst is one timeslot. The midamble is used for both channel equalisation and coherent detection at the receiver. The midamble reduces the user data payload. Table 12.2 shows the burst type I and II structures in detail.

Due to the longer midamble, burst type I is applicable for estimating 16 different uplink channel impulse responses. Burst type II can be used for the downlink independently of the number of active users. If there are fewer than four users within a timeslot, burst type II can also be used for the uplink.

Figure 12.5. Generalised UTRA TDD downlink burst structure. The data fields are separated by a midamble which is used for channel estimation. The transport format combination indicator (TFCI) is used to indicate the combination of used transport channels in the dedicated physical channel (DPCH) and is sent only once per frame. The TFCI uses in-band signalling and has its own coding. The number of TFCI bits is variable and is set at the beginning of the call

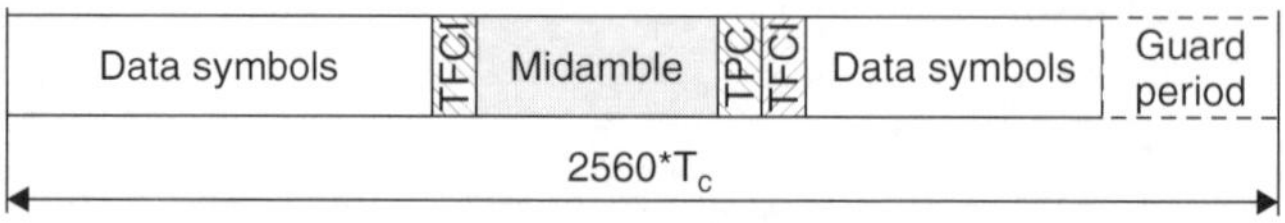

Figure 12.6. Generalised uplink burst structure. Both transmission power control (TPC) and TFCI are present. Both TFCI and TPC are transmitted in the same physical channel and use in-band signalling. The length of the TPC command is one symbol

Table 12.2. Burst type field structures

Burst name	Data field 1 length	Training sequence length	Data field 2 length	Guard period length
Burst type I	976 chips	512 chips	976 chips	96 chips
Burst type II	1104 chips	256 chips	1104 chips	96 chips
Burst type III	976 chips	512 chips	880 chips	192 chips

The midambles, i.e., the training sequences of different users, are time-shifted versions of one periodic basic code. Different cells use different periodic basic codes, i.e., different midamble sets. Due to the generation of midambles from the same periodic basic code, channel estimation of all active users within one timeslot can be performed jointly, for example by one single cyclic correlator. Channel impulse response estimates of different users are obtained sequentially in time at the output of the correlator [6].

In 3GPP Release-99, the downlink uses either a spreading factor of 16 with the possibility of multicode transmission, or a spreading factor of 1 for high bit rate applications in case such a capability is supported by the terminals. In the uplink, orthogonal variable spreading factor (OVSF) codes with spreading factors from 1 to 16 are used. The total number of the burst formats is 20 in the downlink and 90 in the uplink.

The burst Type III is used in the uplink direction only. The has developed for the needs for the PRACH as well as to facilitate handover in cases when timing advance is needed. The guard time of 192 chips (50 µs) equals a cell radius of 7.5 km.

12.2.3.2 Physical Random Access Channel (PRACH)

The logical random access channel (RACH) is mapped to a physical random access (PRACH) channel. Table 12.2 shows the burst type III used with PRACH. and Figure 12.7 illustrate the burst type III structure Spreading factor values of 16 and 8 are used for PRACH. With PRACH there is typically no TPC or TFCI bits used as shown in Figure 12.7.

12.2.3.3 Synchronisation Channel (SCH)

The time division duplex creates some special needs for the synchronisation channel. A capturing problem arises due to the cell synchronisation, i.e. a phenomenon occurring when a stronger signal masks weaker signals. The time misalignment of the different synchronisation channels of different cells would allow for distinguishing several cells within a single timeslot. For this reason a variable time offset (t_{offset}) is allocated between the SCH and the system slot timing. The offset between two consecutive shifts is $71T_c$. There exist two different SCH structures. The SCH can be mapped either to the slot number $k \in \{0\ldots14\}$ or to timeslots k and $k + 8$, $k \in \{0\ldots6\}$. Figure 12.8 shows the latter SCH structure for $k = 0$. This dual-SCH-per-frame structure is intended for cellular use. The position of the SCH can vary on a long-term basis.

The terminal can acquire synchronisation and the coding scheme for the BCCH of the cell in one step and will be able to detect cell messaging instantly. The primary (c_p) and the three secondary (c_s) synchronisation sequences are transmitted simultaneously. Codes are 256 chips long as in the UTRA FDD mode, and the primary code is generated in the same way as in the FDD mode, as a generalised hierarchical Golay sequence. The secondary synchronisation code words (c_s) are chosen from every 16th row of the Hadamard sequence H_8, which is used also in the FDD mode. By doing this there are only 16 possible code

Figure 12.7. UTRA TDD burst type III when used withPRACH

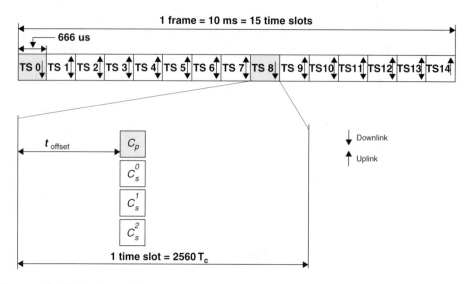

Figure 12.8. UTRA TDD SCH structure. This example has two downlink slots allocated for SCH ($k = 0$). The primary code (c_p) and three QPSK-modulated secondary codes (c_s) are transmitted simultaneously. The time offset (t_{offset}) is introduced to avoid adverse capture effects of the synchronous system. The combined transmission power of the three c_s is equal to the power of c_p

words in comparison to 32 of the FDD mode. The codes are QPSK modulated and the following information is indicated by the SCH:

- Base station code group out of 32 possible alternatives (5 bits)
- Position of the frame in the interleaving period (1 bit)
- Slot position in the frame (1 bit)
- Primary CCPCH locations (3 bits)

With a sequence it is possible to decode the frame synchronisation, the time offset (t_{offset}), the midamble and the spreading code set of the base station as well as the spreading code(s) and location of the broadcast channel (BCCH).

The cell parameters within each code group are cycled over two frames to randomise interference between base stations and to enhance system performance. Also, network planning becomes easier with the averaging property of the parameter cycling.

12.2.3.4 Common Control Physical Channel (CCPCH)

Once the synchronisation has been acquired, the timing and coding of the primary broadcast channel (BCH) are known. The CCPCH can be mapped to any downlink slot(s), including the PSCH slots, and this is pointed by the primary BCH.

The CCPCH is similar to the downlink dedicated physical channel (DPCH). It may be coded with more redundancy than the other channels to simplify acquisition of information.

12.2.3.5 UTRA TDD Shared Channels

The UTRA TDD specification also defines the Downlink Shared Channel (DSCH) and the Uplink Shared Channel (USCH). These channels use exactly the same slot structure as do the dedicated channels. The difference is that they are allocated on a temporary basis.

In the downlink the signalling to indicate which terminals need to decode the channel can be done with TFCI, by detecting midamble in use or by higher layers. In the uplink the USCH uses higher layer signalling and thus is not shared in practice on a frame-by-frame basis.

12.2.3.6 User Data Rates

Table 12.3 shows the UTRA TDD user bit rates with $\frac{1}{2}$-rate channel coding and spreading factor 16. The tail bits, TFCI, TPC or CRC overhead have not been taken into account. Spreading factors other than 16 (from the orthogonal variable spreading scheme) can be seen as subsets of spreading factor 16 (i.e., spreading factor 8 in the uplink corresponds to two parallel codes with spreading factor 16 in the downlink). When the number of needed slots exceeds 7, the corresponding data rate can be provided only for either the uplink or the downlink. The bit rates shown in Table 12.3 are time slot and code limited bit rates, the maximum interference limited bit rate can be lower.

12.2.4   UTRA TDD Physical Layer Procedures

12.2.4.1 Power Control

The purpose of power control is to minimise the interference of separate radio links. Both the uplink and downlink dedicated physical channels (DPCH) and physical random access channel (PRACH) are power controlled. The forward access channel (FACH) may be power controlled. The implementation of advanced receivers such as the joint detector will suppress intra-cell (own cell) interference and reduce the need for fast power control. The optimum multi-user detector is near–far resistant [7] but in practice the limited dynamic range of the suboptimum detector restricts performance. Table 12.4 shows the UTRA TDD power control characteristics.

Table 12.3. UTRA TDD air interface user bit rates

	Number of allocated timeslots		
Number of allocated codes with spreading factor 16	**1**	**4**	**13**
1	13.8 kbps	55.2 kbps	179 kbps
8	110 kbps	441 kbps	1.43 Mbps
16 (or spreading factor 1)	220 kbps	883 kbps	2.87 Mbps

Table 12.4. Power control characteristics of UTRA TDD

	Uplink	**Downlink**
Method	Open loop	SIR-based closed inner loop
Dynamic range	65 dB	30 dB (all the users are within 20 dB in one timeslot)
	Minimum power −44 dBm or less Maximum power 21 dBm	
Step size	1, 2, 3 dB	1, 2, 3 dB
Rate	Variable 1–7 slots delay (2 slot PCCPCH) 1–14 slots delay (1 slot PCCPCH)	From 100 Hz to approx. 750 Hz

In the downlink, closed loop is used after initial transmission. The reciprocity of the channel is used for open loop power control in the uplink. Based on interference level at the base station and on path loss measurements of the downlink, the mobile weights the path loss measurements and sets the transmission power. The interference level and base station transmitter power are broadcast. The transmitter power of the mobile is calculated by the following equation [4]:

$$P_{UE} = \alpha L_{PCCPCH} + (1 - \alpha)L_0 + I_{BTS} + SIR_{TARGET} + C \qquad (12.1)$$

In Equation (12.1) P_{UE} is the transmitter power level in dBm, L_{PCCPCH} is the measured path loss in dB, L_0 is the long-term average of path loss in dB, I_{BTS} is the interference signal power level at the base station receiver in dBm, and α is a weighting parameter which represents the quality of path loss measurements. The α is a function of the time delay between the uplink timeslot and the most recent downlink PCCPCH timeslot. SIR_{TARGET} is the target SNR in dB; this can be adjusted through higher layer outer loop. C is a constant value.

12.2.4.2   Data Detection

UTRA TDD requires that simultaneously active spreading codes within a timeslot are separated by advanced data detection techniques. The usage of conventional detectors, i.e. matched filters or Rake, in the base station requires tight uplink power control, which is difficult to implement in a TDD system since the uplink is not continuously available. Thus advanced data detection techniques should be used to suppress the effect of power differences between users, i.e., the near–far effect. Both inter-symbol interference (ISI) due to multipath propagation and multiple access interference (MAI) between data symbols of different users are present also in downlink. In downlink, the intra-cell interference is suppressed by the orthogonal codes, and the need for advanced detectors is lower than in uplink. In UTRA TDD the number of simultaneously active users is small and the use of relatively short scrambling codes together with spreading make the use of advanced receivers attractive.

The sub-optimal data detection techniques can be categorised as single-user detectors and multi-user detectors (see Section 11.5.2). In UTRA TDD single-user detectors can be applied when all signals pass through the same propagation channel, i.e. they are primarily applied for the downlink [8]. Otherwise, multi-user or joint detection is applied [9, 10].

Single-user detectors first equalise the received data burst to remove the distortion caused by the channel. When perfect equalisation is assumed, the orthogonality of the codes is restored after equalisation. The desired signal can now be separated by code-matched filtering. The advantages of using single-user detectors are that no knowledge of the other user's active codes is required and the computational complexity is low compared to joint detection [8].

To be able to combat both MAI and ISI in UTRA TDD, equalisation based on, for example, zero-forcing (ZF) or minimum mean-square-error (MMSE) can be applied. Both equalisation methods can be applied with or without decision feedback (DF). The computational complexity of the algorithms is essentially the same, but the performance of the MMSE equalisers is better than that of the ZF equalisers [10]. The decision feedback option improves performance (about 3 dB less E_b/N_0 at practical bit error rates) and the MMSE algorithm generally performs better (less than 1 dB difference in E_b/N_0 requirements) than

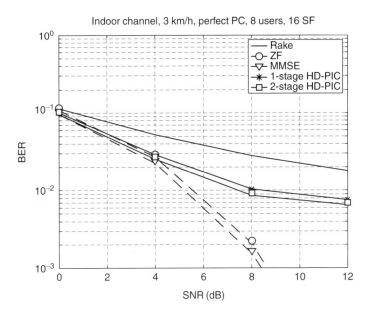

Figure 12.9. Performance of Rake, ZF and MMSE equalisers and one-and two-stage HD-PIC in the
UTRA TDD uplink

zero-forcing. Antenna diversity techniques can be applied with joint detection [11, 12] to
further enhance the performance.

The performance of Rake, ZF equaliser, MMSE equaliser and HD-PIC (hard decision
parallel interference canceller [13]) in the UTRA TDD uplink was studied using Monte
Carlo computer simulations in the UTRA TDD uplink [14]. Eight users with spreading
factor of 16 occupy one timeslot within a 10 ms frame. A two-path channel with tap gains
of 0 dB and −9.7 dB and with a mobile speed of 3 km/h is considered. Channel estimation
and power control are assumed to be ideal and channel coding is omitted. The performance
of Rake, ZF, MMSE, and one- and two-stage HD-PIC are shown in Figure 12.9. The
results show that the advanced base station receivers give a clear gain compared to the
Rake receiver in UTRA TDD even with ideal power control. As the signal-to-noise ratio
(SNR) increases, the performance of ZF and MMSE is better than the performance of
HD-PIC. Channel coding typically increases the differences between the performance of
different detectors. For example, in the operational area of BER = 5–10% the gain from
the advanced receiver structures can be up to 2 dB with perfect power control and even
more with realistic power control. The difference between the presented advanced detectors
is small in this operational area.

12.2.4.3 Timing Advance

To avoid interference between consecutive timeslots in large cells, it is possible to use a
timing advancement scheme to align the separate transmission instants in the base station
receiver. The timing advance is determined by a 6-bit number with an accuracy of four
chips (1.042 μs). The base station measures the required timing advance, and the terminal
adjusts the transmission according to higher layer messaging. The maximum cell range is
9.2 km.

The UTRA TDD cell radius without timing advance can be calculated from the guard period of traffic burst (96 chips = 25 μs), resulting in a range of 3.75 km. This value exceeds practical TDD cell ranges (micro and pico cells) and in practice the timing advance is not likely to be needed.

12.2.4.4 Channel Allocation

The layer 2 medium access control (MAC) entity is responsible timely resource allocation for transmission. Resource unit (RU) allocation (channel frequency, timeslot and code) is done by allocating resources to cells (slow dynamic channel allocation, DCA) and resource allocation to bearer services (fast DCA). Both the terminal and the base station perform periodic monitoring and reporting to support DCA. The termination of the MAC protocol is therefore much more complicated than in the UTRA FDD mode. The fast DCA is always terminated at the base station, but the slow DCA can be terminated at any network entity above the base stations that form the seamless coverage area. In practice this is RNC. Figure 12.10 illustrates the partitioning of different DCA elements in the network.

Slow Dynamic Channel Allocation (DCA)

The slow DCA can be terminated at a network element above the base stations that form a continuous coverage area, i.e. have the same uplink/downlink partitioning. By doing this the network can dynamically allocate resource units based on negotiation between adjacent cells to make handovers between cells fast and to minimise interference. Moreover, the slow DCA algorithm allocates resource units in a cell-related preference list (matrix) for fast DCA to acquire them for different bearers. The cell preference list is updated on a frame-by-frame basis. The slow DCA has the arbitration functionality to solve conflicts between base stations.

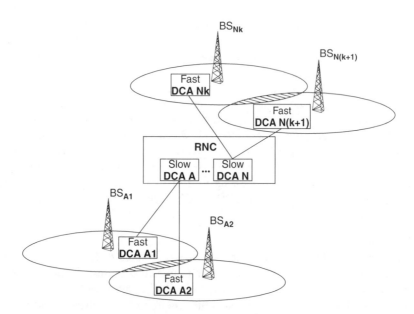

Figure 12.10. Illustration of DCA partitioning between different network elements. The slow DCA is located in RNC and the fast DCA is located in each base station

Fast Dynamic Channel Allocation (DCA)

The fast DCA acquires and releases resource units according to the slow DCA preference list. The multirate services are achieved by pooling resource units in either the code domain (multicode) or the time domain (multislot). A combination of both is also possible. The methods used for fast DCA can be timeslot pooling (as in DECT), frequency pooling or code pooling. The fast DCA can operate on the following strategies:

1. Allocate least interfered resources for traffic (code/timeslot).
2. Allocate several timeslots in order to gain from time diversity.
3. Average inter-cell interference, especially with low to medium data rate users when the cell load is sufficiently low. This is achieved by varying timeslot, code and frequency according to a predetermined scheme.

The number of codes is also dynamic and depends on channel characteristics, environment and system implementation. The fast DCA guidelines are also service dependent. For real-time services channels remain allocated for the whole duration, but the allocated code channel/timeslot (resource unit, or RU) may vary according to the reallocation procedure. For non-real-time services the channels are allocated only for the period of transmission of a data packet by using best effort strategy.

12.2.4.5   Handover

UTRA TDD supports inter-system handovers and intra-system handovers (to UTRA FDD and to GSM). All these handovers are mobile-assisted hard handovers.

UTRA TDD does not use soft handover (or macro diversity). This is a clear difference from UTRA FDD, in which the protocol structure has been designed to support soft handover. The UTRA TDD protocol structure has followed the same architecture as FDD for termination points for maximum commonality above the physical layer. This means, for example, that handover protocols terminate at the same location (RNC) but consist of FDD and TDD mode-specific parameters.

12.2.4.6   UTRA TDD Transmit Diversity

UTRA TDD supports four downlink transmit diversity methods. They are comparable to those in UTRA FDD. For dedicated physical channels Switched Transmitter Diversity (STD) and Transmit Adaptive Antennas (TxAA) methods are supported. The antenna weights are calculated using the reciprocity of the radio link. In order to utilise TxAA method, the required base station receiver and transmitter chain calibration makes the implementation more challenging.

For common channels Time Switched Transmit Diversity (TSTD) is used for PSCH, and Block Space Time Transmit Diversity (Block STTD) is used for primary CCPCH.

For uplink at the base station, the same receiver diversity methods as in FDD are applicable to enhance the performance.

12.3   UTRA TDD Interference Evaluation

In this section we evaluate the effect of interference within the TDD band and between TDD and FDD. TDD–TDD interference is analysed in Section 12.3.1 and the co-existence of TDD and FDD systems in Section 12.3.2.

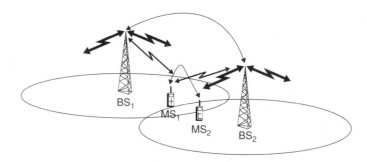

Figure 12.11. Interference between mobiles, between base stations, and between mobile and base station

12.3.1 TDD–TDD Interference

Since both uplink and downlink share the same frequency in TDD, those two transmission directions can interfere with each other. By nature the TDD system is synchronous and this kind of interference occurs if the base stations are not synchronised. It is also present if different asymmetry is used between the uplink and downlink in adjacent cells even if the base stations are frame synchronised. Frame synchronisation requires an accuracy of a few symbols, not an accuracy of chips. The guard period allows more tolerance in synchronisation requirements. Figure 12.11 illustrates possible interference scenarios. The interference within the TDD band is analysed with system simulations in [15].

Interference between uplink and downlink can also occur between adjacent carriers. Therefore, it can also take place between two operators.

In FDD operation the duplex separation prevents interference between uplink and down-link. The interference between a mobile and a base station is the same in both TDD and FDD operation and is not considered in this chapter.

12.3.1.1 Mobile Station to Mobile Station Interference

Mobile-to-mobile interference occurs if mobile MS_2 in Figure 12.12 is transmitting and mobile MS_1 is receiving simultaneously in the same (or adjacent) frequency in adjacent cells. This type of interference is statistical because the locations of the mobiles cannot be controlled. Therefore, it cannot be avoided by network planning. Intra-operator mobile-to-mobile interference occurs especially at cell borders. Inter-operator interference between mobiles can occur anywhere where two operators' mobiles are close to each other and transmitting on fairly high power. Methods to counter mobile-to-mobile interference are:

- DCA and radio resource management
- Power control.

12.3.1.2 Base Station to Base Station Interference

Base station to base station interference occurs if base station BS_1 in Figure 12.12 is transmitting and base station BS_2 is receiving in the same (or adjacent) frequency in adjacent cells. It depends heavily on the path loss between the two base stations and therefore can be controlled by network planning.

Intra-operator interference between base stations depends on the base station locations. Interference between base stations can be especially strong if the path loss is low between

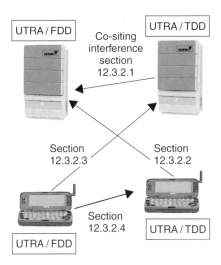

Figure 12.12. Possible interference situations between lower TDD band and FDD uplink band

the base stations. Such cases could occur, for example, in a macro cell, if the base stations are located on masts above rooftops. The best way to avoid this interference is by careful planning to provide sufficient coupling loss between base stations.

The outage probabilities in [15] show that co-operation between TDD operators in network planning is required, or the networks need to be synchronised and the same asymmetry needs to be applied. Sharing base station sites between operators will be very problematic, if not impossible. The situation would change if operators had inter-network synchronisation and identical uplink/downlink split in their systems.

From the synchronisation and co-ordination point of view, the higher the transmission power levels and the larger the intended coverage area, the more difficult will be the co-ordination for interference management. In particular, the locations of antennas of the macro-cell type tend to result in line-of-sight connections between base stations, causing strong interference. Operating TDD in indoor and micro/pico cell environments will mean lower power levels and will reduce the problems illustrated.

12.3.2 TDD and FDD Co-existence

The UTRA FDD and TDD have spectrum allocations that meet at the border at 1920 MHz, and therefore TDD and FDD deployment cannot be considered independently : see Figure 12.13. The regional allocations were shown in Figure 1.2 in Chapter 1. Dynamic channel allocation (DCA) can be used to avoid TDD–TDD interference, but DCA is not effective between TDD and FDD, since FDD has continuous transmission and reception. The possible interference scenarios between TDD and FDD are summarised in Figure 12.12.

12.3.2.1   Co-siting of UTRA FDD and TDD Base Stations

From the network deployment perspective, the co-siting of FDD and TDD base stations looks an interesting alternative. There are, however, problems due to the close proximity of the frequency bands. The lower TDD band, 1900–1920 MHz, is located adjacent to

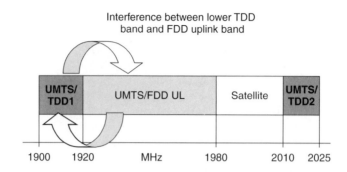

Figure 12.13. Interference between lower TDD band and FDD uplink band

the FDD uplink band, 1920–1980 MHz. The resulting filtering requirements in TDD base stations are expected to be such that co-siting a TDD base station in the 1900–1920 MHz band with an FDD base station is not considered technically and commercially a viable solution. Table 12.5 illustrates the situation. The output power of 24 dBm corresponds to a small pico base station and 43dBm to a macro cell base station.

The required attenuation between TDD macro cell base stations is 78 dB. If we introduce the 5 MHz guard band, with centre frequencies 10 MHz apart, the additional frequency separation of 5 MHz would increase the channel protection by 5 dB. The co-siting (co-located RF parts) is not an attractive alternative with today's technology.

The micro and pico cell environments change the situation, since the TDD base station power level will be reduced to as low as 24 dBm in small pico cells. On the other hand, the assumption of 30 dB antenna-to-antenna separation will not hold if antennas are shared between TDD and FDD systems. Antenna sharing is important to reduce the visual impact of the base station site. Also, if the indoor coverage is provided with shared distributed antenna systems for both FDD and TDD modes, there is no isolation between the antennas. Thus the TDD system should create a separate cell layer in UTRAN. In the pico cell TDD deployment scenario the interference between modes is easier to manage with low RF powers and separate RF parts.

12.3.2.2 Interference from UTRA TDD Mobile to UTRA FDD Base Station

UTRA TDD mobiles can interfere with a UTRA FDD base station. This interference is basically the same as that from a UTRA FDD mobile to a UTRA FDD base station on the adjacent frequency. The interference between UTRA FDD carriers is presented in Section 8.5. There is, however, a difference between these two scenarios: in pure FDD

Table 12.5. Coupling loss analysis between TDD and FDD base station in adjacent frequencies at 1920 MHz

TDD base station output power (pico/macro)	24/43 dBm
Adjacent channel power ratio	−45 dBc
Isolation between antennas (separate antennas for FDD and TDD base stations)	−30 dB
Leakage power into FDD base station receiver	−51/−32 dBm
Allowed leakage power	−110 dBm
Required attenuation	59/78 dB

interference there is always the corresponding downlink interference, while in interference from TDD to FDD there is no downlink interference. In FDD operation the downlink interference will typically be the limiting factor, and therefore uplink interference will not occur. In the interference from a TDD mobile to an FDD base station, the downlink balancing does not exist as between FDD systems, since the interfering TDD mobile does not experience interference from UTRA FDD. This is illustrated in Figure 12.14.

One way to avoid uplink interference problems is to make the base station receiver less sensitive on purpose, i.e. to desensitise the receiver. For small pico cells indoors, base station sensitivity can be degraded without affecting cell size. Another solution is to place the FDD base stations so that the mobile cannot get very close to the base station antenna.

12.3.2.3 Interference from UTRA FDD Mobile to UTRA TDD Base Station

A UTRA FDD mobile operating in 1920–1980 MHz can interfere with the reception of a UTRA TDD base station operating in 1900–1920 MHz. Uplink reception may experience high interference, which is not possible in FDD-only operation. The inter-frequency and inter-system handovers alleviate the problem. The same solutions can be applied here as in Section 12.3.2.2.

12.3.2.4 Interference from UTRA FDD Mobile to UTRA TDD Mobile

A UTRA FDD mobile operating in 1920–1980 MHz can interfere with the reception of a UTRA TDD mobile operating in 1900–1920 MHz. It is not possible to use the solutions of Sections 12.3.2.2 because the locations of the mobiles cannot be controlled. One way to tackle the problem is to use downlink power control in TDD base stations to compensate for the interference from the FDD mobile. The other solution is inter-system/inter-frequency

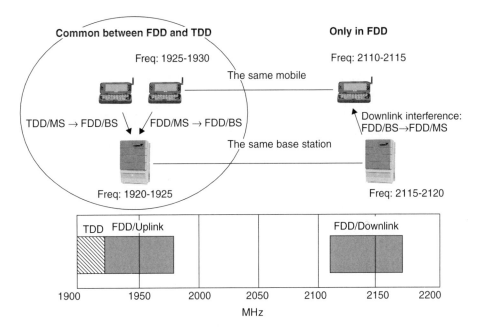

Figure 12.14. Interference from TDD mobile to FDD base station

handover. This type of interference also depends on the transmission power of the FDD mobile. If the FDD mobile is not operating close to its maximum power, the interference to TDD mobiles is reduced. The relative placement of UTRA base stations has effect in the generated interference. Inter-system handover requires multimode FDD/TDD mobiles and this cannot always be assumed.

12.3.3 Unlicensed TDD Operation

Unlicensed operation with UTRA TDD is possible if DCA techniques are applied together with TDMA components. DCA techniques cannot be applied for high bit rates since several timeslots are needed. Therefore, unlicensed operation is restricted to low to medium bit rates if there are several uncoordinated base stations in one geographical area.

12.3.4 Conclusions on UTRA TDD Interference

Sections 12.3.1–12.3.3 considered those UTRA TDD interference issues that are different from UTRA FDD-only operation. The following conclusions emerge:

– Frame-level synchronisation of each operator's UTRA TDD base stations is required.
– Frame-level synchronisation of the base stations of different TDD operators is also recommended if the base stations are close to each other.
– Cell-independent asymmetric capacity allocation between uplink and downlink is not feasible for each cell in the coverage area.
– Dynamic channel allocation is needed to reduce the interference problems within the TDD band.
– Interference between the lower TDD band and the FDD uplink band can occur and cannot be avoided by dynamic channel allocation.
– Inter-system and inter-frequency handovers provide means of reducing and escaping the interference.
– Co-siting of UTRA FDD and TDD macro cell base stations is not feasible, and co-siting of pico base stations sets high requirements for UTRA TDD base station implementation.
– Co-existence of FDD and TDD can affect FDD uplink coverage area and TDD quality of service.
– With proper planning TDD can form a part of the UTRAN where TDD complements FDD.

According to [16], TDD operation should not be prohibited in the FDD uplink band. Based on the interference results in this chapter, there is very little practical sense in such an arrangement, nor is it foreseen to be supported by the equipment offered for the market.

12.4 Concluding Remarks on UTRA TDD

This chapter covered UTRA TDD. The focus was on the physical layer issues, since the higher layer specifications are common, to a large extent, with UTRA FDD. In an actual implementation the algorithms for both the receiver and radio resource management differ between UTRA FDD and TDD, as the physical layers have different parameters to control.

Especially in the TDD base station, advanced receivers are needed, while for mobile stations the required receiver solution will depend on the details of performance requirements.

From the service point of view, both UTRA TDD and FDD can provide both low and high data rate services with similar QoS. The only exception for UTRA TDD is that after a certain point the highest data rates are asymmetric. The coverage of UTRA TDD will be smaller for low and medium data rate services than the comparable UTRA FDD service due to TDMA duty cycle. Also to avoid interference smaller cells provide better starting point. Therefore, UTRA TDD is most suited for small cells and high data rate services.

Interference aspects for UTRA TDD were analysed and will need careful consideration for deployment. With proper planning, UTRA TDD can complement the UTRA FDD network, the biggest benefit being the separate frequency band that can be utilised only with TDD operation.

References

[1]   3GPP Technical Specification 25.221 V3.1.0, Physical Channels and Mapping of Transport Channels onto Physical Channels (TDD).

[2]   3GPP Technical Specification 25.222 V3.1.0, Multiplexing and Channel Coding (TDD).

[3]   3GPP Technical Specification 25.223 V3.1.0, Spreading and Modulation (TDD).

[4]   3GPP Technical Specification 25.224 V3.1.0, Physical Layer Procedures (TDD).

[5]   3GPP Technical Specification 25.102 V3.1.0, UTRA (UE) TDD; Radio Transmission and Reception.

[6]   Steiner, B. and Jung, P., 'Optimum and suboptimum channel estimation for the uplink of CDMA mobile radio systems with joint detection', *European Transactions on Telecommunications and Related Techniques*, Vol. 5, 1994, pp. 39–50.

[7]   Lupas, R. and Verdu, S., 'Near–far resistance of multiuser detectors in asynchronous channels', *IEEE Transactions on Communications*, Vol. 38, no. 4, 1990, pp. 496–508.

[8]   Klein, A., 'Data detection algorithms specially designed for the downlink of CDMA mobile radio systems', in *Proceedings of IEEE Vehicular Technology Conference*, Phoenix, AZ, 1997, pp. 203–207.

[9]   Klein, A. and Baier, P.W., 'Linear unbiased data estimation in mobile radio systems applying CDMA', *IEEE Journal on Selected Areas in Communications*, Vol. 11, no. 7, 1993, pp. 1058–1066.

[10] Klein, A., Kaleh, G.K. and Baier P.W., 'Zero forcing and minimum mean square-error equalization for multiuser detection in code-division multiple-access channels', *IEEE Transactions on Vehicular Technology*, Vol. 45, no. 2, 1996, pp. 276–287.

[11] Jung, P. and Blanz, J.J., 'Joint detection with coherent receiver antenna diversity in CDMA mobile radio systems', *IEEE Transactions on Vehicular Technology*, Vol. 44, 1995, pp. 76–88.

[12] Papathanassiou, A., Haardt, M., Furio, I. and Blanz J.J., 'Multi-user direction of arrival and channel estimation for time-slotted CDMA with joint detection', in *Proceedings of the 1997 13th International Conference on Digital Signal Processing*, Santorini, Greece, 1997, pp. 375–378.

[13] Varanasi, M.K. and Aazhang, B., 'Multistage detection in asynchronous code-division multiple-access communications', *IEEE Transactions on Communications*, Vol. 38, no. 4, 1990, pp. 509–519.

[14] Väätäjä, H., Juntti, M. and Kuosmanen, P., 'Performance of multiuser detection in TD-CDMA uplink', EUSIPCO-2000, Tampere, Finland, 5–8 September 2000, pp. 71–75.

[15] Holma, H., Povey, G. and Toskala, A., 'Evaluation of interference between uplink and downlink in UTRA TDD', *VTC'99/Fall*. Amsterdam, 1999, pp. 2616–2620.

[16] ERC TG1 decision (98)183, February 1999.

13

Multi-Carrier CDMA in IMT-2000

Antti Toskala

13.1 Introduction

As explained in Chapter 4, in addition to the described UTRA FDD and TDD modes in the global ITU-R IMT-2000 CDMA framework, the third mode is the Multi-Carrier (MC) CDMA mode, based on the cdma2000 multi-carrier option being standardised by the 3rd Generation Partnership Project 2 (3GPP2). The key MC mode standards [1–4] scheduled to be completed by the end of 1999 will allow connection to the IS-41 based core network. Later, the necessary extensions are planned to be specified to support connecting the MC mode to GSM-MAP based core networks as well. An overview of the cdma2000 physical layer can be found also in [5].

The MC mode is seen as a natural path of evolution for operators with an existing IS-95 network, especially if a third generation network is to be deployed on the same frequency spectrum as an existing IS-95 network. This kind of spectrum *refarming* approach is foreseen in countries where there is no separate IMT-2000 spectrum, following the North American PCS spectrum allocation.

The name for the MC mode comes from the downlink transmission direction, where instead of a single wideband carrier, multiple (up to 12) parallel narrowband CDMA carriers are transmitted from each base station. Each carrier's chip rate is 1.2288 Mcps, equal to the IS-95 chip rate. The uplink direction is direct spread, very similar to UTRA FDD, with multiple chip rate of 1.2288 Mcps. The first ITU release of cdma2000 will adopt up to three carriers (known as 3X mode) at a chip rate up to 3.6864 Mcps. The term 'MC mode' hereafter will refer to the MC mode (3X) as defined in the cdma2000 standard.

The MC mode has been considered to provide an evolution path for existing IS-95 systems. As illustrated in Figure 13.1, three narrowband IS-95 carriers each with 1.25 MHz are bundled to form a multi-carrier transmission in the downlink with approximately 3.75 MHz (3X) bandwidth in a 5 MHz deployment.

WCDMA for UMTS, edited by Harri Holma and Antti Toskala
© 2001 John Wiley & Sons, Ltd

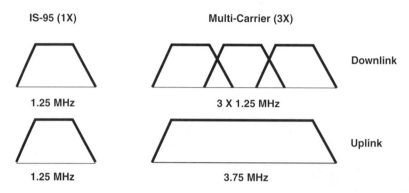

Figure 13.1. Relationship between the MC mode and IS-95 in spectrum usage

In terms of signal bandwidth, there is not much difference between the MC mode's multi-carrier (uplink) chip rate of 3.6864 Mcps and UTRA FDD's 3.84 Mcps. The following sections describe the main characteristics of the physical layer of the MC mode and illustrate the most important differences from UTRA FDD.

With respect to the higher layers, it is worth noting that, although the protocol structures are largely similar, there are differences, such as certain protocols being implemented with less clear solutions through the protocol layers. This means that in practice modifications to the MC mode protocol structure are required when considering interfacing the MC mode with GSM-based networks. This work, which is also known as the development of 'hooks and extensions', is scheduled to be carried out and completed by 3GPP2 during 2000.

In this chapter the focus is on the key principles of the MC mode physical layer; these are also of practical interest if dual-mode terminals between UTRA and the MC mode are to be considered in the future.

13.2   Logical Channels

Corresponding to the UTRA term 'transport channels', which carry data over the air and are mapped directly to the physical channels, the term 'logical channels' is used for cdma2000. The following logical channels are defined in the cdma2000 (Release A) specification which was completed at the end of 1999:

- Dedicated Traffic Channel (f/r-dtch). A point-to-point logical channel that carries data or voice traffic over a dedicated physical channel; it corresponds to the dedicated transport channel in UTRA. As in UTRA, dtch is intended for use by a single terminal.
- Common Control Channels (f/r-cmch control). These are used to carry MAC messages with shared access for several terminals.
- Dedicated Signalling Channel (f/r-dsch). A point-to-point logical channel that carries upper layer signalling traffic over a dedicated physical channel, for a single terminal.
- Common Signalling Channel (f/r-csch). A point-to-multipoint logical channel that carries upper layer signalling traffic over a common physical channel, with shared access for several terminals.

One difference in terminology worth noting is the use of the term 'reverse link' instead of 'uplink', and of 'forward link' instead of 'downlink', in cdma2000 documentation. For convenience and for consistency between the different chapters in this book, this chapter adopts the terms used in UTRA. For example, in physical channel terminology the terms Forward (F) and Reverse (R) link are not used, but are replaced with downlink and uplink respectively.

13.2.1 Physical Channels

The MC mode provides basically the same functionality as does UTRA FDD. Functions such as the Broadcast Channel, random access channel, and so on, are essential to the basic operation of all cellular systems. Also, the paging channel is needed to page the mobiles in the system. The physical layer contains slightly more differences, due to different design philosophies in some areas.

The common channel types more specific to UTRA, such as shared channels, uplink Common Packet Channel, and so on, have no direct counterparts in the MC mode, but the same functionalities are implemented by means of different arrangements.

The corresponding channel for data use is the Supplemental Channel, which in the downlink has similarities with the Downlink Shared Channel (DSCH) in UTRA FDD. The Supplemental Channel also exists in the uplink in the MC mode, but not as an enhancement for the random access channel like the Common Packet Channel in UTRA FDD. The characteristics of the Supplemental Channel are covered in more detail in connection with user data transmission.

The Access Channel differs due to the differences in the higher layer protocols. The typical duration of the MC mode random access message is longer than in UTRA, since in the latter case the change to a dedicated channel takes place earlier. Thus the random access channel transmission may last over several frames. To avoid CDMA near–far problems, a common power control channel may be used to transmit power control information for the MC mode uplink random access procedure.

The MC mode employs a quick paging channel that is used to indicate to the terminal when to listen to the actual paging channel itself. This bears some similarities to the Paging Indicator Channel (PICH) in both UTRA FDD and TDD modes.

13.3 Multi-Carrier Mode Spreading and Modulation

13.3.1 Uplink Spreading and Modulation

Uplink modulation is very similar to that of UTRA FDD in the sense that different channels are provided in either the I or Q branch and then experience a complex valued scrambling operation after spreading to balance the I and Q branch powers. This results in rather similar requirements for amplifier linearity as in UTRA FDD. The multi-code transmission is employed earlier than in UTRA FDD when the data rate increases. When higher data rates are desired with the MC mode, the Supplemental Channel is used in parallel with the Fundamental Channel, which provides only a limited set of possible lower data rates.

The uplink spreading is done with Walsh functions, while in UTRA FDD, OVSF codes are used. Variable rate spreading is not used in the MC mode during the connection, on a frame-by-frame basis, as no rate information is provided in the physical layer signalling.

The uplink long code used for scrambling has a period of $2^{42} - 1$ chips. This is significantly longer than in UTRA FDD, where the code period is 38400 chips for the dedicated channels and the code length is only 256 chips for short scrambling codes. With a period of 38400 chips no degradation is expected, while the code length of 256 chips without advanced receivers usually results in some degradation due to the reduced cross-correlation averaging effect. The Access Channels have a specific scrambling code with a period of 2^{15} chips.

For the adjacent channel attenuation, 40 dBm attenuation in the signal level should be reached outside 4.44 MHz bandwidth. The 3.75 MHz signal bandwidth is not a practical value to be used for frequency planning.

13.3.2 Downlink Spreading and Modulation

The downlink modulation is obviously characterised by its multi-carrier nature. The downlink carriers can be operated independently, or the terminal can demodulate them all. The benefit of receiving on all carriers is the frequency diversity that is improved over a single 1.2288 Mcps carrier. As each carrier contains a pilot channel for channel estimation, they can also be sent from different antennas if desired, to allow additional diversity. This is similar to the transmission diversity methods in UTRA.

The channel on each carrier is spread with Walsh functions using a constant spreading factor during the connection, in a similar way to UTRA with a few exceptions. Like the OVSF codes in UTRA, the Walsh functions separate channels from the same source and have similar orthogonality for transmission from a single source. The spreading factors for data transmission range from 256 down to 4. Downlink modulation consisting of three

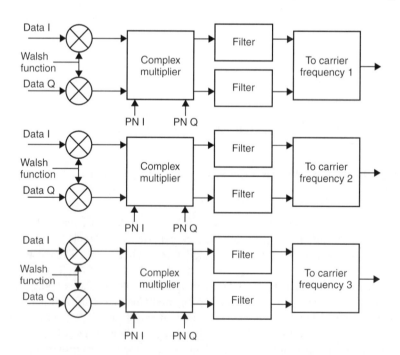

Figure 13.2. Downlink multi-carrier spreading and scrambling

carriers is illustrated in Figure 13.2. Note that PN sequences and Walsh functions on parallel carriers are the same.

Downlink scrambling is characterised by the use of a single code throughout the system. Since the MC mode is operated with synchronised base stations, a single code is used, the different base stations using a different phase of the same code. The number of available phases is 512, corresponding to the number of UTRA FDD primary scrambling codes. In practical networks, the phases with minimum separation are often avoided in order to relax the requirements associated with timing issues in the network planning process.

The MC mode pulse shaping has been specified with exact filter coefficients. Based on these, mean-squared error criteria have been defined that should be met for the filter implementation. Although the single carrier bandwidth discussed has often been 1.25 MHz, the bandwidth that has been defined for a single carrier spectrum mask with 40 dB attenuation for the power level is 1.48 MHz for the base station transmission.

13.4   User Data Transmission

This section describes the key principles of user data transmission in the MC mode, highlighting the main differences from UTRA FDD operation. One general difference from UTRA specifications is that for the MC mode the different data rates have been defined exactly in terms of puncturing or repetition factors, while in UTRA repetition and puncturing rules are given that can generate rate matching for any arbitrary rate. This does not cause practical differences, unless higher layers need to do a lot of padding or other operations to provide the necessary data rate if the physical layer data rate available does not suit the needs of the application. The data rates expected to be added to the MC mode are those needed to support the AMR voice codec used in UTRA and the GSM side, since at low rates the overhead may become significant for implementing the AMR voice codec data rates from the predefined MC mode data rate set.

13.4.1   Uplink Data Transmission

The Fundamental Channel in the MC mode is specified in detail for a given set of data rates, with a maximum data rate of 14.4 kbits/s. It can change the momentary data rate with changes in the repetition, but the symbol rate is not changed. This allows the use of blind rate detection in the base station. The Fundamental Channel and pilot channel structure is illustrated in Figure 13.3, where the pilot channel also contains the power control symbols with a 1.25 ms interval. This allows downlink fast power control at a rate of 800 Hz.

As the data rate increases, larger data rates are not to be introduced on the same channel as in UTRA FDD, but rather with the Supplemental Channel. This is a parallel code channel separated with a different Walsh function from the Fundamental Channel. Uplink transmission may contain one or two Supplemental Channels with data rates ranging from a few kbps up to 1 Mbps, depending on the radio configuration. A typical radio configuration defines 10 or fewer different data rates for a Supplemental Channel. Rates above 14.4 kbps use turbo coding.

Thus, while in UTRA FDD the physical layer control channel maintains constant parameters for the spreading factor, and so on, in the MC mode the parameters for the Fundamental Channel are fixed and then one or more Supplemental Channels can be added (with a fixed

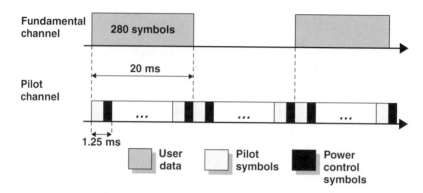

Figure 13.3. Uplink Fundamental Channel structure

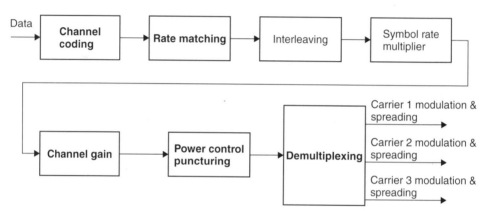

Figure 13.4. User data multiplexing to sub-carriers

spreading factor for a given data rate). There is a natural background to this difference. The MC mode does not contain physical layer control information to inform the receiver of the change of data rate, or more generally of the change of Transport Format Combination (TFI) as termed in UTRA. Thus the in-band signalling on the Fundamental Channel must carry all such information, so the parameters of the Fundamental Channel itself cannot be altered on a frame-by-frame basis.

For user data the radio frame length is 20 ms, while in UTRA it is 10 ms. In any case, speech services in both the MC mode and UTRA use at least 20 ms interleaving, since the AMR speech codec as well as existing GSM speech codecs provide data in 20 ms intervals, so that using 10 ms interleaving does not result in a shorter delay.

In the MC mode, there is no concept corresponding to UTRA's uplink Common Packet Channel (CPCH). However, the enhanced access channel that does the RACH functionality can be used for sending small packets, like the RACH in UTRA. The payload sizes defined for the enhanced access channel range from 172 to 744 information bits. There are three options for the radio frame length: 5,10 or 20 ms. This contrasts with data transmission on the Fundamental or Supplemental Channel where a 20 ms frame is always used.

13.4.2 Downlink Data Transmission

In the downlink direction the MC mode shows a major difference from UTRA FDD. The user data is divided between the three parallel CDMA sub-carriers, each with a chip rate of 1.2288 Mcps. As in the uplink, the lower data rates are implemented with the Fundamental Channel and higher data rates with the Supplemental Channel. A terminal could receive only one of the carriers. However, this would limit the data rate and would not be beneficial from a system capacity point of view, since frequency diversity would be minimised and transmitter antenna diversity unavailable.

The symbol rate for the traffic channels after channel coding and interleaving is multiplied by a factor of three. The power control symbol is then inserted with puncturing and the data demultiplexed to the three different sub-carriers, as shown in Figure 13.4.

The Walsh functions allocated for the Fundamental Channel carry user data with a fixed spreading factor, typically 256 or 128 for the lower data rates. For higher data rates, smaller spreading factors are used.

While the uplink direction has power control information multiplexed with the pilot channel, the downlink direction has only common pilot and the power control information is multiplexed with the data stream by puncturing with the rate of 800 Hz. The power control symbols are transmitted at a constant power level, as indicated in Figure 13.5, and serve as the only power reference for power control operation. The data symbols have a varying power level, since the rate matching is done with repetition or puncturing and the channel symbol rate is kept constant. The common pilot channel used as a phase reference is similar to UTRA CPICH. The power control symbols are not parallel in the time domain for different users in the MC mode downlink. This is to limit the resulting envelope variations by randomising the times when users have their power control symbols, as with DTX only power control symbols are active on the Fundamental Channel.

There are some fundamental differences from UTRA FDD, in addition to the multi-carrier structure. The Fundamental Channel does not carry any pilot symbols or rate information data. This means that blind rate detection with a variable rate connection is necessary. As the different data rates are formed with repetition or puncturing, channel decoding has to be carried out for data rate combinations in order to find out what the transmitted data rate was. This can be done for relatively small data rates, as in the case of the MC mode Fundamental

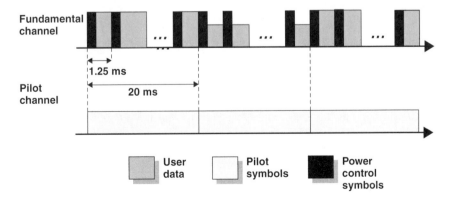

Figure 13.5. Downlink physical layer control multiplexing with user data for one sub-carrier

Channel. The higher data rates are implemented with the Supplemental Channel and higher layer signalling is used to indicate the changes in the data rate of the Supplemental Channel.

13.4.3   Channel Coding for User Data

From the channel coding perspective, the Fundamental Channel always uses convolutional coding, while on the Supplemental Channel Turbo coding is applied. The 8-state Turbo encoder and decoder are identical to the solutions in UTRA, but the turbo interleaving is different, as well as the channel interleaver. The latter would be different in any case, due to the differences in the number of symbols per frame with different spreading factors, and other differences in the frame structure. In the channel encoding there are differences also in the turbo coding rates: rates of $\frac{1}{4}$ and $\frac{1}{2}$ are used in the MC mode, while in UTRA $\frac{1}{4}$ is not applied. Also, in UTRA the coding rate corresponding to $\frac{1}{2}$ rate turbo coding is generated by rate matching from the 1/3 rate code.

Differences in convolutional coding also exist: in addition to UTRA's $\frac{1}{2}$ and 1/3 rates, $\frac{1}{4}$ and 1/6 rate convolutional codes have been specified in the MC mode. The constraint length of 9 is the same as in UTRA.

The differences in the turbo code internal interleaver result from the different kinds of optimisation used in the selection process by the standardisation bodies. The MC mode turbo interleaver was optimised for a fixed set of data rates, while the UTRA interleaver was selected against more generic criteria for a large variety of data rates. The practical differences in performance resulting from this are rather marginal.

13.5   Signalling

The way signalling channels have been defined shows some fundamental differences between UTRA FDD and the MC mode. In the MC mode, all the signalling channels are defined as channels of their own, while UTRA FDD uses the concept of Secondary Common Control Physical Channel (CCPCH), which carries in the physical layer channels such as the paging channel or forward access channel. Another difference is that in UTRA the common channels may use the TFCI for varying the data rate, while the transmission rates in the MC mode are fixed.

13.5.1   Pilot Channel

The MC mode has a separate common pilot channel for each carrier. This pilot channel is used in a similar way as in UTRA FDD; functions such as channel estimation and measurements for handover or cell selection and reselection are also similar. As each of the three carriers has its own pilot channel, they can be sent from separate transmission antennas if desired.

For beam forming, the MC mode always uses beam-specific pilot channels, called Auxiliary Pilot Channels, as there are no other known symbols sent on the dedicated channels to provide the phase reference. To save code space, auxiliary pilots may be generated from the Walsh functions with extended length. The maximum Walsh function length used in the MC mode is 512. The most noticeable difference is when operating with user-specific antenna beams, where UTRA uses pilot symbols on the dedicated channels while in the MC mode a separate pilot channel is provided for each beam.

The MC mode downlink does not use the same kind of transmit diversity as UTRA, but the carriers may be transmitted from different antennas, since each has its own common pilot channel active. For each carrier also Orthogonal Transmit Diversity (OTD) can be used, where the data is 'copied' to two antennas and transmitted from both of them.

13.5.2 Synch Channel

This channel is special to the MC mode. It helps the terminals to acquire initial timing synchronisation. The Synch Channel is a low rate channel with three frames per 80 ms period. The symbol rate is 1.2 kbits/s.

13.5.3 Broadcast Channel

The Broadcast Channel in the MC mode is similar to the UTRA Primary Common Control Channel (PCCCH), which carries the broadcast information in UTRA. Typical information sent on the Broadcast Channel is the availability of Access Channels or Enhanced Access Channels for random access purposes. The MC mode Broadcast Channel is likewise a fixed rate channel with 19.2 ksps. The coding method on the Broadcast Channel is 1/3 rate convolutional coding.

13.5.4 Quick Paging Channel

The Quick Paging Channel in the MC mode is similar to the Paging Indicator Channel in UTRA. It indicates to mobile stations whether they are expected to receive the paging information or information in the Forward Common Control Channel. It is divided into slots and subdivided into paging indicators as well as indicators that show change in the configuration.

13.5.5 Common Power Control Channel

The Common Power Control Channel provides power control information for a number of uplink channels that do not have as a pair a Fundamental Channel providing power control information. Channels that are power controlled in this way are the Reverse Common Control Channel and Enhanced Access Channels.

The Common Power Control Channel provides three different types of common power control groups with 200, 400 or 800 Hz command rates. I and Q branches both provide a command stream of 9.6 kbits/s for power control purposes.

13.5.6 Common and Dedicated Control Channels

The MC mode contains also the concept of common and dedicated control channels in the uplink and downlink directions. These channels are designed to carry higher layer control information for one or more terminals. As in the uplink direction, they are intended to transmit control information for the base station from a single terminal or from multiple terminals when the Reverse Traffic Channel is not used.

The additional higher layer common control channel is the Common Assignment Channel. This carries in the downlink the resource allocation messages for a number of terminals.

13.5.7 Random Access Channel (RACH) for Signalling Transmission

The RACH channel, or Access Channel in the MC mode, performs similar functions as in UTRA, though the detailed RACH procedure differs, with different options depending on whether or not preamble ramping is used. Several access channel frames can be transmitted in the MC mode, while in UTRA only 10 or 20 ms message lengths are used in the RACH. The UTRA Common Packet Channel (CPCH) with a longer message duration corresponds more closely to the MC mode access procedure with closed-loop power control.

For longer messages there is a power control channel that may be used to provide fast power control during the access procedure. The detailed access procedure for the MC mode is covered in the next section. Like the access channel, the enhanced access channel can convey signalling information for random access purposes.

13.6 Physical Layer Procedures

13.6.1 Power Control Procedure

The basic power control procedure is rather similar in the MC mode and UTRA FDD. Fast closed-loop power control is available in both uplink and downlink. Many of the details are different, however. First of all, the power control command rates are different: 1500 Hz with a normal step size of 1 dB in UTRA FDD, and 800 Hz in the MC mode. In the MC mode the fast closed-loop power control does not operate on its own in the uplink, but open-loop power control is also active.

The open-loop power control monitors the received signal strength in the downlink. If threshold values are exceeded it can alter the terminal transmission power. The open-loop power control has a large uncertainty, since terminals operate on a different frequency band for reception and cannot measure the absolute power level very accurately (the open-loop power control needs to compare the terminal transmission power to the received power level in the downlink). The pairing of open-loop power control with closed-loop power control can be turned off by the network; this option is often used in the existing IS-95 networks as well.

For the algorithm in the terminal one difference is caused by the pilot solution. In the MC mode the pilot symbols do not exist on the dedicated channel, thus the only symbols that can be used to aid the SIR estimation are the power control symbols as they preserve the power level unchanged with respect to change in the data rate. The resulting power offset as such does not have a major impact on the downlink peak-to-average due to the large number of parallel transmissions in the downlink. This is also further reduced, as the position for the power control sub-channel is not the same for all users in the slot, so that the envelope variation effects are more averaged over the slot.

13.6.2 Cell Search Procedure

There are essential differences in the cell search procedure between the MC mode and UTRA FDD. As stated earlier, the MC mode uses time-shifted versions of the single scrambling code for all base stations in the network. Upon powering on, the terminal starts searching for the single sequence with a proper receiver, which could be, for example, correlator or matched filter based. The search will continue until one or more code phases have been detected.

As all the cells are synchronised, only a single sequence is needed in the system, and the terminal can search for the different phases of that single sequence. While in UTRA there are 512 different cell-specific scrambling codes, a similar procedure would be too complex or too time-consuming and therefore the search in UTRA starts from the synchronisation code word common to all cells.

As in UTRA, the cell search differs depending on whether an initial search is considered or a search is done for target cells for handover purposes. In the connected mode the terminal will get a list of the neighbouring cells and will perform the search based on the information on the PN-offset of the target cells. The list of neighbouring cells needs to contain the PN-offsets of the cells to be searched, otherwise the terminals in active mode cannot recognise the PN-offset they pick up.

13.6.3   Random Access Procedure

The random access procedure on the Enhanced Access Channel has, like UTRA, a power ramping feature, though there are several different options for the random access operation. The preamble preceding message is of varying length, as is the message itself, with the possibility of trading off preamble length and available base station resources for carrying out the search for the message. Ramping with Enhanced Access Channel Preamble, with the additional preamble active, is illustrated in Figure 13.6. The access probes following the initial access probe are sent with increased power level until the maximum number of access probes is reached or higher layers allow sending of the actual message. If the option of additional preamble is selected, the ramping procedure is ended with an additional preamble to aid channel estimation at the base station.

Once the terminal is allowed to send the actual message, it can have a 5 ms Enhanced Access Header before the Enhanced Access Data, as illustrated in Figure 13.6. The Enhanced Access Data can be operated with 5, 10 or 20 ms frame length.

There also exists a power controlled access mode, where the common power control channel is used together with a random access procedure for controlling the transmission when sending the actual message. Terminals listen to the power control channel; once that

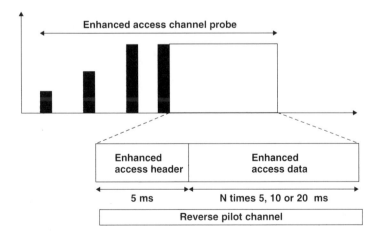

Figure 13.6. MC mode RACH ramping

starts to send down commands, terminals no longer increase the power but follow the power control command stream. This holds the power ramping for the particular channel.

13.6.4   Handover Measurements Procedure

The UTRA FDD handover included many measurements for the purposes of handover to various systems. The MC mode also provides for multimode terminals to hand over to other systems, such as IS-95. Releases later than Release-99 are expected to include the necessary enhancements for measuring and handing over to GSM or UTRA FDD from the physical layer perspective. On the network side, the degree of difficulty will depend on whether one is dealing with a simple voice call only or whether advanced data services are included.

Unlike in UTRA, the MC mode does not offer methods like compressed for inter-frequency measurements. The terminal either has to be dual receiver one or then terminal will do measurements by simply ignoring the data sent on the downlink direction. This aspect is also addressed in [5].

References

[1]   3GPP2 IS-2002.2, Physical Layer Standard for cdma2000 Spread Spectrum Systems.
[2]   3GPP2 IS-2002.3, Medium Access Control (MAC) Standard for cdma2000 Spread Spectrum Systems.
[3]   3GPP2 IS-2002.4, Signaling Layer 2 Standard for cdma2000 Spread Spectrum Systems.
[4]   3GPP2 IS-2002.5, Upper Layer (Layer 3) Signaling Standard for cdma2000 Spread Spectrum Systems.
[5]   Willenegger, S., 'cdma2000 Physical Layer: An Overview', *Journal of Communications and Networks*, Vol.2, No.1, March 2000, pp. 5–17.

Index